Lecture Notes in Mathematics 2082

Editors:
J.-M. Morel, Cachan
B. Teissier, Paris

For further volumes:
http://www.springer.com/series/304

Wolf-Jürgen Beyn
Department of Mathematics
University of Bielefeld
Bielefeld, Germany

Nicola Guglielmi
Dipartimento di Ingegneria, Scienze
 Informatiche Matematica
Università dell'Aquila
L'Aquila, Italy

Jesús María Sanz-Serna
Departamento de Matemática Aplicada
Universidad de Valladolid
Valladolid, Spain

Luca Dieci
School of Mathematics
Georgia Institute of Technology
Atlanta, GE, USA

Ernst Hairer
Section de mathématiques
Université de Genève
Genève, Switzerland

Marino Zennaro
Dipartimento di Matematica e Geoscienze
Università di Trieste
Trieste, Italy

ISBN 978-3-319-01299-5   ISBN 978-3-319-01300-8 (eBook)
DOI 10.1007/978-3-319-01300-8
Springer Cham Heidelberg New York Dordrecht London

Lecture Notes in Mathematics ISSN print edition: 0075-8434
                             ISSN electronic edition: 1617-9692

Library of Congress Control Number: 2013947686

Mathematics Subject Classification (2010): 65C, 65F, 65P, 15A, 34D, 37M

© Springer International Publishing Switzerland 2014

This work is subject to copyright. All rights are reserved by the Publisher, whether the whole or part of the material is concerned, specifically the rights of translation, reprinting, reuse of illustrations, recitation, broadcasting, reproduction on microfilms or in any other physical way, and transmission or information storage and retrieval, electronic adaptation, computer software, or by similar or dissimilar methodology now known or hereafter developed. Exempted from this legal reservation are brief excerpts in connection with reviews or scholarly analysis or material supplied specifically for the purpose of being entered and executed on a computer system, for exclusive use by the purchaser of the work. Duplication of this publication or parts thereof is permitted only under the provisions of the Copyright Law of the Publisher's location, in its current version, and permission for use must always be obtained from Springer. Permissions for use may be obtained through RightsLink at the Copyright Clearance Center. Violations are liable to prosecution under the respective Copyright Law.

The use of general descriptive names, registered names, trademarks, service marks, etc. in this publication does not imply, even in the absence of a specific statement, that such names are exempt from the relevant protective laws and regulations and therefore free for general use.

While the advice and information in this book are believed to be true and accurate at the date of publication, neither the authors nor the editors nor the publisher can accept any legal responsibility for any errors or omissions that may be made. The publisher makes no warranty, express or implied, with respect to the material contained herein.

Printed on acid-free paper

Springer is part of Springer Science+Business Media (www.springer.com)

Studies on Current Challenges in Stability Issues for Numerical Differential Equations

1 Introduction

This volume is the outgrowth of lectures presented during the CIME-EMS Summer School on Applied Mathematics, "Studies on Stability Issues for Numerical Differential Equations", held in Cetraro (Italy) in June 2011. The school was attended by about 50 participants, coming from Belgium, Canada, Germany, Italy, the Netherlands, Poland, Spain, Switzerland, and the USA. The hospitality and generosity of the CIME foundation are gratefully acknowledged.

This volume addresses some current research directions in the general area of stability studies for differential equations. ODEs (Ordinary Differential Equations), PDEs (Partial Differential Equations) and SDEs (Stochastic Differential Equations) are all treated in this monograph.

Although the emphasis of the volume is on issues of concern for numerical studies of differential equations, the arguments treated here will be of interest also to people working on qualitative theory of differential equations.

Above, the word *stability* has to be understood in a broad sense, to indicate some of the many issues which arise in numerical studies of differential equations. These issues range from the need to develop stable and reliable algorithms preserving some qualitative properties of the computed solutions, to the development of tools which are helpful to assess the onset of potential instabilities, to tools which assess the asymptotic properties of the solution or its discretization.

The topics considered in this volume involve the computation of dynamic patterns in evolution PDEs, the long-time integration of conservative ODEs and highly oscillatory systems, the Markov chain Monte Carlo method establishing a connection between SDEs and geometric integration, the continuous decomposition of matrices depending on parameters and the localization of singularities, and the uniform stability analysis of time-dependent linear initial value ODE problems.

In the end, the motivation of the authors remains the one of robust algorithmic development for use in approximation of differential equations. And, in all of the issues considered by the authors, the concerns for reliable approximations

5. The fifth and final chapter of the volume is authored by Nicola Guglielmi and Marino Zennaro and is entitled "Stability of Linear Problems: Joint Spectral Radius of Sets of Matrices".

It is known that the stability analysis of step-by-step numerical methods for differential equations often reduces to the analysis of linear difference equations with variable coefficients. This class of difference equations leads to a family $\mathscr{F}$ of matrices depending on some parameters, and the behaviour of the solutions depends on the convergence properties of the products of the matrices of $\mathscr{F}$. To date, the techniques mainly used in the literature are confined to the search for a suitable norm and for conditions on the parameters such that the matrices of $\mathscr{F}$ are contractive in that norm. In general, the resulting conditions are more restrictive than necessary. An alternative and more rigorous approach is based on the concept of "joint spectral radius" of the family $\mathscr{F}$. It is known that, in analogy with the case of a single matrix, all the products of matrices of $\mathscr{F}$ asymptotically vanish if and only if the joint spectral radius is less than 1. The aim of this chapter is to discuss the main theoretical and computational aspects involved in the analysis of the joint spectral radius and in applying this tool to the stability analysis of the discretizations of differential equations, as well as to other stability problems. It is worth stressing that both theory and numerical methods for the computation of the joint spectral radius still present many open problems.

It is the expectation of the editors that this volume will serve a twofold purpose. On the one hand, the volume will be a valuable entry point into the topics herein considered, with ample and exhaustive references on each subject, and a didactic style of presentation on cutting-edge techniques. On the other hand, this volume will hopefully stimulate researchers to pursue further the topics presented in this volume.

Atlanta, GA, USA Luca Dieci
L'Aquila, Italy Nicola Guglielmi

Contents

Long-Term Stability of Symmetric Partitioned Linear Multistep Methods 1
Paola Console and Ernst Hairer

Markov Chain Monte Carlo and Numerical Differential Equations 39
J.M. Sanz-Serna

Stability and Computation of Dynamic Patterns in PDEs 89
Wolf-Jürgen Beyn, Denny Otten, and Jens Rottmann-Matthes

Continuous Decompositions and Coalescing Eigenvalues for Matrices Depending on Parameters 173
Luca Dieci, Alessandra Papini, Alessandro Pugliese, and Alessandro Spadoni

Stability of Linear Problems: Joint Spectral Radius of Sets of Matrices 265
Nicola Guglielmi and Marino Zennaro

differential equations have been successfully applied to the integration of planetary motion. A theoretical explanation of the observed excellent long-time behavior has been given in [9]. It is based on a backward error analysis, and rigorous estimates for the parasitic solution components are obtained, when the system is Hamiltonian of the form $\ddot{q} = -\nabla U(q)$, and derivative approximations are obtained locally by finite differences.

The main aim of the present contribution is to study to which extend this excellent behavior and its theoretical explanation is valid also in more general situations—separable Hamiltonians $H(p, q) = T(p) + U(q)$ with general functions $T(p)$ and $U(q)$, and problems with position dependent kinetic energy. The presentation of the results is in three parts. In the first part we briefly recall the classical theory of partitioned linear multistep methods (order, zero-stability, convergence) and known results on the long-time behavior of symmetric multistep methods for second order Hamiltonian systems. We also present numerical experiments illustrating an excellent long-time behavior in interesting situations. The theoretical explanation of the long-time behavior is based on a backward error analysis for partitioned multistep methods. Part 2 is devoted to the study of the underlying one-step method. This method is symmetric, and we investigate conditions on the coefficients of the method to achieve good conservation of the Hamiltonian. When using multistep methods one is necessarily confronted with parasitic solution components, because the order of the difference equation is higher than the order of the differential equation. These parasitic terms will be studied in Part 3. On time intervals, where the parasitic terms remain bounded and small, the multistep method essentially behaves like a symmetric one-step method.

1.1 Classical Theory of Partitioned Linear Multistep Methods

Hamiltonian systems are partitioned ordinary differential equations of the form

$$\begin{aligned} \dot{p} &= f(p, q), \ p(0) = p_0, \\ \dot{q} &= g(p, q), \ q(0) = q_0, \end{aligned} \quad (1)$$

where $f(p, q) = -\nabla_q H(p, q)$, $g(p, q) = \nabla_p H(p, q)$, and $H(p, q)$ is a smooth scalar energy function. For their numerical solution we consider partitioned linear multistep methods

$$\begin{aligned} \sum_{j=0}^{k} \alpha_j^p p_{n+j} &= h \sum_{j=0}^{k} \beta_j^p f(p_{n+j}, q_{n+j}) \\ \sum_{j=0}^{k} \alpha_j^q q_{n+j} &= h \sum_{j=0}^{k} \beta_j^q g(p_{n+j}, q_{n+j}), \end{aligned} \quad (2)$$

where the p and q components are discretized by different multistep methods. Following the seminal thesis of Dahlquist, we denote the generating polynomials of the coefficients α_j, β_j of a multistep method by

$$\rho(\zeta) = \sum_{j=0}^{k} \alpha_j \zeta^j, \qquad \sigma(\zeta) = \sum_{j=0}^{k} \beta_j \zeta^j.$$

The generating polynomials of the method (2) are thus $\rho_p(\zeta)$, $\sigma_p(\zeta)$ and $\rho_q(\zeta)$, $\sigma_q(\zeta)$, respectively. In the following we collect some basic properties of linear multistep methods (see e.g., [11]).

Zero-Stability. A linear multistep method is called stable, if the polynomial $\rho(\zeta)$ satisfies the so-called *root condition*, i.e., all zeros of the equation $\rho(\zeta) = 0$ satisfy $|\zeta| \leq 1$, and those on the unit circle are simple.

Order of Consistency. A linear multistep method has order r if

$$\frac{\rho(\zeta)}{\log \zeta} - \sigma(\zeta) = \mathcal{O}\big((\zeta - 1)^r\big) \quad \text{for} \quad \zeta \to 1.$$

For a given polynomial $\rho(\zeta)$ of degree k satisfying $\rho(1) = 0$, there exists a unique $\sigma(\zeta)$ of degree k such that the order of the method is at least $k + 1$; and there exists a unique $\sigma(\zeta)$ of degree $k - 1$ (which yields an explicit method) such that the order of the method is at least k.

Convergence. If both methods of (2) are stable and of order r, then we have convergence of order r. This means that for sufficiently accurate starting approximations and for $t_n = nh \leq T$ we have

$$\|p_n - p(t_n)\| + \|q_n - q(t_n)\| \leq C(T) h^r \quad \text{for} \quad h \to 0. \tag{3}$$

The constant $C(T)$ is independent of n and h. It typically increases exponentially as a function of T.

Symmetry. A multistep method is symmetric if the coefficients satisfy $\alpha_j = -\alpha_{k-j}$ and $\beta_j = \beta_{k-j}$ for all j. In terms of the generating polynomials this reads

$$\rho(\zeta) = -\zeta^k \rho(1/\zeta), \qquad \sigma(\zeta) = \zeta^k \sigma(1/\zeta). \tag{4}$$

If $\alpha_0 = 0$, the number k has to be reduced in this definition. Symmetry together with zero-stability imply that all zeros of $\rho(\zeta)$ have modulus one and are simple.

Remark 1. The idea to use different discretizations for different parts of the differential equation is not new. Already Dahlquist [5, Chap. 7] considers stable combinations of two multistep schemes for the solution of second order differential equations. Often, the vector field is split into a sum of two vector fields (stiff and nonstiff), cf. [2]. In the context of differential-algebraic equations, the

by a finite difference formula as in (7). The aim of this section is to get some insight into the long-time behavior of partitioned linear multistep methods (2) applied to Hamiltonian systems that are more general than (5).

Separable Hamiltonian Systems. Let us first consider separable polynomial Hamiltonians $H(p,q) = T(p) + U(q)$, where

$$T(p) = \sum_{2 \le j+k \le 3} a_{jk}\, p_1^j\, p_2^k + (p_1^4 + p_2^4), \quad U(q) = \sum_{2 \le j+k \le 3} b_{jk}\, q_1^j\, q_2^k + (q_1^4 + q_2^4).$$

The positive definite quartic terms imply that solutions remain in a compact set. We consider the following three situations:

(A) Non-vanishing coefficients are $a_{02} = 1$, $a_{20} = 1$, and $b_{02} = 2$, $b_{20} = 1$, $b_{03} = 1$. Since $T(-p) = T(p)$, the system is reversible with respect to $p \leftrightarrow -p$. Moreover, it is separated into two systems with one degree of freedom.

(B) Non-vanishing coefficients are $a_{02} = 1$, $a_{20} = 1$, $a_{03} = 1$, $a_{30} = -0.5$, and $b_{02} = 2$, $b_{20} = 1$, $b_{03} = 1$. The system is not reversible, but still equivalent to two systems with one degree of freedom.

(C) Non-vanishing coefficients are $a_{02} = 1$, $a_{20} = 1$, and $b_{02} = 2$, $b_{20} = 1$, $b_{12} = -1$, $b_{21} = 2$. The system is reversible, and it is a coupled system with two degrees of freedom.

We consider the following partitioned linear multistep methods:

plmm2 $\quad \rho_p(\zeta) = (\zeta - 1)(\zeta + 1) \quad\quad \sigma_p(\zeta) = 2\zeta$

$\quad\quad\quad \rho_q(\zeta) = (\zeta - 1)(\zeta^2 + 1) \quad \sigma_q(\zeta) = \zeta^2 + \zeta$

plmm4 $\quad \rho_p(\zeta) = \zeta^4 - 1 \quad\quad\quad\quad \sigma_p(\zeta) = \frac{4}{3}(2\zeta^3 - \zeta^2 + 2\zeta)$

$\quad\quad\quad \rho_q(\zeta) = \zeta^5 - 1 \quad\quad\quad\quad \sigma_q(\zeta) = \frac{5}{24}(11\zeta^4 + \zeta^3 + \zeta^2 + 11\zeta)$

Figure 1 shows the numerical Hamiltonian for the second order method "plmm2", and Table 1 presents the qualitative behavior in dependence of time and step size. Looking at Fig. 1, we notice that this partitioned multistep method behaves very similar to (non-symplectic) symmetric one-step methods, as can be seen from the experiments of [13]. For non-reversible problems without any symmetry we have a linear growth in the energy, for reversible problems we observe boundedness for integrable systems and for problems with one degree of freedom, and we observe a random walk behavior of the numerical energy for chaotic solutions. This is illustrated by plotting the numerical Hamiltonian of 4 trajectories with randomly perturbed initial values (perturbation of size $\approx 10^{-15}$) for problem (C).

The intervals considered in the experiments of Fig. 1 are relatively short. What happens on longer time intervals? For problem (A), the numerical energy of the method "plmm2" shows the same regular, bounded, $\mathcal{O}(h^2)$ behavior on intervals as long as 10^7. No secular terms and no influence of parasitic components can be

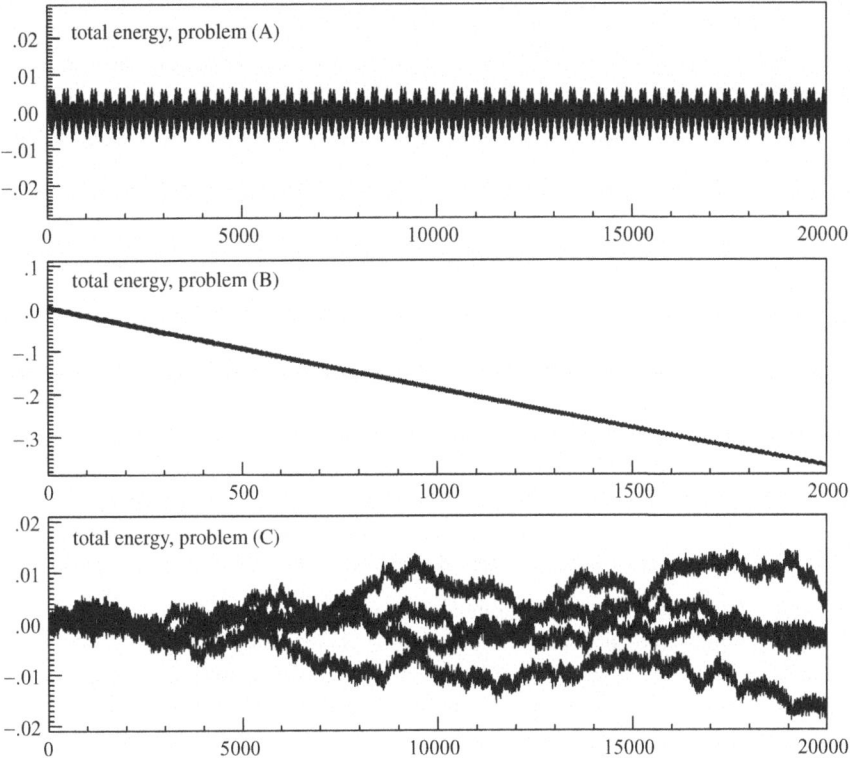

Fig. 1 Numerical Hamiltonian of method "plmm2" applied with step size $h = 0.005$ for problems (A) and (B), and with $h = 0.001$ for problem (C); initial values $q_1(0) = 1$, $q_2(0) = -1.2$, $p_1(0) = 0.2$, $p_2(0) = -0.9$. Starting approximations are computed with high precision

Table 1 Numerical energy behavior on intervals of length $\mathcal{O}(h^{-2})$; t is time, h the step size

Method	Problem (A)	Problem (B)	Problem (C)
plmm2, order 2	$\mathcal{O}(h^2)$	$\mathcal{O}(th^2)$	$\mathcal{O}(\sqrt{t}\,h^2)$
plmm4, order 4	$\mathcal{O}(h^4)$	$\mathcal{O}(th^4)$	$\mathcal{O}(h^4)$
plmm4c, order 4	$\mathcal{O}(h^4)$	$\mathcal{O}(h^4 + th^6)$	$\mathcal{O}(h^4)$

observed. For problem (B) the linear error growth in the energy as $\mathcal{O}(th^2)$ can be observed on intervals of length $\mathcal{O}(h^{-2})$. The behavior for problem (C) is shown in Fig. 2. We observe that after a time that is proportional to h^{-2} (halving the step size increases the length of the interval by a factor four) an exponential error growth is superposed to the random walk behavior of Fig. 1. Such a behavior is not possible for symmetric one-step methods. It will be explained by the presence of parasitic solution components.

We have repeated all experiments with the fourth order partitioned linear multistep method "plmm4" with characteristic polynomials given at the beginning

miltonian of method "plmm2" for problem (C); data as in Fig. 1, but on a

le 1 shows the behavior on intervals of length $\mathcal{O}(h^{-2})$. Whereas roblems (A) and (B) is expected, we cannot observe a random roblem (C). On very long time intervals, the energy error remains size $\mathcal{O}(h^4)$ for the problem (A). For the problems (B) and (C), ential error growth like $\delta \exp(ch^2 t)$ with small δ is superposed, isible after an interval of length $\mathcal{O}(h^{-2})$. Consequently, the e length of the interval is not related to the order of the method.

Triple Pendulum. For non-separable Hamiltonians, symmetric and/or symplectic one-step methods are in general implicit. It is therefore of interest to study the behavior of explicit symmetric multistep methods applied to such systems. We consider the motion of a triple pendulum, which leads to a Hamiltonian system with

$$H(p,q) = \frac{1}{2} p^T M(q)^{-1} p + U(q),$$

where $U(q) = -3\cos q_1 - 2\cos q_2 - \cos q_3$ and

$$M(q) = \begin{pmatrix} 3 & 2\cos(q_2 - q_1) & \cos(q_3 - q_1) \\ 2\cos(q_2 - q_1) & 2 & \cos(q_3 - q_2) \\ \cos(q_3 - q_1) & \cos(q_3 - q_2) & 1 \end{pmatrix}.$$

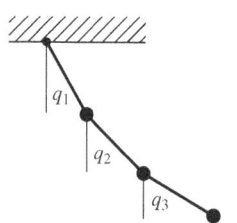

This matrix is positive definite with det $M(q) = 4 - 2\cos^2(q_2 - q_1) - \cos^2(q_3 - q_2)$. We have experimented with both partitioned multistep methods (order 2 and order 4) and we observed that the methods give excellent results when the angles are not to large, and the motion is not too chaotic.

with boundary conditions $p_{k+N} = p_k$ and $q_{k+N} = q_k$. This system can be written in the non-canonical Hamiltonian form

$$\dot{p} = -D(p,q)\nabla_q H(p,q), \qquad \dot{q} = D(p,q)\nabla_p H(p,q),$$

where $D(p,q)$ is the diagonal matrix with entries $d_k(p,q) = \frac{1}{\Delta x}(1 + \Delta x^2 (p_k^2 + q_k^2))$, and the Hamiltonian is given by

$$H(p,q) = \frac{1}{\Delta x}\sum_{k=1}^{N}(p_k p_{k-1} + q_k q_{k-1}) - \frac{1}{\Delta x^3}\sum_{k=1}^{N}\ln(1 + \Delta x^2(p_k^2 + q_k^2)). \quad (9)$$

Furthermore, the expression

$$I(p,q) = \frac{1}{\Delta x}\sum_{k=1}^{N}(p_k p_{k-1} + q_k q_{k-1}) \quad (10)$$

is a first integral of the system (8). Since the system is completely integrable, there are in addition $N-2$ other independent first integrals.

Since we are confronted with a Poisson system with non-separable Hamiltonian, there do not exist symplectic and/or symmetric integrators that are explicit. It is therefore of high interest to study the performance of explicit partitioned linear multistep methods, when applied to the system (8). Notice that the system is

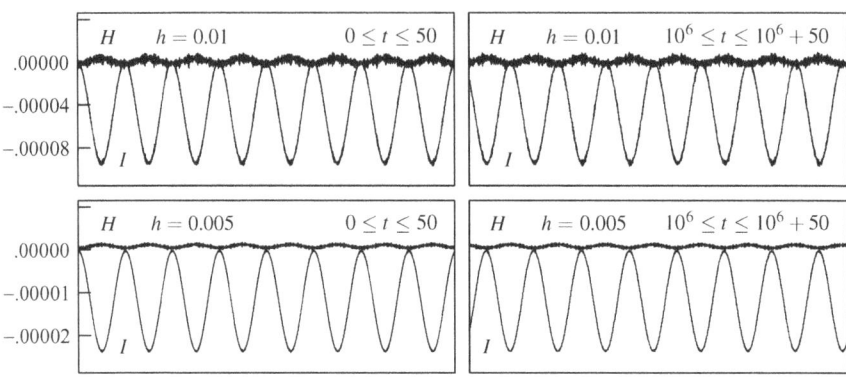

Fig. 3 Numerical preservation of the invariants H and I, defined in (9) and (10), with the method "plmm2" applied with step sizes $h = 0.01$ and $h = 0.005$; initial data are that of (11)

reversible with respect to the symmetries $p \leftrightarrow -p$ and $q \leftrightarrow -q$. Following [10, 16], we consider initial values

$$p_k(0) = \frac{1}{2}\big(1 - \varepsilon \cos(bx_k)\big), \qquad q_k(0) = 0, \qquad (11)$$

where $x_k = -L/2 + (k-1)\Delta x$, $\Delta x = L/N$, $b = 2\pi/L$ with $L = 2\pi\sqrt{2}$, and $\varepsilon = 0.01$. We apply the second order method "plmm2" to the system with $N = 16$, and we use various time step sizes for integrations over long time intervals. Figure 3 shows the error in both first integrals, H and I, to the left on the first subinterval of length 50, and to the right on the final subinterval starting at $t = 10^6$. We observe that halving the step size decreases the error by a factor of $4 = 2^2$, which is in accordance with a second order integrator. Similar to an integration with a symplectic scheme, the partitioned multistep method behaves very well over long times and no drift in the invariants can be seen. Comparing the results for different step sizes at the final interval, we notice a time shift in the numerical solution, but amplitude and shape of the oscillations are not affected. We also observe that the errors are a superposition of a slowly varying function scaled with h^2, and of high oscillations that decrease faster than with a factor 4, when the step size is halved.

The same qualitative behavior can be observed with the 4th order, explicit, partitioned multistep method "plmm4" for step sizes smaller than $h = 0.005$. As expected, the error decreases by a factor of $16 = 2^4$ when having the step size. For larger values of ε, say $\varepsilon \geq 0.05$ the behavior of the partitioned multistep method is less regular.

Further numerical experiments can be found in [7]. Excellent long-time behavior of partitioned linear multistep methods is reported for the Kepler problem and for a test problem in molecular dynamics simulation (frozen Argon crystal). Exponentially fitted partitioned linear multistep methods are considered in [18] for the long-term integration of N-body problems.

2 Long-Time Analysis of the Underlying One-Step Method

For one-step methods, the long-time behavior of numerical approximations is easier to analyze than for multistep methods. Whereas the notions of symplecticity and energy preservation are straightforward for one-step methods, this is not the case for multistep methods. It has been shown by Kirchgraber [14] that the numerical solution of strictly stable[2] linear multistep methods essentially behaves like that of a one-step method, which we call *underlying one-step method*. For a fixed step size h and a differential equation $\dot y = f(y)$, it is defined as the mapping $\Phi_h(y)$, such that the sequence defined by $y_{n+1} = \Phi_h(y_n)$ satisfies the multistep formula. This means that for starting approximations given by $y_j = \Phi_h^j(y_0)$ for $j = 0, 1, \ldots, k-1$, the numerical approximations obtained by the multistep formula coincides with that of the underlying one-step method (neglecting round-off effects).

For symmetric linear multistep methods, which cannot be strictly stable, such an underlying one-step method exists as a formal series in powers of h (see [6, p. 274] and [12, Sect. XV.2.2]). Despite its non-convergence, it can give much insight into the long-time behavior of the method.

2.1 Analysis for the Harmonic Oscillator

Consider a harmonic oscillator, written as a first order Hamiltonian system,

$$\dot p = -\omega q, \; p(0) = p_0,$$
$$\dot q = \omega p, \; q(0) = q_0.$$

Applying the partitioned linear multistep method (2) to this system yields the difference equations

$$\rho_p(E)\, p_n = -\omega h\, \sigma_p(E)\, q_n, \qquad \rho_q(E)\, q_n = \omega h\, \sigma_q(E)\, p_n, \qquad (12)$$

where we have made use of the shift operator $E y_n = y_{n+1}$. Looking for solutions of the form $p_n = a\zeta^n$, $q_n = b\zeta^n$ we are led to the 2-dimensional linear system

$$R(\omega h, \zeta)\begin{pmatrix} a \\ b \end{pmatrix} = 0 \quad \text{with} \quad R(\omega h, \zeta) = \begin{pmatrix} \rho_p(\zeta) & \omega h\, \sigma_p(\zeta) \\ -\omega h\, \sigma_q(\zeta) & \rho_q(\zeta) \end{pmatrix}. \qquad (13)$$

It has a nontrivial solution if and only if $\det R(\omega h, \zeta) = 0$. For small values of ωh the roots of this equation are close to the zeros of the polynomials $\rho_p(\zeta)$ and $\rho_q(\zeta)$. By consistency we have two roots close to 1, they are conjugate to each other, and

---

[2] A linear multistep is called strictly stable, if $\zeta_1 = 1$ is a simple zero of the ρ polynomial, and all other zeros have modulus strictly smaller than one.

they satisfy $\zeta_0 = \zeta_0(\omega h) = 1 + i\omega h + \mathcal{O}(h^2)$ and $\overline{\zeta}_0 = \overline{\zeta}_0(\omega h) = 1 - i\omega h + \mathcal{O}(h^2)$ (principal roots). They lead to approximations to the exact solution, which is a linear combination of $e^{i\omega t}$ and $e^{-i\omega t}$. The other roots lead to parasitic terms in the numerical approximations. The general solution $(p_n, q_n)^\top$ of the difference equation (12) is in fact a linear combination of $\zeta^n(a,b)^\top$, where ζ is a root of $\det R(\omega h, \zeta) = 0$, and the vector $(a,b)^\top$ satisfies the linear system (13).

Underlying One-Step Method. We consider a numerical solution of (12) that is built only on linear combinations of ζ_0^n and $\overline{\zeta}_0^n$. It has to be of the form

$$\begin{pmatrix} p_n \\ q_n \end{pmatrix} = \Phi_n \begin{pmatrix} p_0 \\ q_0 \end{pmatrix}, \qquad \Phi_n = \frac{1}{2}\left(\zeta_0^n + \overline{\zeta}_0^n\right) I + \frac{1}{2i}\left(\zeta_0^n - \overline{\zeta}_0^n\right) C, \qquad (14)$$

where the matrix C satisfies $R_0(I - iC) = 0$ and $\overline{R}_0(I + iC) = 0$, so that the vectors multiplying ζ_0^n and $\overline{\zeta}_0^n$ satisfy the relation (13) with $R_0 = R(\omega h, \zeta_0)$. It follows from the consistency of the method that for small but nonzero ωh the real and imaginary parts of the matrix R_0 are invertible. This permits us to compute the real matrix $C = -i(R_0 + \overline{R}_0)^{-1}(R_0 - \overline{R}_0)$. As a consequence of $R_0 = \frac{1}{2}(R_0 + \overline{R}_0)(I + iC)$ and $\det R_0 = 0$ we have $\det C = 1$ and trace $C = 0$, which implies $C^2 = -I$. The matrix Φ_n of (14) thus satisfies $\Phi_{n+1} = \Phi_n \Phi_1$, and consequently $\Phi_n = \Phi_1^n$, so that the underlying one-step method is seen to be given by

$$\begin{pmatrix} p_{n+1} \\ q_{n+1} \end{pmatrix} = \Phi(\omega h) \begin{pmatrix} p_n \\ q_n \end{pmatrix}, \qquad \Phi(\omega h) = \frac{1}{2}\left(\zeta_0 + \overline{\zeta}_0\right) I + \frac{1}{2i}\left(\zeta_0 - \overline{\zeta}_0\right) C. \qquad (15)$$

Notice that $\Phi(\omega h)$ is not an analytic function of ωh.

Properties of the Underlying One-Step Method. The above derivation is valid for all partitioned multistep methods. If the method is symmetric, also the coefficients of the polynomial $\det R(h\omega, \zeta)$ are symmetric, so that with $\zeta_0 = \zeta_0(\omega h)$ also its inverse is a solution of $\det R(h\omega, \zeta) = 0$. This implies $\zeta_0^{-1} = \overline{\zeta}_0$, and hence also $|\zeta_0| = 1$. Similarly, the symmetry of the methods (ρ_p, σ_p) and (ρ_q, σ_q) imply that $C(-\omega h) = C(\omega h)$. Consequently, we have $\Phi(-\omega h)\Phi(\omega h) = I$, which proves the *symmetry* of the underlying one-step method.

Furthermore, the mapping defined by the matrix $\Phi(\omega h)$ is *symplectic*:

$$\Phi(\omega h)^\top J \, \Phi(\omega h) = J \qquad \text{with} \qquad J = \begin{pmatrix} 0 & 1 \\ -1 & 0 \end{pmatrix}. \qquad (16)$$

This follows from the relations $C^\top J + JC = 0$ and $C^\top J C = J$, which are a consequence of $\det C = 1$ and trace $C = 0$.

Since the eigenvalues of C are $\pm i$, we have

$$T C T^{-1} = \begin{pmatrix} 0 & 1 \\ -1 & 0 \end{pmatrix} \qquad \text{with} \qquad T = \begin{pmatrix} 1 & 0 \\ a & b \end{pmatrix},$$

where (a, b) is the first row of the matrix C. Notice that we have $a = \mathcal{O}((\omega h)^2)$ and $b = 1 + \mathcal{O}((\omega h)^2)$. This transformation implies that $T\Phi(\omega h)T^{-1}$ is an orthogonal matrix, so that

$$\frac{\omega}{2}\left\|T\begin{pmatrix}p_n\\q_n\end{pmatrix}\right\|^2 = \frac{\omega}{2}\left(p_n^2 + (ap_n + bq_n)^2\right)$$

is a conserved quantity that is $\mathcal{O}(h^2)$ close to the true Hamiltonian.

Parasitic Solution Components. The complete solution of the difference equation (12) is given by

$$\begin{pmatrix}p_n\\q_n\end{pmatrix} = \Phi_1(\omega h)^n \begin{pmatrix}a\\b\end{pmatrix} + \sum_{l=1}^{2k-2} \zeta_l(\omega h)^n \begin{pmatrix}a_l\\b_l\end{pmatrix},$$

where $\zeta_l(\omega h)$ are the roots of $\det R(\omega h, \zeta) = 0$ which are different from the principal roots $\zeta_0(\omega h)$ and $\overline{\zeta}_0(\omega h)$. They are called parasitic roots of the method. Initial approximations (p_j, q_j) for $j = 0, 1, \ldots, k-1$ uniquely determine the vectors (a, b) and (a_l, b_l), recalling that (a_l, b_l) has to satisfy the relation (13).

If the starting values (p_j, q_j) approximate for $j = 0, 1, \ldots, k-1$ the exact solution $(p(t_0 + jh), q(t_0 + jh))$ up to an error of size $\mathcal{O}(h^{\nu+1})$ with $\nu \le r$, then we have

$$\begin{pmatrix}a\\b\end{pmatrix} = \begin{pmatrix}p_0\\q_0\end{pmatrix} + \mathcal{O}(h^{\nu+1}), \qquad \begin{pmatrix}a_l\\b_l\end{pmatrix} = \mathcal{O}(h^{\nu+1}) \qquad \text{for all } l.$$

For zero-stable multistep methods, all roots of $\det R(\omega h, \zeta) = 0$ can be bounded by $|\zeta_l(\omega h)| \le 1 + \gamma \omega h$ (here $\gamma > 0$ and $\omega > 0$). This implies that $|\zeta_l(\omega h)^n| \le e^{\gamma \omega T}$ for $nh \le T$, and the parasitic solution components remain small of size $\mathcal{O}(h^{\nu+1})$ on intervals of fixed length. To have a similar estimate on arbitrarily long intervals, the roots $\zeta_l(\omega h)$ have to be bounded by 1.

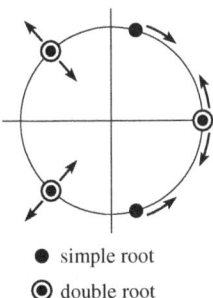

● simple root
◉ double root

In general, we do not have a control on the modulus of ζ_l. However, for symmetric methods we know that with ζ_l not only the complex conjugate $\overline{\zeta}_l$, but also the inverse ζ_l^{-1} are roots of $\det R(\omega h, \zeta) = 0$. Furthermore, the roots $\zeta_l(\omega h)$

depend continuously on its argument. If $\zeta_l(0)$ is a double root of $\det R(0, \zeta) = 0$, then it is possible that it splits for $\omega h > 0$ into a pair of roots, one of which has modulus larger than 1, and one smaller than 1 (see the figure). If $\zeta_l(0)$ is a simple root, then we must have $\overline{\zeta}_l(\omega h) = \zeta_l(\omega h)^{-1}$, implying $|\zeta_l(\omega h)| = 1$ for sufficiently small $\omega h > 0$.

Consequently, if apart from the double root at 1, all roots of $\det R(0, \zeta) = 0$ are simple (i.e., with the exception of 1, all zeros of $\rho_p(\zeta)$ are different from those of $\rho_q(\zeta)$), the parasitic solution components remain bounded of size $\mathcal{O}(h^{\nu+1})$ independent of the length of the integration interval.

Linear Change of Coordinates. Partitioned linear multistep methods are invariant with respect to linear transformations of the form $\tilde{p} = T_p \, p$, $\tilde{q} = T_q \, q$. However, care has to be taken when p and q components are mixed. Suppose, for example, that after such a transformation the harmonic oscillator reduces to a Hamiltonian system with (we put $\omega = 1$ for convenience)

$$H(p,q) = \frac{1}{2}(p^2 + 2\varepsilon \, p\, q + q^2),$$

where $\varepsilon \neq 0$ is a small parameter. An application of the partitioned multistep method yields the difference equation

$$\begin{aligned}\rho_p(E)\, p_n &= -h\left(\varepsilon \, \sigma_p(E)\, p_n + \sigma_p(E)\, q_n\right),\\ \rho_q(E)\, q_n &= h\left(\sigma_q(E)\, p_n + \varepsilon \, \sigma_q(E)\, q_n\right).\end{aligned} \quad (17)$$

Instead to (13) we are led this time to the system

$$R(h,\zeta)\begin{pmatrix}a\\b\end{pmatrix} = 0 \quad \text{with} \quad R(h,\zeta) = \begin{pmatrix}\rho_p(\zeta)+\varepsilon h\sigma_p(\zeta) & h\,\sigma_p(\zeta)\\ -h\,\sigma_q(\zeta) & \rho_q(\zeta)-\varepsilon h\sigma_q(\zeta)\end{pmatrix}.$$

Even if we only consider symmetric partitioned linear multistep methods, the coefficients of the polynomial $\det R(h, \zeta)$ are no longer symmetric, so that the modulus of its zeros is in general not equal to one. A straightforward computation shows that for simple roots of $R(0, \zeta) = 0$ (for example if we have $\rho_p(\zeta_l) = 0$ but $\rho_q(\zeta_l) \neq 0$), the continuous continuation satisfies

$$\zeta_l(h) = \zeta_l\bigl(1 - \mu_l \varepsilon h + \mathcal{O}(h^2)\bigr), \qquad \mu_l = \frac{\sigma_p(\zeta_l)}{\zeta_l \, \rho'_p(\zeta_l)}.$$

From the symmetry of the method it follows that μ_l is a real number. It is called growth parameter. We conclude from this asymptotic formula that $|\zeta_l(h)| > 1$ for small h, if the product $\mu_l \varepsilon$ is negative. In such a situation parasitic solution components grow exponentially with time, and the numerical solution becomes meaningless on integration intervals whose length T is such that $h^{\nu+1} e^{-\mu_l \varepsilon T} \geq 1$.

2.2 Backward Error Analysis (Smooth Numerical Solution)

An important tool for the study of the long-time behavior of numerical approximations is "backward error analysis". The idea is to interpret the numerical solution of a one-step method as the exact solution of a modified differential equation (for details see Chap. IX of [12]). For linear multistep methods, it is in principle possible to construct the underlying one-step method as a formal series in powers of the step size h, and then to apply the well-established techniques. Here, we follow the approach of [7,9], where the modified differential equation is directly obtained from the multistep schemes without passing explicitly through the underlying one-step method.

Theorem 1 (modified differential equation). *Consider a consistent, partitioned linear multistep method (2), applied to a partitioned system (1). There then exist h-independent functions $f_j(p,q)$, $g_j(p,q)$, such that for every truncation index N every solution $p_h(t), q_h(t)$ of the system*

$$\dot{p} = f(p,q) + hf_1(p,q) + \ldots + h^{N-1}f_{N-1}(p,q)$$
$$\dot{q} = g(p,q) + hg_1(p,q) + \ldots + h^{N-1}g_{N-1}(p,q) \tag{18}$$

satisfies the multistep formula up to a defect of size $\mathcal{O}(h^{N+1})$, i.e.,

$$\sum_{j=0}^{k} \alpha_j^p p_h(t+jh) = h \sum_{j=0}^{k} \beta_j^p f\big(p_h(t+jh), q_h(t+jh)\big) + \mathcal{O}(h^{N+1})$$
$$\sum_{j=0}^{k} \alpha_j^q q_h(t+jh) = h \sum_{j=0}^{k} \beta_j^q g\big(p_h(t+jh), q_h(t+jh)\big) + \mathcal{O}(h^{N+1}). \tag{19}$$

The constant symbolized by $\mathcal{O}$ is independent of h, but depends on the truncation index N. It also depends smoothly on t. If the method is of order r, then we have $f_j(p,q) = g_j(p,q) = 0$ for $1 \leq j < r$.

Proof. We closely follow the proof for second order equations in [9]. Denoting time differentiation by D, the Taylor series expansion of a function can be written as $y(t+h) = e^{hD} y(t)$. Equation (19) thus becomes

$$\rho_p(e^{hD}) p_h(t) = h\, \sigma_p(e^{hD}) f\big(p_h(t), q_h(t)\big) + \mathcal{O}(h^{N+1})$$
$$\rho_q(e^{hD}) q_h(t) = h\, \sigma_q(e^{hD}) g\big(p_h(t), q_h(t)\big) + \mathcal{O}(h^{N+1}). \tag{20}$$

With the coefficients of the expansions

$$\frac{x\, \sigma_p(e^x)}{\rho_p(e^x)} = 1 + \mu_1^p x + \mu_2^p x^2 + \ldots, \qquad \frac{x\, \sigma_q(e^x)}{\rho_q(e^x)} = 1 + \mu_1^q x + \mu_2^q x^2 + \ldots, \tag{21}$$

this becomes equivalent to (omitting the argument t)

$$\dot{p}_h = \left(1 + \mu_1^p h D + \mu_2^p h^2 D^2 + \ldots\right) f(p_h, q_h) + \mathcal{O}(h^N)$$
$$\dot{q}_h = \left(1 + \mu_1^q h D + \mu_2^q h^2 D^2 + \ldots\right) g(p_h, q_h) + \mathcal{O}(h^N). \tag{22}$$

For a function $\Psi(p, q)$, we have

$$D\Psi(p_h, q_h) = \partial_p \Psi(p_h, q_h) f_h(p_h, q_h) + \partial_q \Psi(p_h, q_h) g_h(p_h, q_h),$$

where the functions $f_h(p,q)$ and $g_h(p,q)$ are an abbreviation for the right-hand side of (18). Applying this formula iteratively to the expressions in (22) and collecting equal powers of h, a comparison of (18) and (22) determines recursively the functions $f_j(p,q)$ and $g_j(p,q)$. □

The flow of the modified differential equation (18) depends on the parameter h. If we denote this flow by $\varphi_t^{[h]}(p,q)$, then the underlying one-step method of the partitioned linear multistep method is given by $\Phi_h(p,q) = \varphi_h^{[h]}(p,q)$ up to an error of size $\mathcal{O}(h^{N+1})$.

Corollary 1. *Assume that the partitioned linear multistep method is symmetric, i.e., both multistep schemes satisfy the symmetry relations (4). We then have:*

(a) *The expansion of the vector field of the modified differential equation (18) is in even powers of h.*
(b) *If the differential equation (1) is reversible, i.e., $f(-p,q) = f(p,q)$ and $g(-p,q) = -g(p,q)$, then the modified differential equation (18) is also reversible.*

Proof. The symmetry relations (4) imply that the expressions of (21) are even functions of x. This proves statement (a).

If $(f_h(p,q), g_h(p,q))$ is a reversible vector field, then the function $D^2\Psi(p,q)$ has the same parity in p as the function $\Psi(p,q)$. As a consequence of the recursive construction of the modified differential equation, and of the fact that only even powers of D appear in (22), this observation proves the statement (b). □

Theorem 1 tells us that the solution of the truncated modified differential equation (18) satisfies the multistep formulas up to a defect of size $\mathcal{O}(h^{N+1})$. Consequently, the classical analysis shows that on intervals of length $T = \mathcal{O}(1)$,

$$\|p_n - p_h(nh)\| + \|q_n - q_h(nh)\| \leq C(T) h^N.$$

2.3 Near Energy Preservation

Whereas the analysis of the previous Sect. 2.2 is valid for general partitioned differential equations, we assume here that the vector field is Hamiltonian and given by

$$f(p,q) = -\nabla_q H(p,q), \qquad g(p,q) = \nabla_p H(p,q). \tag{23}$$

In this situation the exact solution satisfies $H(p(t), q(t)) = \text{const}$, and it is of interest to study whether numerical approximations of partitioned linear multistep methods (nearly) preserve the energy $H(p,q)$ over long times. Recall that in this chapter we consider only "smooth" numerical solutions, which are given by the flow of the modified differential equation (18) up to an arbitrarily small error of size $\mathcal{O}(h^N)$. We therefore have to investigate the near preservation of $H(p_h(t), q_h(t))$.

The solution of the truncated modified equation satisfies (20). Instead of dividing by the ρ polynomial, which led us to the construction of the modified differential equation, we divide the relation by the σ polynomial. This leads to

$$\begin{aligned}(1 + \lambda_1^p h D + \lambda_2^p h^2 D^2 + \ldots)\dot{p}_h &= -\nabla_q H(p_h, q_h) + \mathcal{O}(h^N) \\ (1 + \lambda_1^q h D + \lambda_2^q h^2 D^2 + \ldots)\dot{q}_h &= \nabla_p H(p_h, q_h) + \mathcal{O}(h^N),\end{aligned} \tag{24}$$

where the coefficients in the expansion are given by

$$\frac{\rho_p(e^x)}{x\,\sigma_p(e^x)} = 1 + \lambda_1^p x + \lambda_2^p x^2 + \ldots, \qquad \frac{\rho_q(e^x)}{x\,\sigma_q(e^x)} = 1 + \lambda_1^q x + \lambda_2^q x^2 + \ldots. \tag{25}$$

For symmetric methods, we are concerned with even functions of x, so that the expansions in (24) are in even powers of h. In this situation we multiply the first relation of (24) with $\dot{q}_h$, the second one with $\dot{p}_h$, and we subtract both so that the right-hand side becomes a total differential. This yields

$$\dot{q}_h^T(1 + \lambda_2^p h^2 D^2 + \ldots)\dot{p}_h - \dot{p}_h^T(1 + \lambda_2^q h^2 D^2 + \ldots)\dot{q}_h + \frac{d}{dt} H(p_h, q_h) = \mathcal{O}(h^N). \tag{26}$$

The main ingredient for a further simplification is the fact that

$$\dot{q}_h^T p_h^{(2j+1)} - \dot{p}_h^T q_h^{(2j+1)} = \frac{d}{dt}\left(\sum_{l=1}^{2j}(-1)^{l+1} q_h^{(l)T} p_h^{(2j+1-l)}\right) \tag{27}$$

is also a total differential. We now distinguish the following situations:

Case A: Both Multistep Methods are Identical. This case has been treated in Sect. XV.4.3 of [12]. We have $\lambda_j^p = \lambda_j^q$ for all j, and it follows from (27) that the entire left-hand side of (26) is a total differential. Using the modified differential equation (18), first and higher derivatives of p_h and q_h can be substituted with expressions depending only on p_h and q_h. This proves the existence of functions $H_{2j}(p,q)$, such that after integration of (26)

$$H(p_h, q_h) + h^2 H_2(p_h, q_h) + h^4 H_4(p_h, q_h) + \ldots = \text{const} + \mathcal{O}(th^N). \tag{28}$$

As long as the solution of the modified differential equation (i.e., the numerical solution) remains in a compact set, we thus have $H(p_h, q_h) = \text{const} + \mathcal{O}(h^r) + \mathcal{O}(th^N)$, where r is the order of the method and N can be chosen arbitrarily large.

This is a nice result, but of limited interest. If the p and q components are discretized by the same multistep method, parasitic components are usually not under control and they destroy the long-time behavior of the underlying one-step method.

Case B: Separable Hamiltonian with Quadratic Kinetic Energy. This situation is treated in [9]. For a Hamiltonian of the form $H(p, q) = \frac{1}{2} p^\mathsf{T} M^{-1} p + U(q)$ (without loss of generality we assume $M = I = $ identity) we have $\nabla_p H(p, q) = p$. The second relation of (24) therefore permits to express p_h as a linear combination of odd derivatives of q_h. Inserted into (26), this gives rise to a linear combination of terms $q_h^{(m)\mathsf{T}} q_h^{(2j+1-m)}$, which all can be written as total differentials because of

$$2 q_h^{(m)\mathsf{T}} q_h^{(2j+1-m)} = \frac{\mathrm{d}}{\mathrm{d}t}\left(\sum_{l=m}^{2j-m} (-1)^{l-m} q_h^{(l)\mathsf{T}} q_h^{(2j-l)} \right). \tag{29}$$

Without any assumptions on the coefficients λ_j^p and λ_j^q, a modified Hamiltonian satisfying (28) can be obtained as in Case (A). This is an important result, because the parasitic components can be shown to remain bounded and small (see [9] and Chap. 3 below).

Case C: Additional Order Conditions. If both multistep schemes are of order r, then $\lambda_j^p = \lambda_j^q = 0$ holds for $1 \le j < r$. Can we construct schemes, where the polynomials $\rho_p(\zeta)$ and $\rho_q(\zeta)$ have no common zeros other than $\zeta = 1$, such that $\lambda_j^p = \lambda_j^q$ also for $j = r$ (and possibly also for larger j)?

The class of explicit, symmetric 3-step methods of order $r = 2$ is given by

$$\rho(\zeta) = (\zeta - 1)(\zeta^2 + 2a\zeta + 1), \qquad \sigma(\zeta) = (a + 1)(\zeta^2 + \zeta),$$

where $|a| < 1$ by stability (for $a = 1$ it is reducible and equivalent to the 2-step explicit midpoint rule). The coefficient λ_2 in the expansion (25) is $\lambda_2 = \frac{1}{2}\left(\frac{1}{a+1} - \frac{1}{6}\right)$, and it is not possible to have the same λ_2 for different values of a.

Symmetric 5-step methods of order $r = 4$ are given by

$$\rho(\zeta) = (\zeta - 1)(\zeta^2 + 2a_1\zeta + 1)(\zeta^2 + 2a_2\zeta + 1),$$

where $|a_1| < 1$ and $|a_2| < 1$ (one of these coefficients is allowed to be equal to 1, but then the method reduces to a 4-stage method). The polynomial $\sigma(\zeta)$ is uniquely determined by assuming the method to be explicit and of order 4. In this case, the coefficient

$$\lambda_4 = \frac{131 - 19(a_1 + a_2) + 11 a_1 a_2}{720 (1 + a_1)(1 + a_2)}$$

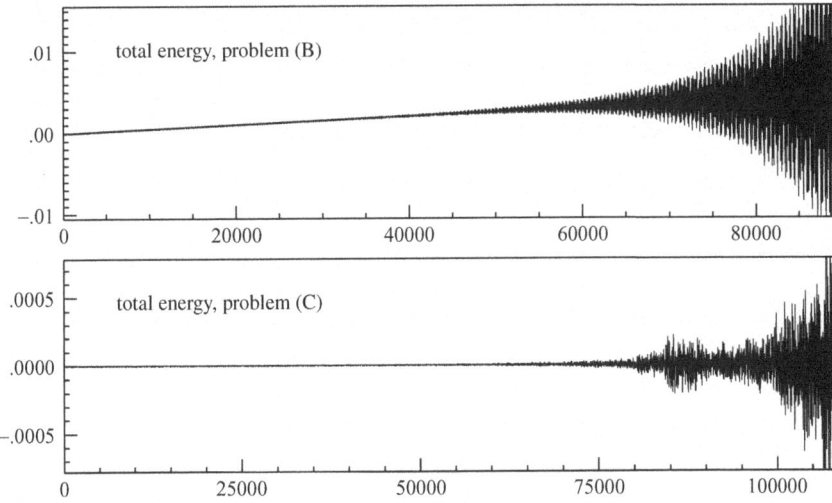

Fig. 4 Numerical Hamiltonian of method "plmm4c" applied with step size $h = 0.005$ for problem (B), and with $h = 0.001$ for problem (C); initial values and starting approximations as in Fig. 1

in (25) depends on two parameters, and it is possible to construct different methods with the same value of λ_4. This happens, for example, when the coefficients $a_j^p = \cos\theta_j^p$ and $a_j^q = \cos\theta_j^q$ for the two ρ polynomials are given by

$$\text{plmm4c} \quad \begin{aligned} \rho_p(\zeta) &: \theta_1^p = \pi/8 \quad \theta_2^p = 3\pi/4 \\ \rho_q(\zeta) &: \theta_1^q = 3\pi/8 \quad \theta_2^q \approx 0.68\,\pi \end{aligned}$$

(here, θ_1^p, θ_2^p, θ_1^q are arbitrarily fixed, and θ_2^q is computed to satisfy $\lambda_4^p = \lambda_4^q$). We apply this method to the three problems with separable Hamiltonian of Sect. 1.3. For problem (A) there is no difference to the behavior of methods plmm2 and plmm4. The error in the Hamiltonian is of size $\mathcal{O}(h^4)$ and no drift can be observed. Numerical results for problems (B) and (C) are presented in Fig. 4. For problem (B) we expect that the dominant error term in the Hamiltonian remains bounded. In fact, experiments with many different values of the step size h indicate that the error in the Hamiltonian is bounded by $\mathcal{O}(h^4) + \mathcal{O}(th^6)$ on intervals of length $\mathcal{O}(h^{-2})$. Similarly, also for problem (C) the dominant error term remains bounded. In this case we expect the error to behave like $\mathcal{O}(h^4) + \mathcal{O}(\sqrt{t}h^6)$. The second term is invisible on intervals of length $\mathcal{O}(h^{-2})$, see also Table 1. Beyond such an interval, Fig. 4 shows that for both problems, (B) and (C), the error behaves like $\delta \exp(ch^2 t)$ with a small constant δ. This undesirable exponential error growth will be explained by studying parasitic solution components in Chap. 3.

2.4 Near Preservation of Quadratic First Integrals

We again consider general differential equations (1) and we assume the existence of a quadratic first integral of the form $L(p,q) = p^\mathsf{T} E q$, i.e.,

$$f(p,q)^\mathsf{T} E q + p^\mathsf{T} E g(p,q) = 0 \quad \text{for all } p \text{ and } q. \tag{30}$$

The exact solution satisfies $L(p(t), q(t)) = \text{const}$, and we are interested to know if the numerical approximation can mimic this behavior. As in the previous section we consider only smooth numerical approximations, which are formally equal to the values of the solution $(p_h(t), q_h(t))$ at $t = nh$ of the modified differential equation. We therefore have to study the evolution of $L(p_h(t), q_h(t))$.

Dividing the relations in (20) by h times the σ polynomials, the solution of the modified differential equation is seen to verify

$$\begin{aligned}(1 + \lambda_1^p h D + \lambda_2^p h^2 D^2 + \ldots) \dot{p}_h &= f(p_h, q_h) + \mathcal{O}(h^N) \\ (1 + \lambda_1^q h D + \lambda_2^q h^2 D^2 + \ldots) \dot{q}_h &= g(p_h, q_h) + \mathcal{O}(h^N),\end{aligned} \tag{31}$$

where the coefficients λ_j^p and λ_j^q are given by (25). We restrict our considerations to symmetric methods, so that the series are in even powers of h. We multiply the transposed first relation of (31) with $E q_h$ from the right, and the second one with $p_h^\mathsf{T} E$ from the left, and we add both so that by (30) the right-hand side becomes an expression of size $\mathcal{O}(h^N)$. We thus obtain

$$\bigl((1 + \lambda_2^p h^2 D^2 + \ldots)\dot{p}_h\bigr)^\mathsf{T} E q_h + p_h^\mathsf{T} E (1 + \lambda_2^q h^2 D^2 + \ldots)\dot{q}_h = \mathcal{O}(h^N). \tag{32}$$

An important simplification can be achieved by using the identity

$$(p_h^{2j+1})^\mathsf{T} E q_h + p_h^\mathsf{T} E q_h^{(2j+1)} = \frac{\mathrm{d}}{\mathrm{d}t}\Bigl(\sum_{l=0}^{2j}(-1)^l (p_h^{(2j-l)})^\mathsf{T} E q_h^{(l)}\Bigr) \tag{33}$$

As in the previous section we now distinguish the following situations:

Case A: Both Multistep Methods are Identical. This is the case considered in Sect. XV.4.4 of [12]. We have $\lambda_j^p = \lambda_j^q$ for all j, and it follows from (33) that the expression in (32) is a total differential. As in Sect. 2.3, first and higher derivatives of p_h and q_h can be substituted with expressions depending only on p_h and q_h. Hence, there exist functions $L_{2j}(p,q)$ with $L_0(p,q) = L(p,q) = p^\mathsf{T} E q$, such that after integration of (32)

$$L(p_h, q_h) + h^2 L_2(p_h, q_h) + h^4 L_4(p_h, q_h) + \ldots = \text{const} + \mathcal{O}(th^N). \tag{34}$$

As long as the solution of the modified differential equation (i.e., the numerical solution) remains in a compact set, we thus have $L(p_h, q_h) = \text{const} + \mathcal{O}(h^r) + \mathcal{O}(th^N)$, where r is the order of the method and N can be chosen arbitrarily large.

Note that such a result is not true in general for symmetric one-step methods. However, it is of limited interest, because parasitic components are usually not under control for the situation, where both multistep methods are identical.

Case B: Special Form of the Differential Equation. We consider problems of the form

$$\dot{p} = f(q), \qquad \dot{q} = M^{-1} p,$$

which are equivalent to second order differential equations $\ddot{q} = M^{-1} f(q)$. This corresponds to the situation treated in [9]. Without loss of generality we assume in the following that $M = I$ = identity). For such special differential equations the condition (30) splits into two conditions

$$f(q)^\mathsf{T} E q = 0, \qquad p^\mathsf{T} E p = 0 \qquad \text{for all } p \text{ and } q,$$

which implies that E is a skew-symmetric matrix. Moreover, because of $g(p, q) = p$, the second relation of (31) permits to express p_h as a linear combination of odd derivatives of q_h. Inserted into (32), this gives rise to a linear combination of terms $q_h^{(2m+1)\mathsf{T}} E q_h^{(2j-2m+1)}$, which can be written as total differentials because

$$q_h^{(2m+1)\mathsf{T}} E q_h^{(2j-2m+1)} = \frac{d}{dt}\left(\sum_{l=2m+1}^{j} (-1)^{l-1} q_h^{(l)\mathsf{T}} E q_h^{(2j-l+1)} \right).$$

Without any assumptions on the coefficients λ_j^p and λ_j^q, a formal first integral of the form (34) is obtained that is $\mathcal{O}(h^r)$-close to the invariant $L(p, q) = p^\mathsf{T} E q$ of the differential equation. This result is important, because the parasitic components will be shown to remain bounded and small (see also [9]).

Case C: Additional Order Conditions. If the partitioned multistep method is of order r, we have $\lambda_j^p = \lambda_j^q = 0$ for $1 \le j < r$. If the coefficients of the method are constructed such that $\lambda_j^p = \lambda_j^q$ also for $j = r$, we can apply the computation of case (A) to the leading error term. In this way an improved near conservation of quadratic first integrals can be achieved, similar to the near energy conservation in the previous section.

2.5 Symplecticity and Conjugate Symplecticity

In the numerical solution of Hamiltonian systems it is unavoidable to speak also about symplecticity. Together with the differential equation whose flow we denote by $\varphi_t(p_0, q_0)$, we consider the variational differential equation

$$\dot{P} = -\nabla^2_{qp} H(p,q) P - \nabla^2_{qq} H(p,q) Q, \qquad (36)$$
$$\dot{Q} = \nabla^2_{pp} H(p,q) P + \nabla^2_{pq} H(p,q) Q,$$

where we use the notation $\nabla^2_{qp} H(p,q) = \left(\frac{\partial^2 H}{\partial q_i \partial p_j}\right)$. Here, $P(t)$ and $Q(t)$ are the derivatives with respect to initial values,

$$P(t) = \left(\frac{\partial p(t)}{\partial p_0}, \frac{\partial p(t)}{\partial q_0}\right), \quad Q(t) = \left(\frac{\partial q(t)}{\partial p_0}, \frac{\partial q(t)}{\partial q_0}\right) \quad \text{and} \quad \varphi'_t(p_0, q_0) = \begin{pmatrix} P(t) \\ Q(t) \end{pmatrix}.$$

The flow map $\varphi_t(p_0, q_0)$ of (35) is a symplectic transformation, see e.g., [12, VI.2]. This means, by definition, that its Jacobian matrix satisfies

$$\varphi'_t(p_0, q_0)^T J \, \varphi'_t(p_0, q_0) = J \quad \text{or equivalently} \quad P(t)^T Q(t) - Q(t)^T P(t) = J,$$

where J is the canonical structure matrix already encountered in (16). The important observation is that symplecticity just means that $P^T Q - Q^T P$ is a quadratic first integral of the combined system (35)–(36).

The smooth numerical solution of a partitioned multistep method is formally equal to the exact solution of the modified differential equation of Theorem 1. We therefore call the multistep method *symplectic*, if the derivative $(P_h(t), Q_h(t))$ (with respect to initial values) of the solution $(p_h(t), q_h(t))$ of the modified differential equation (18) satisfies

$$P_h(t)^T Q_h(t) - Q_h(t)^T P_h(t) = J.$$

Unfortunately, this is never satisfied unless for some trivial exceptions (implicit mid-point rule, symplectic Euler method, and the Störmer-Verlet scheme) which are partitioned linear multistep methods and one-step methods at the same time. Intuitively this is clear from the considerations of Sect. 2.4, because we did not encounter any result on the exact preservation of quadratic first integrals. A rigorous proof of this negative result has first been given by Tang [17] (see also [12, Sect. XV.4]).

In view of this negative result, it is natural to consider a weaker property than symplecticity, which nevertheless retains the same qualitative long-time behavior. We call a matrix-valued mapping $\Phi_h : (p,q) \mapsto (P, Q)$ *conjugate symplectic*, if there exists a global change of coordinates $(\hat{p}, \hat{q}) = \chi_h(p, q)$ that is $\mathcal{O}(h^r)$-close to the identity, such that the mapping is symplectic in the new coordinates, i.e., the mapping $\hat{\Phi}_h = \chi_h \circ \Phi_h \circ \chi_h^{-1}$ is a symplectic transformation. Since

$$\hat{\Phi}'_h(\hat{p},\hat{q}) = \Phi'_h(p,q) + h^r K_r(p,q) + h^{r+1} K_{r+1}(p,q) + \dots,$$

the symplecticity of $\hat{\Phi}_h$ yields the existence of functions $L_j(p,q)$ such that

$$\Phi'_h(p,q)^\mathsf{T} J \, \Phi'_h(p,q) + h^r L_r(p,q) + h^{r+1} L_{r+1}(p,q) + \dots = J. \qquad (37)$$

This means that for a method that is conjugate symplectic, there exists a modified first integral (as a formal series in powers of h) of the modified differential equation which is $\mathcal{O}(h^r)$-close to $P_h^\mathsf{T} Q_h - Q_h^\mathsf{T} P_h = (\Phi'_h)^\mathsf{T} J \, \Phi'_h$.

If Φ_h represents the underlying one-step method of a partitioned multistep method, we know from Sect. 2.4 that under suitable assumptions there exist functions $L_j(p,q)$ such that (37) holds. Does this imply that the method Φ_h is conjugate symplectic? That this is indeed the case follows from results of Chartier et al. [4], see also [12, Sect. XV.4.4]. We do not pursue this question in the present work.

3 Long-Term Stability of Parasitic Solution Components

We consider the partitioned linear multistep method (2) applied to the differential equation (1). We assume that both multistep methods are symmetric and stable, so that the zeros of the polynomials $\rho_p(\zeta)$ and $\rho_q(\zeta)$ are all on the unit circle. We denote these zeros by $\zeta_0 = 1$, and ζ_j, $\zeta_{-j} = \bar{\zeta}_j$ for $j = 1, \dots, \kappa$ (if -1 is such a zero, we let $\zeta_{-\kappa} = \zeta_\kappa = -1$). Furthermore, we consider finite products of the zeros of the ρ-polynomials, which we again denote by ζ_j and $\zeta_{-j} = \bar{\zeta}_j$. The resulting index set is denoted by $\mathscr{I}$, so that

$$\{\zeta_l\}_{l \in \mathscr{I}} = \{\zeta = \zeta_1^{m_1} \cdot \ldots \cdot \zeta_\kappa^{m_\kappa} \, ; \, m_j \in \{0, \pm 1, \pm 2, \dots\}\}.$$

The index set can be finite (if all zeros of the ρ-polynomials are roots of unity) or it can be infinite. It is convenient to denote $\mathscr{I}^* = \mathscr{I} \setminus \{0\}$.

Our aim is to write the numerical solution of (2) in the form

$$\begin{pmatrix} p_n \\ q_n \end{pmatrix} = \begin{pmatrix} p(t_n) \\ q(t_n) \end{pmatrix} + \sum_{l \in \mathscr{I}^*} \zeta_l^n \begin{pmatrix} u_l(t_n) \\ v_l(t_n) \end{pmatrix}, \qquad (38)$$

where $t_n = nh$. Here, $(p(t), q(t))$ is an h-dependent approximation to the exact solution of (1), called *principal solution component*. To avoid any confusion, we denote in this chapter the exact solution of (1) as $(p_{exact}(t), q_{exact}(t))$. The functions $(u_l(t), v_l(t))$ also depend on the step size h, and they are called *parasitic solution components*. This chapter is devoted to get bounds on these parasitic solution components and to investigate the length of time intervals, where the parasitic components do not significantly perturb the principal solution component.

A similar representation of the numerical solution has been encountered when discussing the numerical solution for the harmonic oscillator in Sect. 2.1. There, only zeros of the ρ-polynomials are present in the sum. The appearance of products of such zeros in (38) is due to the nonlinearity of the vector field in (1).

3.1 Modified Differential Equation (Full System)

We first study the existence of the coefficient functions in the representation (38). This is an extension of the backward error analysis of the smooth numerical solution as discussed in Sect. 2.2. It follows closely the presentation of [12, Sect. XV.3.2]. In the following we use the notations $y(t) = (p(t), q(t))$, $z_l(t) = (u_l(t), v_l(t))$, and we collect in the vector $\mathbf{z}(t)$ the components $u_l(t)$ ($l \neq 0$) for which $\rho_p(\zeta_l) = 0$ and the components $v_l(t)$ ($l \neq 0$) for which $\rho_q(\zeta_l) = 0$.

Theorem 2. *Consider a consistent, symmetric, partitioned linear multistep method (2), applied to the differential equation (1). Then, there exist h-independent functions $f_j(p, q, \mathbf{z})$, $g_j(p, q, \mathbf{z})$, and $f_{l,j}(p, q, \mathbf{z})$, $g_{l,j}(p, q, \mathbf{z})$, such that for an arbitrarily chosen truncation index N and for every solution $p(t), q(t), u_l(t), v_l(t)$ of the system*

$$
\begin{aligned}
\dot{p} &= f(p,q) + h f_1(p,q,\mathbf{z}) + \ldots + h^{N-1} f_{N-1}(p,q,\mathbf{z}) \\
\dot{q} &= g(p,q) + h g_1(p,q,\mathbf{z}) + \ldots + h^{N-1} g_{N-1}(p,q,\mathbf{z}) \\
\dot{u}_l &= f_{l,0}(p,q,\mathbf{z}) + h f_{l,1}(p,q,\mathbf{z}) + \ldots + h^{N-1} f_{l,N-1}(p,q,\mathbf{z}) \quad \text{if } \rho_p(\zeta_l) = 0 \\
\dot{v}_l &= g_{l,0}(p,q,\mathbf{z}) + h g_{l,1}(p,q,\mathbf{z}) + \ldots + h^{N-1} g_{l,N-1}(p,q,\mathbf{z}) \quad \text{if } \rho_q(\zeta_l) = 0 \\
u_l &= h f_{l,1}(p,q,\mathbf{z}) + \ldots + h^N f_{l,N}(p,q,\mathbf{z}) \quad \text{if } \rho_p(\zeta_l) \neq 0 \\
v_l &= h g_{l,1}(p,q,\mathbf{z}) + \ldots + h^N g_{l,N}(p,q,\mathbf{z}) \quad \text{if } \rho_q(\zeta_l) \neq 0 \\
u_l &= 0, \; v_l = 0 \quad \text{if } \zeta_l \neq \zeta_1^{m_1} \cdot \ldots \cdot \zeta_\kappa^{m_\kappa} \text{ with } |m_1| + \ldots + |m_\kappa| < N,
\end{aligned}
$$
(39)

with initial values $\mathbf{z}(0) = \mathcal{O}(h)$, the function (with $n = t/h$)

$$\begin{pmatrix} p_h(t) \\ q_h(t) \end{pmatrix} = \begin{pmatrix} p(t) \\ q(t) \end{pmatrix} + \sum_{l \in \mathscr{I}^*} \zeta_l^n \begin{pmatrix} u_l(t) \\ v_l(t) \end{pmatrix}, \tag{40}$$

satisfies the multistep formula up to a defect of size $\mathcal{O}(h^{N+1})$, i.e.,

$$
\begin{aligned}
\sum_{j=0}^{k} \alpha_j^p p_h(t+jh) &= h \sum_{j=0}^{k} \beta_j^p f\big(p_h(t+jh), q_h(t+jh)\big) + \mathcal{O}(h^{N+1}) \\
\sum_{j=0}^{k} \alpha_j^q q_h(t+jh) &= h \sum_{j=0}^{k} \beta_j^q g\big(p_h(t+jh), q_h(t+jh)\big) + \mathcal{O}(h^{N+1})
\end{aligned}
$$
(41)

as long as $(p(t), q(t))$ remain in a compact set, and $\|\mathbf{z}(t)\| \le Ch$. The constant symbolized by $\mathcal{O}$ is independent of h, but depends on the truncation index N. It also depends smoothly on t. If the partitioned multistep method is of order r, then we have $f_l(p, q) = g_l(p, q) = 0$ for $1 \le l < r$.

Remark 2. Because of the last line in (39), the sum in (40) is always finite. Substituting $\mathbf{z} = 0$ in the upper two equations of (39) yields the modified differential equation (18) of Sect. 2.2. The solution of the system (39) satisfies $u_{-l}(t) = \overline{u}_l(t)$, $v_{-l}(t) = \overline{v}_l(t)$, whenever these relations hold for the initial values.

Proof. The proof is very similar to that of Theorem 1, and we highlight here only the main differences. We insert the finite sum (40) into (41), we expand the nonlinearities around $(p(t), q(t))$, which we also denote by $(u_0(t), v_0(t))$, and we compare the coefficients of ζ_j^n. This yields, recalling that $y(t) = (p(t), q(t)) = (u_0(t), v_0(t))$ and $z_l(t) = (u_l(t), v_l(t))$, and omitting the argument t,

$$\rho_p(\zeta_l e^{hD}) u_l = h \sigma_p(\zeta_l e^{hD}) \sum_{m \ge 0} \frac{1}{m!} \sum_{\zeta_{l_1} \cdots \zeta_{l_m} = \zeta_l} f^{(m)}(y)(z_{l_1}, \ldots, z_{l_m}) + \mathcal{O}(h^{N+1}),$$

$$\rho_q(\zeta_l e^{hD}) v_l = h \sigma_q(\zeta_l e^{hD}) \sum_{m \ge 0} \frac{1}{m!} \sum_{\zeta_{l_1} \cdots \zeta_{l_m} = \zeta_l} g^{(m)}(y)(z_{l_1}, \ldots, z_{l_m}) + \mathcal{O}(h^{N+1}),$$

(42)

where the second sum is over indices $l_1 \ne 0, \ldots, l_m \ne 0$. The summand for $m = 0$, which is $f(y(t))$, resp. $g(y(t))$, is present only for $l = 0$, i.e., for $\zeta_l = 1$. Notice further that for $l = 0$ the summand for $m = 1$ vanishes, because we always have $\zeta_{l_1} \ne \zeta_0$. In view of an inversion of the operators $\rho_p(\zeta_l e^{hD})$ and $\rho_q(\zeta_l e^{hD})$ we introduce the coefficients of the expansions (cf. (21) for $\zeta_0 = 1$)

$$\frac{x \sigma_p(\zeta_l e^x)}{\rho_p(\zeta_l e^x)} = \mu_{l0}^p + \mu_{l1}^p x + \mu_{l2}^p x^2 + \ldots, \qquad \frac{x \sigma_q(\zeta_l e^x)}{\rho_q(\zeta_l e^x)} = \mu_{l0}^q + \mu_{l1}^q x + \mu_{l2}^q x^2 + \ldots$$

(43)

If $\rho_p(\zeta_l) \ne 0$, we have $\mu_{l0}^p = 0$. If $\rho_p(\zeta_l) = 0$, the expansion exists because ζ_l is a simple zero, and we have $\mu_{l0}^p \ne 0$ because $\sigma_p(\zeta_l) \ne 0$ as a consequence of the irreducibility of the method. The same statements hold for the second method. We therefore obtain the differential equations

$$\dot{u}_l = \left(\mu_{l0}^p + \mu_{l1}^p hD + \ldots \right) \sum_{m \ge 0} \frac{1}{m!} \sum_{\zeta_{l_1} \cdots \zeta_{l_m} = \zeta_l} f^{(m)}(y)(z_{l_1}, \ldots, z_{l_m}) + \mathcal{O}(h^N),$$
$$\text{if } \rho_p(\zeta_l) = 0,$$

$$\dot{v}_l = \left(\mu_{l0}^q + \mu_{l1}^q hD + \ldots \right) \sum_{m \ge 0} \frac{1}{m!} \sum_{\zeta_{l_1} \cdots \zeta_{l_m} = \zeta_l} g^{(m)}(y)(z_{l_1}, \ldots, z_{l_m}) + \mathcal{O}(h^N),$$
$$\text{if } \rho_q(\zeta_l) = 0,$$

(44)

and the algebraic relations

$$u_l = \left(\mu_{l1}^p hD + \mu_{l2}^p h^2 D^2 + \ldots\right) \sum_{m \geq 1} \frac{1}{m!} \sum_{\zeta_{l_1} \cdots \zeta_{l_m} = \zeta_l} f^{(m)}(y)(z_{l_1}, \ldots, z_{l_m}) + \mathcal{O}(h^{N+1}),$$
$$\text{if } \rho_p(\zeta_l) \neq 0,$$

$$v_l = \left(\mu_{l1}^q hD + \mu_{l2}^q h^2 D^2 + \ldots\right) \sum_{m \geq 1} \frac{1}{m!} \sum_{\zeta_{l_1} \cdots \zeta_{l_m} = \zeta_l} g^{(m)}(y)(z_{l_1}, \ldots, z_{l_m}) + \mathcal{O}(h^{N+1}),$$
$$\text{if } \rho_q(\zeta_l) \neq 0.$$
(45)

As in the proof of Theorem 1 we use (44) to recursively eliminate first and higher derivatives of u_l if $\rho_p(\zeta_l) = 0$ and of v_l if $\rho_q(\zeta_l) = 0$. Similarly, we use (45) to recursively eliminate u_l and its derivatives if $\rho_p(\zeta_l) \neq 0$ and of v_l and its derivatives if $\rho_q(\zeta_l) \neq 0$. Collecting equal powers of h yields the functions $f_j(p, q, \mathbf{z})$, $g_j(p, q, \mathbf{z})$, and $f_{l,j}(p, q, \mathbf{z})$, $g_{l,j}(p, q, \mathbf{z})$.

If $\zeta_l \neq \zeta_1^{m_1} \cdots \zeta_K^{m_K}$ with $|m_1| + \ldots + |m_K| < N$, the right-hand side of (45) contains at least N factors of components of $\mathbf{z}$. By our assumption $\|\mathbf{z}(t)\| \leq Ch$, this implies $u_l = \mathcal{O}(h^{N+1})$ and $v_l = \mathcal{O}(h^{N+1})$, so that these functions can be included in the remainder term. This justifies the last line of (39) and concludes the proof of the theorem. □

Initial Values for the System (39). For an application of the multistep formula (2), starting approximations (p_j, q_j) for $j = 0, \ldots, k-1$ have to be provided. We assume that they satisfy (with $0 \leq \nu \leq r$)

$$p_j - p_{exact}(jh) = \mathcal{O}(h^{\nu+1}), \quad q_j - q_{exact}(jh) = \mathcal{O}(h^{\nu+1}), \quad j = 0, \ldots, k-1.$$
(46)

Initial values for the differential equation (39) have to be such that

$$\begin{pmatrix} p_j \\ q_j \end{pmatrix} = \begin{pmatrix} p(jh) \\ q(jh) \end{pmatrix} + \sum_{l \in \mathscr{I}^*} \zeta_l^j \begin{pmatrix} u_l(jh) \\ v_l(jh) \end{pmatrix}, \quad j = 0, \ldots, k-1. \quad (47)$$

The solution of (39) is uniquely determined by the initial values $y(0), \mathbf{z}(0)$ (for the notation of y and $\mathbf{z}$ see the beginning of Sect. 3.1), so that the system (47) can be written as $F(y(0), \mathbf{z}(0), h) = 0$. For $h = 0$, it represents a linear Vandermonde system for $y(0), \mathbf{z}(0)$, which gives a unique solution. The Implicit Function Theorem thus proves the local existence of a solution of $F(y(0), \mathbf{z}(0), h) = 0$ for sufficiently small step sizes h. Note that the initial values depend smoothly on h. Under the assumption (46) we have $p(0) = p_{exact}(0) + \mathcal{O}(h^{\nu+1})$, $q(0) = q_{exact}(0) + \mathcal{O}(h^{\nu+1})$, and $\mathbf{z}(0) = \mathcal{O}(h^{\nu+1})$.

3.2 Growth Parameters

Before attacking the question of bounding rigorously the parasitic solution components, we try to get a feeling of the solution of the system (39). This system is equivalent to (44) and (45). Our aim is to have small parasitic solution components. We therefore neglect all terms that are at least quadratic in **z**.

Equation (44) for $l = 0$ (principal solution components) become equivalent to the modified equation already studied in Chap. 2. If we consider only the leading (h-independent) term in the expansion (45), we get zero functions. All that remains are (44) with $l \neq 0$ which, for $h = 0$, are as follows:

- if ζ_l is a common zero of $\rho_p(\zeta)$ and $\rho_q(\zeta)$, we have

$$\dot{u}_l = \mu_{l0}^p \big(f_p(p(t), q(t)) u_l + f_q(p(t), q(t)) v_l \big)$$
$$\dot{v}_l = \mu_{l0}^q \big(g_p(p(t), q(t)) u_l + g_q(p(t), q(t)) v_l \big), \tag{48}$$

- if ζ_l is a zero of $\rho_p(\zeta)$, but $\rho_q(\zeta_l) \neq 0$, we have

$$\dot{u}_l = \mu_{l0}^p f_p(p(t), q(t)) u_l, \tag{49}$$

- if ζ_l is a zero of $\rho_q(\zeta)$, but $\rho_p(\zeta_l) \neq 0$, we have

$$\dot{v}_l = \mu_{l0}^q g_q(p(t), q(t)) v_l. \tag{50}$$

The coefficient $\mu_l = \mu_{l0}$ is called *growth parameter* of a multistep method with generating polynomials $\rho(\zeta)$ and $\sigma(\zeta)$. It is defined by (43) for the limit $x \to 0$, and can be computed from

$$\mu_l = \frac{\sigma(\zeta_l)}{\zeta_l \rho'(\zeta_l)}.$$

We remark that for a symmetric linear multistep method the growth parameter is always real. This follows from $\sigma(1/\zeta_l) = \zeta_l^k \sigma(\zeta_l)$ and $-\zeta_l^{-2} \rho'(\zeta_l) = \zeta_l^k \rho'(\zeta_l)$, which is obtained by differentiation of the relation $\rho(1/\zeta) = \zeta^k \rho(\zeta)$.

Already when we use for $(p(t), q(t))$ the exact solution of the original problem (48)–(50) give much insight into the behavior of the multistep method. For example, if we consider the harmonic oscillator, for which $f(p, q) = -q$, $g(p, q) = p$, the differential equation (48) gives bounded solutions only if the product of the growth parameters of both methods satisfy $\mu_l^p \mu_l^q > 0$ for all l. For nonlinear problems, the differential equation (48) has bounded solutions only in very exceptional cases.

If the polynomials $\rho_p(\zeta)$ and $\rho_q(\zeta)$ do not have common zeros with the exception of $\zeta_0 = 1$, the situation with (48) cannot arise. Therefore, only (49) and (50) are relevant. There are many interesting situations, where the solutions of these equations are bounded, e.g., if $f(p,q)$ only depends on q and $g(p,q)$ only depends on p, what is the case for Hamiltonian systems with separable Hamiltonian.

3.3 Bounds for the Parasitic Solution Components

We study the system (39) of modified differential equations. We continue to use the notation $y = (p,q)$ and, as in Sect. 3.1, we denote by $\mathbf{z}(t)$ the vector whose components are $u_l(t)$ ($l \neq 0$) for which $\rho_p(\zeta_l) = 0$ and $v_l(t)$ ($l \neq 0$) for which $\rho_q(\zeta_l) = 0$. The system (39) can then be written in compact notations as

$$\begin{aligned} \dot{y} &= F_{h,N}(y) + G_{h,N}(y, \mathbf{z}) \\ \dot{\mathbf{z}} &= A_{h,N}(y)\,\mathbf{z} + B_{h,N}(y, \mathbf{z}), \end{aligned} \qquad (51)$$

where $G_{h,N}(y, \mathbf{z})$ and $B_{h,N}(y, \mathbf{z})$ collect those terms that are quadratic or of higher order in $\mathbf{z}$. Note that, by the construction via the system (44), the differential equation for y does not contain any linear term in $\mathbf{z}$.

We consider a compact subset K_0 of the $y = (p, q)$ phase space, and for a small positive parameter δ we define

$$K = \{(y, \mathbf{z})\,;\, y \in K_0, \|\mathbf{z}\| \le \delta\}. \qquad (52)$$

Regularity of the (original) differential equation implies that there exists a constant L such that

$$\|G_{h,N}(y, \mathbf{z})\| \le L\,\|\mathbf{z}\|^2, \qquad \|B_{h,N}(y, \mathbf{z})\| \le L\,\|\mathbf{z}\|^2 \qquad \text{for} \quad (y, \mathbf{z}) \in K. \qquad (53)$$

Our aim is to get bounds on the parasitic solution components $\mathbf{z}(t)$, which then allow to get information on the long-time behavior of partitioned linear multistep methods. To this end, we consider the simplified system

$$\begin{aligned} \dot{y} &= F_{h,N}(y), \\ \dot{\mathbf{z}} &= A_{h,N}(y)\,\mathbf{z}, \end{aligned} \qquad (54)$$

where quadratic and higher order terms of $\mathbf{z}$ have been removed from (51). The differential equation for y is precisely the modified differential equation for the smooth numerical solution (Sect. 2.2). The differential equation for $\mathbf{z}$ is linear with coefficients depending on time t through the solution $y(t)$. Its dominant h-independent term is the differential equation studied in Sect. 3.2.

In the case of linear multistep methods for second order Hamiltonian systems, a formal invariant of the full system (51) has been found that is close to $\|\mathbf{z}\|$ (see [9] or [12, Sect. XV.5.3]; the ideas are closely connected to the study of adiabatic invariants in highly oscillatory differential equations [8]). This was the key for getting bounds of the parasitic solution components on time intervals that are much longer than the natural time scale of the system (54). Here, we include the existence of such a formal invariant in an assumption ("S" for stability and "I" for invariant), and we later discuss situations, where it is satisfied.

Stability Assumption (SI). *We say that a partitioned linear multistep method (2) applied to a partitioned differential equation (1) satisfies the stability assumption (SI), if there exists a smooth function $I_{h,N}(y, \mathbf{z})$ such that, for $0 < h \leq h_0$,*

- *the invariance property*

$$I_{h,N}(y(h), \mathbf{z}(h)) = I_{h,N}(y(0), \mathbf{z}(0)) + \mathcal{O}(h^{M+1}\|z(0)\|^2)$$

holds for solutions of the differential equation (54), for which $(y(t), \mathbf{z}(t)) \in K$ for t in the interval $0 \leq t \leq h$;
- *there exists a constant $C \geq 1$, such that*

$$I_{h,N}(y, \mathbf{z}) \leq \|\mathbf{z}\|^2 \leq C\, I_{h,N}(y, \mathbf{z}) \quad \text{for} \quad (y, \mathbf{z}) \in K.$$

We are interested in situations, where the stability assumption (SI) is satisfied with $M > 0$, and we obviously focus on situations which admit a large M.

Lemma 1. *Under the stability assumption (SI) we have, for $0 < h \leq h_0$,*

$$I_{h,N}(y(h), \mathbf{z}(h)) = I_{h,N}(y(0), \mathbf{z}(0)) + \mathcal{O}(h^{M+1}\|\mathbf{z}(0)\|^2) + \mathcal{O}(h\,\delta\,\|\mathbf{z}(0)\|^2)$$

along solutions of the complete system (51) of modified differential equations, provided that they stay in the compact set K for $0 \leq t \leq h$.

Proof. The defect of the solution $(y(t), \mathbf{z}(t))$ of (51), when inserted into (54), is bounded by $\mathcal{O}(\|\mathbf{z}(0)\|^2)$. An application of the Gronwall Lemma therefore proves that the difference of the solutions of the two systems with identical initial values is bounded by $\mathcal{O}(h\|\mathbf{z}(0)\|^2)$. The statement then follows from the mean value theorem applied to the function $I_{h,N}(y, \mathbf{z})$ and from the fact that the derivative still contains a factor of $\mathbf{z}$. $\square$

We are now able to state and prove the main result of this chapter. It tells us the length of the integration interval, on which the parasitic solution components do not destroy the long-time behavior of the underlying one-step method.

Theorem 3. *In addition to the stability assumption (SI) we require that*

(A1) the partitioned linear multistep method (2) is symmetric, of order r, and the generating polynomials $\rho_p(\zeta)$ and $\rho_q(\zeta)$ do not have common zeros with the exception of $\zeta = 1$;

(A2) the vector field of (1) is defined and analytic in an open neighborhood of a compact set K_1;

(A3) the numerical solution $y_n = (p_n, q_n)$ stays for all n with $0 \le nh \le T_0$ in a compact set $K_0 \subset K_1$ which has positive distance from the boundary of K_1;

(A4) the starting approximations (p_j, q_j), $j = 0, \ldots, k-1$ are such that the initial values for the full modified differential equation (51) satisfy $y(0) \in K_0$, and $\|\mathbf{z}(0)\| \le \delta/\sqrt{2eC}$ with C from the stability assumption (SI) and $\delta = \mathcal{O}(h)$.

For sufficiently small h and δ and for a fixed truncation index N, chosen large enough such that $h^N \le \max(h^M \delta, \delta^2)$, there exist constants c_1, c_2 and functions $y(t), z_l(t)$ on an interval of length

$$T = \min(T_0, c_1 \delta^{-1}, c_2 h^{-M}), \tag{55}$$

such that

- *the numerical solution satisfies $y_n = y(nh) + \sum_{l \in \mathscr{I}^*} \zeta_l^n z_l(nh)$ for $0 \le nh \le T$;*
- *on every subinterval $[mh, (m+1)h)$, the functions $y(t), z_l(t)$ are a solution of the system (51);*
- *at the time instants $t_m = mh$ the functions $y(t), z_l(t)$ have jump discontinuities of size $O(h^{N+1})$;*
- *the parasitic solution components are bounded: $\|\mathbf{z}(t)\| \le \delta$ for $0 \le nh \le T$.*

Proof. The proof closely follows that of Theorem 8 in the publication [9], see also [12, Sect. XV.5.3]. We separate the integration interval into subintervals of length h. On a subinterval $[mh, (m+1)h)$ we define the functions $y(t) = (p(t), q(t))$ and $z_l(t) = (u_l(t), v_l(t))$ as the solution of the system (39) with initial values such that (47) holds with $j = m, m+1, \ldots, m+k-1$. It follows from Theorem 2, formula (41), that $y_{m+k} - y(t_{m+k}) = \mathcal{O}(h^{N+1})$. Consequently, the construction of initial values for the next subinterval $[(m+1)h, (m+2)h)$ yields for the functions $y(t)$ and $z_l(t)$ a jump discontinuity at t_{m+1} that is bounded by $\mathcal{O}(h^{N+1})$.

We now study how well the expression $I_{h,N}(y(t), \mathbf{z}(t))$ is preserved on long time intervals. Lemma 1 gives a bound on the maximal deviation within a subinterval of length h. Together with the $\mathcal{O}(h^{N+1})$ bound on the jump discontinuities at t_m this proves for $I_m = I_{h,N}(y(t_m), \mathbf{z}(t_m))$ the estimate

$$I_{m+1} = I_m(1 + C_1 h^{M+1} + C_2 h \delta) + C_3 h^{N+1} \delta$$

as long as $(y(t), \mathbf{z}(t))$ remains in K. With $\gamma = C_1 h^M + C_2 \delta$ the discrete Gronwall Lemma thus yields

$$I_m = I_0 (1 + \gamma h)^m + \frac{(1 + \gamma h)^m - 1}{\gamma h} C_3 h^{N+1} \delta,$$

which, for $\gamma t_m \le 1$, gives the estimate $I_m \le I_0 e + C_3 (e-1) h^N \delta t_m$. This implies

$$\|\mathbf{z}(t)\|^2 \le C e \|\mathbf{z}(0)\|^2 + C_4 h^N \delta t,$$

so that $\|z(t)\| \leq \delta$ for times t subject to $\gamma t \leq 1$, if the truncation index N is chosen sufficiently large. □

It is straight-forward to construct partitioned linear multistep methods of high order satisfying (A1). The assumption (A2) is satisfied for many important differential equations. The assumption (A3) can be checked a posteriori. If the method is of order r and if the starting approximations are computed with very high precision, then assumption (A4) is fulfilled with $\delta = \mathcal{O}(h^{r+1})$. This follows from the construction of the initial values for the system (39) as explained in the end of Sect. 3.1. The difficult task is the verification of the stability assumption (SI).

3.4 Near Energy Conservation

Combining our results on the long-time behavior of smooth numerical solutions with the bounded-ness of parasitic solution components we obtain the desired statements on the preservation of energy and of quadratic first integrals.

The near energy preservation has been studied analytically in Sect. 2.3 for smooth numerical solutions of symmetric partitioned multistep methods. We consider methods which, when applied to Hamiltonian systems, have a modified energy

$$H_h(p,q) = H(p,q) + h^r H_r(p,q) + \ldots + h^{N-1} H_{N-1}(p,q), \quad (56)$$

where r is the order of the method and $N > r$, such that

$$H_h(p_h, q_h) = \text{const} + \mathcal{O}(th^N) \quad (57)$$

along solutions of the modified differential equation (18). There are situations (cases (A) and (B) of Sect. 2.3), where N is arbitrarily large. This is the best behavior we can hope for. In the case (C) of Sect. 2.3 we achieve $N = r + 2$. The worst behavior is when $N = r$, in which case a linear drift for the numerical Hamiltonian is present from the beginning. This behavior of smooth numerical solutions carries over to the general situations as follows:

Theorem 4. *Consider a partitioned linear multistep method (2) of order r, applied to a Hamiltonian system (23). Assume that there exists a modified energy (56) such that (57) holds for smooth numerical solutions.*

Under the assumptions of Theorem 3 with $\delta = \mathcal{O}(h^r)$, the numerical solution satisfies

$$H(p_n, q_n) = \text{const} + \mathcal{O}(h^r) \quad \text{for} \quad nh \leq T,$$

where the length of the time interval T is limited by (55) and by $T \leq \mathcal{O}(h^{r-N})$.

Proof. Let $y(t) = (p(t), q(t))$ and $z_l(t)$ (for $t_m \le t \le t_{m+1}$, $t_m = mh$) be a solution of the complete system (51) as in the statement of Theorem 3. Applying the proof of Lemma 1 to the near invariant $H_h(p,q)$ yields

$$H_h\big(p(t_{m+1}), q(t_{m+1})\big) = H_h\big(p(t_m), q(t_m)\big) + \mathcal{O}(h\delta^2) + \mathcal{O}(h^{N+1}).$$

Since the jump discontinuities at the grid points t_m can be neglected, we obtain by following the proof of Theorem 3 that

$$H_h(p_n, q_n) = H_h(p_0, q_0) + \mathcal{O}(t_n \delta^2) + \mathcal{O}(t_n h^N),$$

so that the statement follows from (56) and the requirement $\delta = \mathcal{O}(h^r)$. □

Analogous statements are obtained for the near conservation of quadratic first integrals. In this case the results of Sect. 2.4 have to be combined with the boundedness of the parasitic solution components (Theorem 3).

3.5 Verification of the Stability Assumption (SI)

It remains to study the stability assumption (SI), and to investigate how large the number M in the invariance property can be. The nice feature is that we only have to consider the simplified system (54), where the subsystem for the principle solution component y is separated from the parasitic solution components. Therefore, the differential equation for z is a linear differential equation with coefficients depending on t via the principle solution $y(t)$. Another nice feature is that we are concerned only with a local result (estimates on an interval of length h which is the step size of the integrator).

The linear system $\dot{z} = A_{h,N}(y(t))z$ is obtained from (42), where terms are neglected that are either at least quadratic in z or contain a sufficiently high power of h. We consider $\zeta_l \ne 1$ satisfying $\rho_p(\zeta_l) = 0$ and $\rho_q(\zeta_l) \ne 0$. By irreducibility of the method we then have $\sigma_p(\zeta_l) \ne 0$. For ease of presentation, we assume[3] that also $\sigma_q(\zeta_l) \ne 0$. We then can apply the inverse of the operators $\sigma_p(\zeta_l e^{hD})$ and $\sigma_q(\zeta_l e^{hD})$ to both sides of (42) and thus obtain

$$\begin{aligned}\left(\frac{\rho_p}{\sigma_p}\right)(\zeta_l e^{hD})u_l &= h \sum_{m \ge 1} \frac{1}{m!} \sum_{\zeta_{l_1} \cdots \zeta_{l_m} = \zeta_l} f^{(m)}(y)(z_{l_1}, \ldots, z_{l_m}) + \mathcal{O}(h^{N+1}), \\ \left(\frac{\rho_q}{\sigma_q}\right)(\zeta_l e^{hD})v_l &= h \sum_{m \ge 1} \frac{1}{m!} \sum_{\zeta_{l_1} \cdots \zeta_{l_m} = \zeta_l} g^{(m)}(y)(z_{l_1}, \ldots, z_{l_m}) + \mathcal{O}(h^{N+1}).\end{aligned} \quad (58)$$

---

[3]The case $\sigma_q(\zeta_l) = 0$ needs special attention, see the end of Sect. 3.5 or [9] for the special case of second order differential equations.

Expanding the left-hand side into powers of h leads to the consideration of the series

$$\mathrm{i}\frac{\rho_p(\zeta_l\,\mathrm{e}^{\mathrm{i}x})}{\sigma_p(\zeta_l\,\mathrm{e}^{\mathrm{i}x})} = \lambda^p_{l0}+\lambda^p_{l1}x+\lambda^p_{l2}x^2+\dots, \qquad \mathrm{i}\frac{\rho_q(\zeta_l\,\mathrm{e}^{\mathrm{i}x})}{\sigma_q(\zeta_l\,\mathrm{e}^{\mathrm{i}x})} = \lambda^q_{l0}+\lambda^q_{l1}x+\lambda^q_{l2}x^2+\dots$$

(note that $\lambda^p_{l0} = 0$ if $\rho_p(\zeta_l) = 0$). The symmetry of the methods implies that the coefficients λ^p_{lj} and λ^q_{lj} are real. For the conjugate root $\zeta_{-l} = \overline{\zeta_l}$ we have

$$\lambda^p_{-l,j} = (-1)^{j+1}\lambda^p_{l,j}, \qquad \lambda^q_{-l,j} = (-1)^{j+1}\lambda^q_{l,j}. \tag{59}$$

Removing in (58) the terms with $m \geq 2$, we thus obtain

$$\dots + \lambda^p_{l2}(-\mathrm{i}h)^2\,\ddot{u}_l + \lambda^p_{l1}(-\mathrm{i}h)\,\dot{u}_l = \mathrm{i}h\left(f_p(p,q)\,u_l + f_q(p,q)\,v_l\right)$$
$$\dots + \lambda^q_{l2}(-\mathrm{i}h)^2\,\ddot{v}_l + \lambda^q_{l1}(-\mathrm{i}h)\,\dot{v}_l + \lambda^q_{l0}\,v_l = \mathrm{i}h\left(g_p(p,q)\,u_l + g_q(p,q)\,v_l\right) \tag{60}$$

and the same relations for l replaced by $-l$. An important ingredient for a further study is the fact that

$$\Re\left(\overline{z}^\mathsf{T} z^{(2m+1)}\right) = \frac{1}{2}\frac{d}{dt}\left(\sum_{j=0}^{2m}(-1)^j\,(\overline{z}^{(j)})^\mathsf{T} z^{(2m-j)}\right)$$
$$\Im\left(\overline{z}^\mathsf{T} z^{(2m)}\right) = \frac{1}{2\mathrm{i}}\frac{d}{dt}\left(\sum_{j=0}^{2m-1}(-1)^j\,(\overline{z}^{(j)})^\mathsf{T} z^{(2m-j-1)}\right) \tag{61}$$

are total differentials. We first put the main result of [9] on the long-time behavior of parasitic solution components into the context of the present investigation.

Second Order Hamiltonian Systems. We consider partitioned systems

$$\dot{p} = -\nabla U(q), \qquad \dot{q} = p,$$

which are equivalent to second order differential equations $\ddot{q} = -\nabla U(q)$. In this case we have $g_q(p,q) = 0$ and $g_p(p,q) = I$, so that from the lower line of (60) the expression $\mathrm{i}hu_l$ is seen to be a linear combination of derivatives of v_l. Inserted into the upper relation of (60) this gives

$$\dots - \lambda_{l3}(-\mathrm{i}h)^2\,v_l^{(3)} - \lambda_{l2}(-\mathrm{i}h)\,\ddot{v}_l - \lambda_{l1}\,\dot{v}_l = -\mathrm{i}h\,\nabla^2 U(q)\,v_l, \tag{62}$$

where $\lambda_{l1} = \lambda^p_{l1}\lambda^q_{l0}$, $\lambda_{l2} = \lambda^p_{l2}\lambda^q_{l0}+\lambda^p_{l1}\lambda^q_{l1}$, etc. are real coefficients. It follows from the symmetry of the Hessian matrix $\nabla^2 U(q)$ that $\Im(\overline{v}_l^\mathsf{T}\nabla^2 U(q)\,v_l) = 0$. Taking the scalar product of (62) with $\overline{v}_l^\mathsf{T}$ and considering its real part, we thus obtain

$$\dots + h^2\lambda_{l3}\,\Re(\overline{v}_l^\mathsf{T} v_l^{(3)}) - h\lambda_{l2}\,\Im(\overline{v}_l^\mathsf{T}\ddot{v}_l) - \lambda_{l1}\,\Re(\overline{v}_l^\mathsf{T}\dot{v}_l) = 0.$$

The magic formulas (61) show that the left-hand expression is a total differential. Its dominant term is the derivative of $-\lambda_{l1}\frac{1}{2}\|v_l\|^2$. The other terms are the derivative expressions containing higher derivatives of v_l. These can be eliminated with the help of the simplified modified differential equation. Because of $\lambda_{l1} \neq 0$, we thus get a formal invariant (a near invariant if the series is truncated) of the system (60), which is of the form

$$\ldots + h^2 I_{l2}(y,\mathbf{z}) + h\, I_{l1}(y,\mathbf{z}) + \|v_l\|^2 = I_l(y,\mathbf{z}).$$

Since all functions $I_{lj}(y,\mathbf{z})$ are bounded by a constant times $\|\mathbf{z}\|^2$ and since we obtain such a formal invariant for all components of $\mathbf{z}$, the stability assumption (SI) is proved with $C = 1 + \mathcal{O}(h)$ and for arbitrarily large M.

Remark 3. This derivation of a near invariant that is close to $\|v_l\|^2$ essentially relies on the fact that the polynomials $\rho_p(\zeta)$ and $\rho_q(\zeta)$ do not have common roots other than $\zeta = 1$. If, in addition to $\rho_p(\zeta_l) = 0$, also $\rho_q(\zeta_l) = 0$ would be satisfied, then the coefficient λ_{l0}^q would be zero. This would imply $\lambda_{l1} = 0$, so that the formal invariant does not contain the term $\|v_l\|^2$.

Separable Hamiltonian Systems. We next consider a Hamiltonian system with

$$H(p,q) = T(p) + U(q).$$

We still consider partitioned linear multistep methods (2), where the ρ-polynomials do not have common zeros with the exception of $\zeta = 1$. In the situation of (60) the vector v_l contains a factor h. Since $f_p(p,q) = 0$ for a separable Hamiltonian system, the differential equation for u_l contains an additional factor h. Consequently, the differential equation (54) for $\mathbf{z}$ is in fact of the form $\dot{\mathbf{z}} = h\, A_{h,N}^0(y)\mathbf{z}$. Therefore we have $\|\mathbf{z}(h)\| \leq \|\mathbf{z}(0)\|(1 + \mathcal{O}(h^2))$, so that the stability assumption (SI) is satisfied with $M = 1$.

Discussion of the Examples of Sect. 1.3. In the numerical experiments of Sect. 1.3 we have seen situations, where the parasitic solution components remain bounded on intervals of length $\mathcal{O}(h^{-2})$. According to our Theorem 3 this requires the stability assumption to be satisfied for $M = 2$. The system (60) is of the form

$$\begin{aligned}\lambda_{l1}^p \dot{u}_l &= \nabla^2 U(q) v_l + \mathcal{O}(h^2 \|\mathbf{z}\|) \\ \lambda_{l0}^q v_l &= ih\nabla^2 T(p) u_l + \mathcal{O}(h^2 \|\mathbf{z}\|)\end{aligned} \quad (63)$$

which yields the differential equation

$$\dot{u}_l = i\, h\, \lambda\, \nabla^2 U(q) \nabla^2 T(p)\, u_l + \mathcal{O}(h^2 \|\mathbf{z}\|)$$

with $\lambda = \lambda_{l0}^q / \lambda_{l1}^p$. If the product of the two Hessian matrices is symmetric or, equivalently, if their commutator vanishes, i.e.,

$$[\nabla^2 U(q), \nabla^2 T(p)] = \nabla^2 U(q)\nabla^2 T(p) - \nabla^2 T(p)\nabla^2 U(q) = 0, \quad (64)$$

we can multiply the differential equation with $\bar{u}_l^\mathsf{T}$ and we obtain

$$\|u_l(h)\|^2 = \|u_l(0)\|^2 + \mathcal{O}(h^3 \|\mathbf{z}(0)\|^2)$$

as a consequence of $\Im(\bar{u}_l^\mathsf{T} \nabla^2 U(q)\nabla^2 T(p) u_l) = 0$. This proves the validity of the stability assumption (SI) with $M = 2$. Unfortunately, the commutativity of the two Hessian matrices is a strong requirement and not often satisfied.

The examples (A) and (B) of Sect. 1.3 are separable Hamiltonian equations, which split into independent subsystems having one degree of freedom. The condition (64) is therefore trivially satisfied.

For the example (C) the condition (64) is not satisfied, so that we do not have better than $M = 1$ in the stability assumption (SI). Let us explain the behavior observed in Fig. 2. The parasitic roots of method "plmm2" are $\zeta_1 = i$, $\zeta_{-1} = -i$, and $\zeta_2 = -1$.

We have $\sigma_q(\zeta_2) = 0$, so that the division by $\sigma_q(\zeta_2 e^{hD})$ is not permitted in (58). We thus go back to formula (42), which shows that for $\rho_q(\zeta_l) \neq 0$ and $\sigma_q(\zeta_l) = 0$ the vector v_l is an expression multiplied by h^2. Inserted into the first equation of (63) we see that the right-hand side of the differential equation for u_2 contains the factor h^2, so that $\|u_2(h)\|^2 = \|u_2(0)\|^2 + \mathcal{O}(h^3 \|\mathbf{z}(0)\|^2)$.

For the root $\zeta_1 = i$ we study numerically the dominant term of the parasitic solution component. We have $\lambda_{l0}^p = -1$ and $\lambda_{l1}^q = 2$ for the method "plmm2", so that the differential equation for v_l becomes

$$\dot{v}_1 = -\frac{ih}{2} \nabla^2 T(p)\nabla^2 U(q)\, v_1 + \mathcal{O}(h^2 \|\mathbf{z}\|).$$

We neglect the $\mathcal{O}(h^2 \|\mathbf{z}\|)$ term and solve the linear differential equation for v_1 numerically with the code DOPRI5 of [11]. Since the problem is chaotic, care has to be taken about the credibility of the results. We therefore solve the problem with a high accuracy requirement of $tol = 10^{-12}$ and with many different initial values of norm $\|v_1(0)\| = 1$. The result is qualitatively the same for all runs, and we plot in Fig. 5 one such parasitic solution.

If the starting approximations for the partitioned multistep method are computed with high accuracy (what is the case for all our numerical experiments), the initial values of the parasitic solution components are of size $\mathcal{O}(h^{r+1})$ (where r denotes the order of the method). Consequently, the functions shown in Fig. 5 have to be scaled with a factor $\mathcal{O}(h^{r+1})$. A comparison with Fig. 2 shows that this solution, where we have removed quadratic and higher order terms in $\mathbf{z}$ as well a linear terms in $\mathbf{z}$ with a factor of at least h^2, cannot be the reason of the exponential divergence in Fig. 2. It must be a consequence of the next term having a factor h^2. This nicely explains why the parasitic solution components remain small and bounded on intervals of length $\mathcal{O}(h^{-2})$.

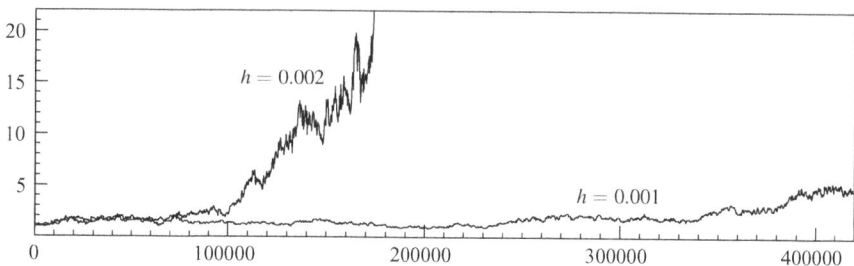

Fig. 5 Euclidean norm of the parasitic solution component v_1; data for the Hamiltonian system are as in Fig. 1, problem (C); initial data for the parasitic component are normalized to $\|v_1(0)\| = 1$

Conclusion

We have studied the long-time behavior of partitioned linear multistep methods applied to Hamiltonian systems. These are methods, where the momenta p and the positions q of the system are treated by two different multistep formula. It turns out that the following two properties are essential for a qualitative correct simulation over long times:

- both multistep schemes have to be symmetric;
- the generating polynomials $\rho_p(\zeta)$ and $\rho_q(\zeta)$ of the two methods are not allowed to have common zeros with the exception of $\zeta = 1$.

The study is motivated by the analysis of [9] for special multistep methods and Hamiltonian systems of the form $\ddot{q} = -\nabla U(q)$. We have extended the techniques of proof to a more general situation.

The positive insight of our investigation is that for problems having symmetries and a regular solution behavior, the numerical results concerning long-time preservation of energy and quadratic first integrals are excellent. This is remarkable, because the considered methods are explicit, of arbitrarily high order, and can be implemented very efficiently. We expect that this excellent long-time behavior is typical for all nearly integrable systems. A more thorough investigation of this question is outside the scope of the present work.

For separable Hamiltonian systems with chaotic solution, we observed that the "smooth" numerical solution behaves exactly like a symmetric (non-symplectic) one-step method. The parasitic solution components are typically bounded on a time interval of length $\mathcal{O}(h^{-2})$, but usually not on longer time intervals. This observation is independent of the order of the method.

Recently we have extended our numerical experiments and also the theoretical investigations to constrained Hamiltonian systems, which are differential-algebraic equations of index 3. Preliminary results are very encouraging and we expect to obtain a new efficient class of methods for such problems.

Acknowledgements We are grateful to Luca Dieci and Nicola Guglielmi, organizers of the CIME-EMS summer school on "Current challenges in stability issues for numerical differential equations", for including our research in their program. We acknowledge financial support from the Fonds National Suisse, Project No. 200020-126638 and 200021-129485.

References

1. M.J. Ablowitz, J.F. Ladik, A nonlinear difference scheme and inverse scattering. Stud. Appl. Math. **55**, 213–229 (1976)
2. G. Akrivis, M. Crouzeix, C. Makridakis, Implicit-explicit multistep methods for quasilinear parabolic equations. Numer. Math. **82**(4), 521–541 (1999)
3. C. Arévalo, G. Söderlind, Convergence of multistep discretizations of DAEs. BIT **35**(2), 143–168 (1995)
4. P. Chartier, E. Faou, A. Murua, An algebraic approach to invariant preserving integrators: The case of quadratic and Hamiltonian invariants. Numer. Math. **103**, 575–590 (2006)
5. G. Dahlquist, Stability and error bounds in the numerical integration of ordinary differential equations. *Transactions of the Royal Institute of Technology*, vol. 130, Stockholm (1959)
6. K. Feng, *Collected Works II* (National Defense Industry Press, Beijing, 1995)
7. E. Hairer, Backward error analysis for multistep methods. Numer. Math. **84**, 199–232 (1999)
8. E. Hairer, C. Lubich, Long-time energy conservation of numerical methods for oscillatory differential equations. SIAM J. Numer. Anal. **38**, 414–441 (2001)
9. E. Hairer, C. Lubich, Symmetric multistep methods over long times. Numer. Math. **97**, 699–723 (2004)
10. E. Hairer, C. Schober, Corrigendum to: Symplectic integrators for the Ablowitz-Ladik discrete nonlinear Schrödinger equation [Phys. Lett. A 259 (1999), 140–151]. Phys. Lett. A **272**(5–6), 421–422 (2000)
11. E. Hairer, S.P. Nørsett, G. Wanner, *Solving Ordinary Differential Equations I. Nonstiff Problems*. Springer Series in Computational Mathematics, vol. 8, 2nd edn. (Springer, Berlin, 1993)
12. E. Hairer, C. Lubich, G. Wanner, *Geometric Numerical Integration. Structure-Preserving Algorithms for Ordinary Differential Equations*. Springer Series in Computational Mathematics, vol. 31, 2nd edn. (Springer, Berlin, 2006)
13. E. Hairer, R. McLachlan, R. Skeel, On energy conservation of the simplified Takahashi–Imada method. ESAIM Math. Model. Numer. Anal. **43**(4), 631–644 (2009)
14. U. Kirchgraber, Multi-step methods are essentially one-step methods. Numer. Math. **48**, 85–90 (1986)
15. G.D. Quinlan, S. Tremaine, Symmetric multistep methods for the numerical integration of planetary orbits. Astron. J. **100**, 1694–1700 (1990)
16. C.M. Schober, Symplectic integrators for the Ablowitz-Ladik discrete nonlinear Schrödinger equation. Phys. Lett. A **259**, 140–151 (1999)
17. Y.-F. Tang, The symplecticity of multi-step methods. Comput. Math. Appl. **25**, 83–90 (1993)
18. D.S. Vlachos, T.E. Simos, Partitioned linear multistep method for long term integration of the N-body problem. Appl. Numer. Anal. Comput. Math. **1**(3), 540–546 (2004)

Markov Chain Monte Carlo and Numerical Differential Equations

J.M. Sanz-Serna

Abstract The aim of this contribution is to provide a readable account of Markov Chain Monte Carlo methods, with particular emphasis on their relations with the numerical integration of deterministic and stochastic differential equations. The exposition is largely based on numerical experiments and avoids mathematical technicalities. The presentation is largely self-contained and includes tutorial sections on stochastic processes, Markov chains, stochastic differential equations and Hamiltonian dynamics. The Metropolis Random-Walk algorithm, Metropolis adjusted Langevin algorithm and Hybrid Monte Carlo are discussed in detail, including some recent results.

1 Introduction

This contribution presents a—hopefully readable—introduction to Markov Chain Monte Carlo methods with particular emphasis on their combination with ideas from deterministic or stochastic numerical differential equations. Markov Chain Monte Carlo algorithms are widely used in many sciences, including physics, chemistry and statistics; their importance is comparable to those of the Gaussian Elimination or the Fast Fourier Transform. We have tried to keep the presentation as self-contained as it has been feasible. A basic knowledge of applied mathematics and probability[1] is assumed, but there are tutorial sections devoted to the necessary prerequisites

---

[1] We assume notions such as discrete and continuous random variables, expectation, variance, conditional probability and independence.

J.M. Sanz-Serna (✉)
Departamento de Matemática Aplicada e IMUVA, Facultad de Ciencias, Universidad de Valladolid, Valladolid, Spain
e-mail: sanzsern@mac.uva.es

in stochastic processes (Sect. 2), Markov chains (Sect. 3), stochastic differential equations (Sect. 6) and Hamiltonian dynamics/statistical physics (Sect. 8). The basic Random Walk Metropolis algorithm for discrete or continuous distributions is presented in Sects. 4 and 5. Sections 7 and 9 are respectively devoted to MALA, an algorithm based on stochastic differential equations proposals, and to the Hybrid Monte Carlo method, founded on ideas from Hamiltonian mechanics.

We have avoided throughout mathematical technicalities (that in the study of continuous-time stochastic processes may be overwhelming). We have rather followed the style of presentation taken by D. Higham in his tutorial paper on stochastic differential equations [18] and aimed at an exposition based on computer experiments; we believe that this approach may provide much insight and be a very useful entry point to the study of the issues considered here.

2 Stochastic Processes

We begin with a few introductory definitions and some useful examples.

2.1 Preliminaries

The definition of stochastic process ([15], Chap. 8) is simple:

Definition 1. Let T be a set of indices. *A stochastic process* is a family $\{X_t\}_{t \in T}$ of random variables defined on a common probability space $(\Omega, \mathscr{A}, \mathbb{P})$.[2]

In the applications, the variable t often corresponds to time. If $T = \mathbb{R}$, $T = [0, \infty)$ or $T = [a, b]$ the process is said to occur in *continuous time*. If $T = \{0, 1, \dots\}$ the process takes place in *discrete time* and we write $\{X_n\}_{n \geq 0}$.

The variables X_t may take values in a *continuous state space* like $\mathbb{R}^d$ or in a finite or infinite *discrete state space* E. In the latter case and without loss of generality, one may assume that E has been identified with a subset of $\mathbb{Z}$.

By definition, X_t may be seen as a function of two arguments: t and ω.[3] For a given value of t, X_t is a function of ω (the chance), so that the value of X_t will be different in different instances of the random experiment. For a given draw of the

---

[2] Recall that (1) Ω is a set and each point $\omega \in \Omega$ corresponds to a possible outcome of a random experiment, (2) $\mathscr{A}$ (a σ-algebra) is the family of those subsets $A \subseteq \Omega$ called *events* to which a probability $\mathbb{P}(A)$ is assigned, (3) $\mathbb{P}$ is a probability measure, $\mathbb{P} : \mathscr{A} \to [0, 1]$. The probability space plays very little explicit role in the study of the process; this is carried out in terms of the distributions of the X_t (see the examples in this section).

[3] Often the dependence of X_t on ω is not incorporated explicitly to the notation.

chance ω, the value of X_t changes with time. In this way there are two different, complementary ways to study a given process $\{X_t\}_{t \in T}$:

- By studying the distribution of the variable X_t for each $t \in T$ and (since in all interesting cases the X_t's are not mutually independent) the distribution of the pair (X_{t_1}, X_{t_2}) for each $t_1, t_2 \in T, \ldots$, the distribution of $(X_{t_1}, \ldots, X_{t_n})$ for each $t_1, \ldots, t_n \in T, \ldots$
- By drawing ω from Ω and studying the map $t \mapsto X_t(\omega)$: *a trajectory or path or realization* of the process.

These considerations will hopefully become clearer with the examples that follow.

2.2 Some Simple Stochastic Processes

Let us examine three well-known, useful examples of time-discrete, discrete state space processes.

2.2.1 The Symmetric Random Walk

At each time $n = 1, 2, \ldots$, Mary and Paul toss a fair coin and bet one euro. Let X_n be Mary's accumulated gain before the $(n+1)$-st toss ($X_0 = 0$). Here the state space is $E = \mathbb{Z}$.

The distributions of the first few X_n are easily found:

$$\mathbb{P}(X_0 = 0) = 1;$$

$$\mathbb{P}(X_1 = -1) = 1/2, \quad \mathbb{P}(X_1 = 1) = 1/2;$$

$$\mathbb{P}(X_2 = -2) = 1/4, \quad \mathbb{P}(X_2 = 0) = 1/2, \quad \mathbb{P}(X_2 = 2) = 1/4;$$

....

The joint distribution of any pair (X_m, X_n) may also be determined readily. For instance for (X_3, X_4) we compute:

$$\mathbb{P}(X_3 = 1, X_4 = 0) = 3/16,$$

$$\mathbb{P}(X_3 = 1, X_4 = 1) = 0,$$

$$\mathbb{P}(X_3 = 0, X_4 = 1) = 0,$$

....

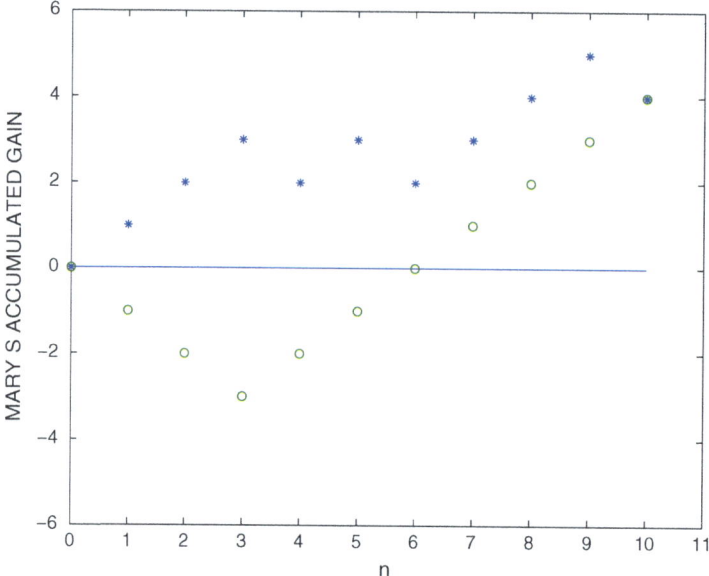

Fig. 1 Two possible trajectories of the symmetric random walk, $0 \leq n \leq 10$

For (X_1, X_2, X_3):

$$\mathbb{P}(X_1 = 1, X_2 = 2, X_3 = 1) = 1/8,$$

....

Sets of four, five, ... X_n's are dealt with in the same way.

Note that $X_{n+1} = X_n + Z_n$,[4] where the variables Z_n (gain in toss $n + 1$) are mutually independent and take values ± 1 with probability $1/2$ each. This leads to the formulae $\mathbb{E}(X_n) = 0$, $\mathrm{Var}(X_n) = n$ for the expectation and variance.

Figure 1 shows, for $0 \leq n \leq 10$, two possible trajectories of the process. A computer-generated, longer trajectory may be seen in Fig. 2, where we note a few remarkable facts. (A complete study of the symmetric random walk using elementary means may be found in [9], Chap. 3.)

- The vertical axis covers only a small range slightly larger than $[-100, 100]$, in spite of the fact that Mary's gains might in principle have been in the range $-10{,}000 \leq X_n \leq 10{,}000$. This happens because the standard deviation $\sigma(X_n)$ equals $\sqrt{n}$.

---

[4]In general, a process $\{X_n\}_{n \geq 0}$ is a *random walk* if $X_{n+1} = X_n + Z_n$, where Z_n is independent of $X_n, \ldots, X_0$.

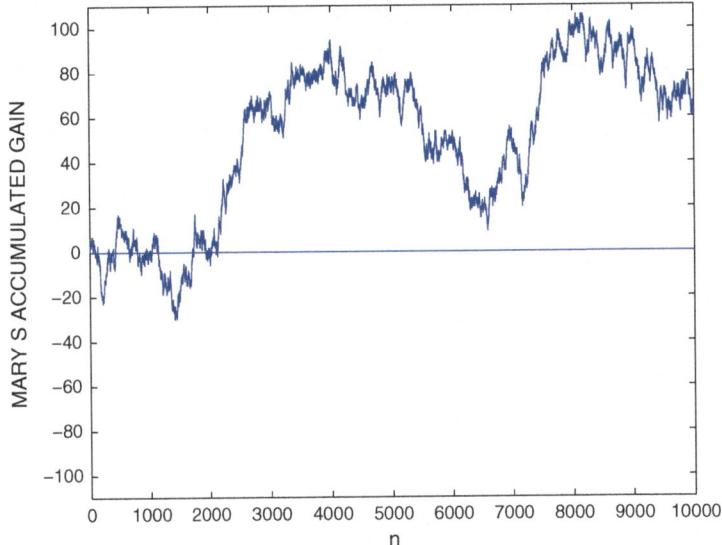

Fig. 2 A typical trajectory $0 \le n \le 10{,}000$ of the symmetric random walk

- While the game is "fair" i.e. $\mathbb{E}(X_n) = 0$, Mary has been winning most of the time. This is not a peculiarity of the particular trajectory shown: typically either Mary is ahead most of the time or Paul is ahead most of the time.
- In the first two or three thousand tosses Mary and Paul tied ($X_n = 0$) a few times. Since after a tie the game restarts afresh—the coin has no memory—one would have expected that similar ties would keep happening after, say, $n = 3{,}000$. Clearly this has not been the case. Our intuition suggests that, if T_i is the number of tosses between consecutive ties, then the average $A_n = (T_1 + \cdots + T_n)/n$ should converge to a limit as $n \to \infty$. However it may be proved that the size of A_n grows proportionally to n, so that $T_1 + \cdots + T_n$ grows like n^2. In fact it is likely that one among $T_1, \ldots, T_n$ be of size proportional to n^2.

2.2.2 The Non-symmetric Random Walk

Everything is as before but Mary's chance of winning an individual bet is now $p \ne 1/2$ so that Paul's is $q = 1 - p$.

For the distributions of the X_n we find now:

$$\mathbb{P}(X_0 = 0) = 1,$$
$$\mathbb{P}(X_1 = -1) = q, \quad \mathbb{P}(X_1 = 1) = p,$$
$$\mathbb{P}(X_2 = -2) = q^2, \quad \mathbb{P}(X_2 = 0) = 2pq, \quad \mathbb{P}(X_2 = 2) = p^2,$$
$$\ldots\ldots\ldots\ldots\ldots\ldots\ldots\ldots\ldots\ldots\ldots\ldots\ldots\ldots\ldots\ldots$$

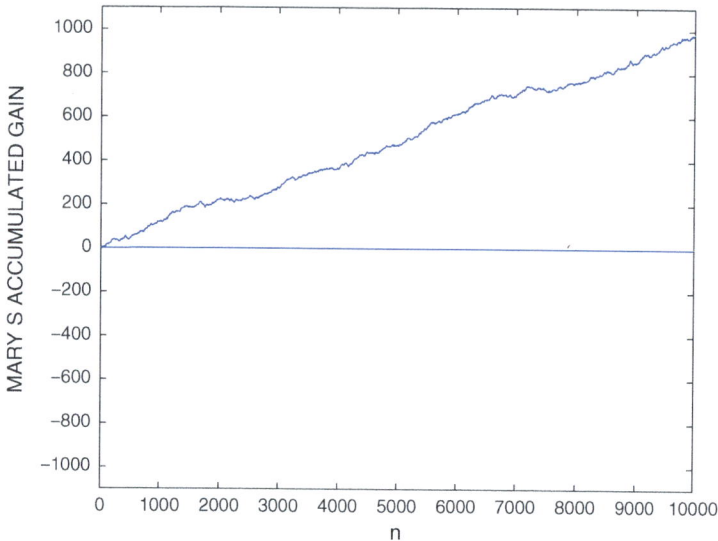

Fig. 3 Non-symmetric random walk, $p = 0.55$, $0 \le n \le 10{,}000$. The expectation of $X_{10{,}000}$ is 1,000 and its typical deviation ≈ 100. Drift offsets fluctuations due to chance

From $X_{n+1} = X_n + Z_n$ we compute $\mathbb{E}(X_n) = n(p - q)$ and $\text{Var}(X_n) = 4npq$. Since the expectation grows like n and the standard deviation only like $\sqrt{n}$, the *drift* arising from the lack of fairness of the coin, $p \ne q$, will in the long run dominate the *fluctuations* due to chance, even if $|p - q|$ is very small. This is borne out in Fig. 3, where $p = 0.55$.

2.2.3 The Ehrenfest Diffusion Model

This was proposed in 1907 by P. Ehrenfest (1880–1933) to illustrate the second law of thermodynamics. Two containers, left and right, are adjacent to each other and contain gas that may move between them through a small aperture. There are in total M molecules. At each time n, a molecule, randomly chosen among the M, moves to the other container. Let X_n be the number of molecules in the left box before the $(n + 1)$-st move. We assume that initially all molecules are in the left container, $X_0 = M$. The state space is $\{0, 1, \ldots, M\}$ and the distributions are:

$$\mathbb{P}(X_1 = M - 1) = 1,$$

$$\mathbb{P}(X_2 = M - 2) = (M - 1)/M, \quad \mathbb{P}(X_2 = M) = 1/M,$$

....

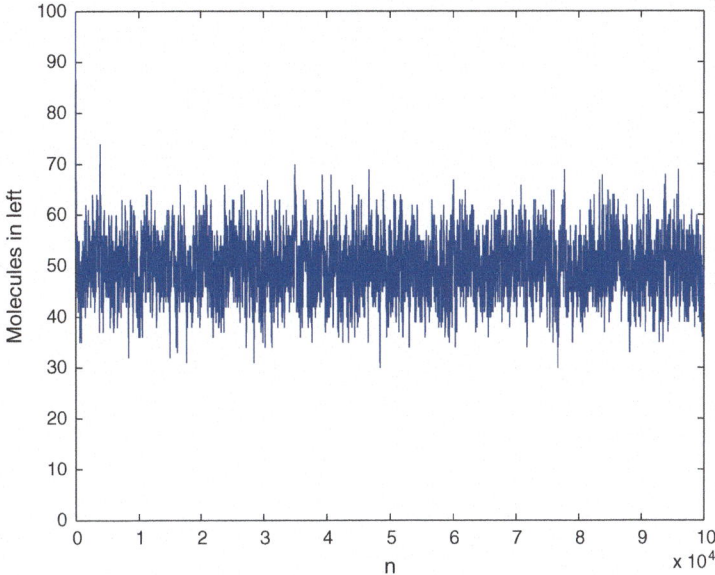

Fig. 4 A trajectory of the Ehrenfest model, 100,000 steps for $M = 100$ molecules. Outside the initial transient X_n has not left the interval $[25, 75]$

A typical trajectory may be seen in Fig. 4, where we observe that, after an initial transient, X_n has not left the interval $[25, 75]$, i.e. the molecules distribute themselves more or less evenly between both containers. It may be shown ([9], Chap. XVII, Example 7.c) that, with $M = 10^6$ the probability of finding, for n outside the initial transient, more than $505,000$ molecules in one container (i.e. of finding fluctuations larger than 1 % around the break-even situation $X_n = M/2$) is of the order of 10^{-23}. In statistical physics M is of course much, much larger and the size of the fluctuations around $M/2$ correspondingly smaller: for all practical purposes the gas, driven by sheer chance, will remain in the maximum entropy state $X_n = M/2$.

3 Discrete State Space Markov Chains

A Markov chain (MC) is a process $\{X_n\}_{n \geq 0}$ where the distribution of X_{n+1} conditional on $X_0, \ldots, X_n$ coincides with the distribution of X_{n+1} conditional on X_n. This is sometimes expressed by saying "in order to know the future, the knowledge of the past does not add anything to the knowledge of the present". The precise definition is:

Definition 2. A discrete-time process $\{X_n\}_{n\geq 0}$ with values in a countable space E is a *Markov chain* if for all $n \geq 0$ and all states $i_0, i_1, \ldots, i_{n-1}, i, j \in E$

$$\mathbb{P}(X_{n+1} = j \mid X_n = i, X_{n-1} = i_{n-1}, \ldots, X_0 = i_0) = \mathbb{P}(X_{n+1} = j \mid X_n = i)$$

(whenever both sides are well defined[5]).

Markov chains are named after A. Markov (1856–1922). Cases with infinite, countable state spaces where first considered by A. Kolmogorov in 1936.

If, for each pair of states $i, j \in E$, $\mathbb{P}(X_{n+1} = j \mid X_n = i)$ is independent of n, then the MC is called *homogeneous*; only homogeneous MC are considered in this paper.

The symmetric and non-symmetric random walks in Sects. 2.2.1 and 2.2.2 are (homogeneous) MC: the structure $X_{n+1} = X_n + Z_n$ noted before makes it clear that the knowledge of the values of $X_0, \ldots, X_{n-1}$ adds nothing to the knowledge of the value of X_n (in both trajectories in Fig. 1 $X_{10} = 4$ and the distribution of X_{11} conditional on the past is the same: $X_{11} = 5$ or $X_{11} = 3$ with probability $1/2$ each). In the symmetric random walk, the so-called transition probabilities are

$$\mathbb{P}(X_{n+1} = i + 1 \mid X_n = i) = \mathbb{P}(Z_n = 1) = 1/2,$$
$$\mathbb{P}(X_{n+1} = i - 1 \mid X_n = i) = \mathbb{P}(Z_n = -1) = 1/2,$$
$$\mathbb{P}(X_{n+1} = j \mid X_n = i) = \mathbb{P}(Z_n \neq \pm 1) = 0, \quad j \neq i \pm 1.$$

The Ehrenfest model (Sect. 2.2.3) is another example of MC. The transition probabilities are:

$$\mathbb{P}(X_{n+1} = i + 1 \mid X_n = i) = (M - i)/M, \quad i < M,$$
$$\mathbb{P}(X_{n+1} = i - 1 \mid X_n = i) = i/M, \quad i > 0$$

(other transitions are impossible).

3.1 The Transition Matrix

The *transition probabilities* of a MC are defined by

$$p_{ij} := \mathbb{P}(X_{n+1} = j \mid X_n = i).$$

---

[5]Note that e.g. $\mathbb{P}(X_{n+1} = j \mid X_n = i)$ does not make sense if $\mathbb{P}(X_n = i) = 0$. Here we shall not pay attention to the difficulties created by probabilities conditioned to events $\{X_n = i\}$ of 0 probability. These difficulties are easily avoided if, as in [9], the Markov chain is defined in the first place by means of the transition probabilities rather than in terms of the variables X_n.

Since, as j ranges in E with i fixed, the p_{ij} are a probability distribution, we may write

$$p_{ij} \geq 0, \qquad \sum_{j \in E} p_{ij} = 1,$$

so that the *transition matrix* $P = (p_{ij})_{(i,j) \in E \times E}$ is a *stochastic matrix*.[6] The i-th row of P provides the distribution of X_{n+1} conditional on $X_n = i$.

A transition over two steps from i to k must be accomplished through an intermediate visit to some state j and therefore we may write the *Chapman-Kolmogorov equation*

$$\mathbb{P}(X_{n+2} = k \mid X_n = i) = \sum_j \mathbb{P}(X_{n+1} = j \mid X_n = i)\, \mathbb{P}(X_{n+2} = k \mid X_{n+1} = j) = \sum_j p_{ij} p_{jk}.$$

In this way $\mathbb{P}(X_{n+2} = k \mid X_n = i)$ is given by the (i,k) entry of P^2. The entries of higher powers $P^3, P^4, \ldots$, give similarly the probabilities of transitions in $3, 4, \ldots$ steps.

The (unconditional) distribution of each X_n is determined by the distribution of X_0 together with the transition matrix P:

$$\mathbb{P}(X_n = \ell) = \sum_i \mathbb{P}(X_0 = i)\, \mathbb{P}(X_n = \ell \mid X_0 = i) = \sum_i \mathbb{P}(X_0 = i)\, (P^n)_{i\ell}.$$

It is customary to collect in a column vector $\mu^{(n)}$ the probabilities $\mathbb{P}(X_n = \ell)$, $\ell \in E$, and then the preceding formula may be rewritten as

$$\mu^{(n)T} = \mu^{(0)T} P^n.$$

Equivalently one has the following expression for the evolution of the distributions $\mu^{(n)}$:

$$\mu^{(n+1)T} = \mu^{(n)T} P, \qquad n = 0, 1, \ldots \tag{1}$$

The joint distributions of (X_m, X_n), (X_ℓ, X_m, X_n), ... are also easily determined once $\mu^{(0)}$ and P are known. In practice it is customary to describe a MC by specifying the transition matrix together with the initial distribution.[7] Although strictly speaking the MC is the sequence $\{X_n\}$, in practice we often speak as though the chain were the matrix P together with the initial distribution $\mu^{(0)}$ or even the matrix P with an undetermined $\mu^{(0)}$.

---

[6] If E comprises an infinite number of states, this "matrix" will of course have infinitely many rows/columns. Sums like $\sum_j p_{ij} p_{jk}$ that we shall find below have a finite value if P is stochastic.

[7] See [5], Theorem 8.1 for the construction of the X_n and the underlying probability space.

3.2 Classifying Markov Chains

The following concept is needed in order to define persistent states:

Definition 3. For each state $i \in E$ define the *return time* T_i (a random variable) by:

$$T_i := \inf\{n \geq 1 : X_n = i\} \in [0, \infty].$$

Here it is understood that the inf of the empty set is ∞. Note that $n \geq 1$ so that, in particular $X_0 = i$ does not imply $T_i = 0$.

Definition 4. A state $i \in E$ is *persistent or recurrent* if $\mathbb{P}(T_i < \infty \mid X_0 = i) = 1$. Otherwise it is called transient.

A recurrent state $i \in E$ is *positive* if $\mathbb{E}(T_i \mid X_0 = i) < \infty$. Otherwise it is called *null*.

If i is persistent and the chain is started at i, then the number of visits to i (i.e. the number of values of n with $X_n = i$) is infinite with probability 1 (see [5], Theorem 8.2).

For the chain

$$P = \begin{bmatrix} 1/2 & 1/2 & 0 \\ 0 & 1/2 & 1/2 \\ 0 & 1/2 & 1/2 \end{bmatrix} \qquad (2)$$

the second state is persistent. In fact, if started at 2, the chain returns to 2 in one move with probability $1/2$, in two moves with probability $1/4$, etc. The third state is persistent for the same reason. The first state is transient: conditional to $X_0 = 1$, T_1 only takes the values 1 (with probability $1/2$) and ∞.

The matrix (2) is certainly special: states 2 and 3, on their own, would make up a MC. This matrix provides an example of reducibility in the sense of the next definition because moves from 2 to 1 or from 3 to 1 in n steps are impossible for each $n \geq 1$.

Definition 5. A MC is *irreducible* if for each ordered pair of states $i, j \in E$ there exists a number $n = 1, 2, \ldots$ such that

$$\mathbb{P}(X_n = j \mid X_0 = i) > 0.$$

The following important result holds ([6], Chap. 3, Sect. 1.3):

Theorem 1. *For an irreducible chain, one of the following three possibilities holds true:*

- *All states are positive recurrent.*
- *All states are null recurrent.*
- *All states are transient.*

If, in addition, E is finite, then the chain is necessarily positive recurrent.

The Ehrenfest model, the symmetric random walk, and the non-symmetric random walk are irreducible and provide examples of the three possibilities in the theorem ([6], Chap. 3, Example 1.2). For the symmetric random walk the fact that the expected waiting time for the first tie $X_n = 0$ is infinite is related to some of the counterintuitive features we saw in Fig. 2; for instance to the fact that the average A_n of the first n times between successive returns to equilibrium does not approach a finite limit. For the Ehrenfest model positive recurrence implies that, except for a set of trajectories with probability 0, in each trajectory there are infinitely many n_r such that $X_{n_r} = M$: all the molecules will be back in the left container infinitely many times! There is no contradiction with Fig. 4: the expectation $\mathbb{E}(T_M \mid X_0 = M)$ is positive but exponentially small as we shall see.

3.3 Stationary Distributions

The following concept is extremely important.

Definition 6. A probability distribution μ on E ($\mu_i \geq 0, i \in E, \sum_i \mu_i = 1$) is called a *stationary or invariant or equilibrium* distribution of the MC with transition matrix P if

$$\sum_i \mu_i p_{ij} = \mu_j, \qquad j \in E,$$

or, in matrix notation,

$$\mu^T P = \mu^T. \qquad (3)$$

From (1) it follows that if X_0 possesses the distribution μ and μ is invariant, then all the X_n, $n = 0, 1, 2, \ldots$ share the same distribution; we then say that the chain is at *stationarity*. Note that at stationarity the X_n are identically distributed but, except for trivial cases,[8] *not independent*. It is easy to see that the symmetric and non-symmetric random walk do not possess invariant probability distributions. For the Ehrenfest model the distribution

$$\mu_i = \binom{M}{i} \frac{1}{2^M} \qquad (4)$$

is readily seen to be invariant. Note that μ_i coincides with the probability that, when each of the M molecules is randomly assigned to the left or right container (with probability $1/2$ each), then the left container receives i molecules.

---

[8] If all the rows of P are equal, X_{n+1} is independent of X_n.

As a further example consider the *doubly stochastic matrix* ($\sum_j p_{ij} = \sum_i p_{ij} = 1$, $p_{ij} \geq 0$):

$$P = \begin{bmatrix} 1/4 & 1/2 & 1/4 \\ 1/4 & 1/4 & 1/2 \\ 1/2 & 1/4 & 1/4 \end{bmatrix}. \tag{5}$$

The invariant distribution is $[1/3, 1/3, 1/3]$: at stationarity all states have the same probability, something that happens with all doubly stochastic transition matrices.

The following general result holds ([6], Chap. 3, Theorems 3.1, 3.2):

Theorem 2. *Assume the chain to be irreducible. Then it is positive recurrent if and only if there is a stationary distribution. The stationary distribution μ, if it exists is unique and $\mu_i = 1/\mathbb{E}(T_i \mid X_0 = i) > 0$ for each $i \in E$.*

Since from (4) $\mu_M = 2^{-M}$, the theorem shows that in Fig. 4 the expected number of interchanges for all 100 molecules to return to the left container is $2^{100} \approx 1.27 \times 10^{30}$.

3.4 Reversibility

Consider a MC with a stationary distribution such that $\mu_i > 0$ for each $i \in E$ (a particular case is given by an irreducible, positive recurrent chain, see Theorem 2). The matrix Q with entries $q_{ij} = \mu_j \, p_{ji}/\mu_i$ is stochastic,

$$\sum_j q_{ij} = \sum_j \frac{\mu_j \, p_{ji}}{\mu_i} = \frac{\mu_i}{\mu_i} = 1,$$

and also has μ_i as an invariant distribution

$$\sum_i \mu_i q_{ij} = \sum_i \mu_j p_{ji} = \mu_j \sum_i p_{ji} = \mu_j.$$

What is the meaning of Q? Assume that the initial distribution $\mu^{(0)}$ coincides with μ (i.e. the chain is at stationarity) then:

$$\mathbb{P}(X_n = j \mid X_{n+1} = i) = \frac{\mathbb{P}(X_{n+1} = i \mid X_n = j)\,\mathbb{P}(X_n = j)}{\mathbb{P}(X_{n+1} = i)} = \frac{p_{ji}\,\mu_j}{\mu_i} = q_{ij}.$$

Thus, Q is the transition matrix of a chain where the "arrow of time" has been reversed, because n and $n+1$ have interchanged their roles. As a simple example consider the chain (5) for which $Q = P^T$, since the stationary distribution is $\mu_1 = \mu_2 = \mu_3 = 1/3$. If P is at stationarity the three events $\{X_n = 1, X_{n+1} = 2\}$, $\{X_n = 2, X_{n+1} = 3\}$, $\{X_n = 3, X_{n+1} = 1\}$ have probability $1/6$ each, while the six events $\{X_n = 1, X_{n+1} = 1\}$, $\{X_n = 1, X_{n+1} = 3\}$, etc. have probability $1/12$ each. For Q it is the events $\{X_n = 2, X_{n+1} = 1\}$, $\{X_n = 3, X_{n+1} = 2\}$, $\{X_n = 1, X_{n+1} = 1/3\}$ that have probability $1/6$.

Definition 7. A probability distribution $\mu > 0$ and a transition matrix P satisfy the *detailed balance* condition if:

$$\forall i, j \in E, \qquad \mu_i \, p_{ij} = \mu_j \, p_{ji}. \qquad (6)$$

Of course $\mu_i \, p_{ij}$ is the probability of the event $(X_n = i, X_{n+1} = j)$. For the example (5) the detailed balance condition does not hold: the event $(X_n = 1, X_{n+1} = 2)$ is more likely than the event $(X_n = 2, X_{n+1} = 1)$.

Since (6) implies

$$\sum_i \mu_i \, p_{ij} = \sum_i \mu_j \, p_{ji} = \mu_j,$$

we may conclude:

Theorem 3. *Under the assumption of detailed balance (6):*

- *μ is an invariant distribution with respect to P.*
- *At stationarity, the reversed matrix Q coincides with P and therefore the chain and its time-reversal are statistically the same.*

The chain is then called reversible with respect to μ.

The Ehrenfest chain is in detailed balance with the distribution (4) and hence reversible with respect to it.

3.5 Ergodicity

The ergodic theorem ([6], Chap. 3, Theorem 4.1) is the foundation of all our later work:

Theorem 4. *Let $\{X_n\}_{n \geq 0}$ be an irreducible, positive recurrent MC and denote by μ its stationary distribution as in Theorem 2. For any function $f : E \to \mathbb{R}$ such that*

$$\sum_{i \in E} |f(i)| \mu_i < \infty$$

and any initial distribution $\pi^{(0)}$, $\mathbb{P}_{\mu^{(0)}}$ almost sure:

$$\lim_{N \to \infty} \frac{1}{N+1} \sum_{k=0}^{N} f(X_k) = \sum_{i \in E} f(i) \mu_i. \qquad (7)$$

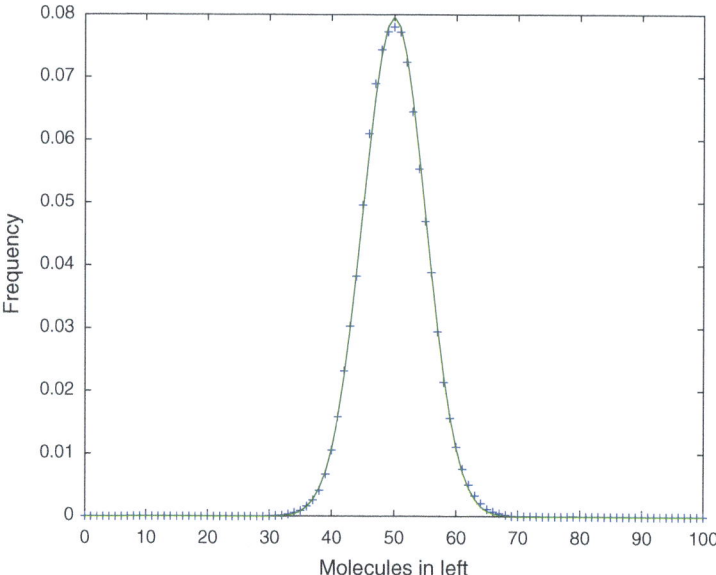

Fig. 5 Ehrenfest's model: an histogram of the trajectory in Fig. 4 (+ signs) and the invariant distribution (4) (*solid line*)

The last sum is of course the expectation of $f(X)$ when X has the distribution μ. Figure 5 shows the ergodic theorem at work in Ehrenfest's model. The + signs measure the frequency with which the $M+1$ states $0,\ldots,M$ have been occupied along the single trajectory of Fig. 4 and the solid line represents the equilibrium probabilities (4). For each state i, the probability in (4) is of course the expectation of the function f such that $f(i) = 1$ and $f(j) = 0$ for $j \neq i$ (the indicator of the set $\{i\}$) and the frequency of occupation is the average of f along the trajectory; we see in Fig. 5 that both very approximately coincide, in agreement with the ergodic theorem. In statistical physics one considers an *ensemble*, i.e. a very large number of Ehrenfest experiments running in parallel, independently of one another, so that, at any fixed n, the distribution of the number of molecules in the left containers across the experiments coincides with the distribution of the random variable X_n. The ergodic theorem implies that the behavior of a trajectory in a single experiment started e.g. from $X_0 = M$ coincides with the behavior of the ensemble at any fixed time n *when the initial distribution across the ensemble is given by (4), i.e. if the chain is at stationarity*.

Ergodicity makes it possible to compute expectations with respect to the stationary distribution by computing averages along trajectories. This is the basis of Monte Carlo algorithms that we shall study in the next sections.

3.6 Convergence to Steady State

With $M = 2$ molecules, the Ehrenfest transition matrix is

$$P = \begin{bmatrix} 0 & 1 & 0 \\ 1/2 & 0 & 1/2 \\ 0 & 1 & 0 \end{bmatrix};$$

it is easy to check that returns to the initial state $X_n = X_0$ are only possible if n is even. This motivates the following definition.

Definition 8. *The period $d(i)$ of a state i is the greatest common divisor of all the numbers n such that a return to i in n steps is possible. If $d(i) = 1$, the state i is said to be* aperiodic. *If $d(i) > 1$ the state i is said to be* periodic *with period $d(i)$.*

It turns out that for an irreducible chain either all states are aperiodic or all states are periodic and share the same period ([15], Sect. 6.3). One then says that the chain is aperiodic or periodic respectively. The Ehrenfest chain, regardless of the number of molecules M, is periodic with period 2 and so are the symmetric and non-symmetric random walks of Sects. 2.2.1 and 2.2.2.[9]

The result below ([6], Theorem 2.1) shows that, if we exclude periodic chains, the distribution $\mu^{(n)}$ of X_n in an irreducible, positive recurrent chain will approach as $n \uparrow \infty$ the stationary distribution, regardless of the choice of the distribution of X_0. This fact is sometimes expressed by saying that an irreducible, aperiodic and positive recurrent chain is *asymptotically stationary*; for n large the chain "forgets its origin."

Theorem 5. *Assume that a MC is irreducible, positive recurrent and aperiodic and let μ be the corresponding invariant distribution as in Theorem 2. For each choice of the distribution*[10] *$\mu^{(0)}$ of X_0*

$$\lim_{n \to \infty} | \mu^{(0)T} P^n - \mu^T | = \lim_{n \to \infty} | \mu^{(n)T} - \mu^T | = 0.$$

It is perhaps useful to stress that the ergodic Theorem 4 holds for both periodic and aperiodic (irreducible, positive recurrent) chains (after all Fig. 5 illustrates this theorem at work in the periodic Ehrenfest chain). However, if the chain is irreducible, positive recurrent and aperiodic, the combination of Theorems 4 and 5 implies that the average in the left-hand side of (7) approximately coincides with

---

[9]This should not lead to the conclusion that period 2 is the rule for MCs. The three examples in the last section are not typical in this respect and were chosen in view of the fact that are very easily described—in each of them transitions may only occur between state i and states $i \pm 1$. Any chain where the diagonal elements of P are all $\neq 0$ only contains aperiodic states.

[10]If λ, ν are distributions the notation $| \lambda^T - \nu^T |$ means $\sum_i |\lambda_i - \nu_i|$.

the expectation of f with respect to the distribution $\mu^{(n)}$ of X_n, $n \gg 1$, *regardless of the choice of the distribution* $\mu^{(0)}$.

In this connection consider an ensemble of many couples, Marys and Pauls, with each couple tossing a coin and betting repeatedly ($X_0 = 0$).[11] At any fixed n, the distribution of X_n is symmetric and there will be approximately as many Marys ahead as Pauls ahead—this is a property of the *ensemble*. However, as illustrated in Fig. 2, in the typical trajectory of a single couple either Mary or Paul will be ahead most of the time. Here what is typical for a single trajectory is markedly different from what happens in the ensemble; this difference explains why some people find Fig. 2 disconcerting.

4 Sampling from a Target Distribution by Monte Carlo Algorithms: The Discrete Case

Markov Chain Monte Carlo (MCMC) algorithms [28] are aimed at computing expectations[12] with respect to a given *target distribution* μ on a state space E, that for the time being we assume discrete. These algorithms construct a MC $\{X_n\}_{n\geq 0}$ for which μ is an invariant distribution and, as pointed out before, invoke ergodicity to approximate expectations by empirical means along a trajectory as in (7).

MCMC is a very popular technique in computational chemistry and physics; examples will be considered in later. Bayesian statistics ([30], Sect. 1.3) is another important field of application. There a *prior* probability distribution μ_0 is postulated on the (discrete) set Θ of possible values of a parameter θ appearing in a probabilistic model (note that Θ is now the state space and that the different states are the different values of θ). Then data y are collected and "incorporated" into the model via Bayes's theorem to define a *posterior* distribution for θ

$$\mu(\theta \mid y) = \frac{\phi(y \mid \theta)\mu_0(\theta)}{\sum_{\zeta \in \Theta} \phi(y \mid \zeta)\mu_0(\zeta)} \tag{8}$$

($\phi(y \mid \theta)$ is the probability of y when the parameter value is θ.) As a rule, the posterior is neither one of the familiar distributions nor tractable analytically and in order to compute expectations

$$\mathbb{E}(f(\theta) \mid y) = \sum_{\theta \in \Theta} f(\theta)\mu(\theta \mid y)$$

it is necessary to resort to Monte Carlo techniques.

---

[11]Note that there is no invariant probability distribution, since the chain is null recurrent.

[12]Of course computing the probability of an event A is equivalent to computing the expectation of its indicator, i.e. of the random variable that takes the value 1 if $\omega \in A$ and 0 if $\omega \notin A$.

The idea behind MCMC methods [26] was suggested by Metropolis and his coworkers in 1953. The problem to be addressed is essentially[13] how to construct a MC $\{X_n\}_{n\geq 0}$ in a given state space E having the target μ as an invariant distribution or, more precisely, how to compute trajectories of that chain in order to be able to use the empirical average in the left-hand side of (7) as an approximation to the expectation in the right-hand side. Metropolis algorithms use two ingredients to obtain realizations from $\{X_n\}_{n\geq 0}$:

1. Realizations $u_0, u_1 \ldots$ of a sequence of mutually independent random variables $U_0, U_1, \ldots$ with uniform distribution in the unit interval. These realizations are of course readily available on any computing system.
2. Samples from the distribution of Y_{n+1} conditional on Y_n in an *auxiliary* MC $\{Y_n\}_{n\geq 0}$ in the same state space E (not in the MC $\{X_n\}_{n\geq 0}$ we wish to construct!). At this stage the only requirement we impose on $\{Y_n\}_{n\geq 0}$ is that the transition probabilities p_{ij}^* satisfy the symmetry requirement

$$\forall i, j, \qquad p_{ij}^* = p_{ji}^*. \qquad (9)$$

For example, if $E = \mathbb{Z}$ then $\{Y_n\}_{n\geq 0}$ may be defined through $Y_{n+1} = Y_n + Z_n$, where the Z_n are mutually independent, integer-valued and with a symmetric distribution (i.e. $\mathbb{P}(Z_n = i) = \mathbb{P}(Z_n = -i)$ for each $i \in \mathbb{Z}$). To generate a sample of $Y_{n+1} \mid Y_n = y_n$ just set $y_{n+1} = y_n + z_n$ where z_n is a sample of Z_n. If $Z_n = \pm 1$ with probability $1/2$ each, then $\{Y_n\}_{n\geq 0}$ is of course the symmetric random walk in Sect. 2.2.1.

The algorithm is as follows:

- Choose a value i_0 for X_0 (randomly or e.g. $i_0 = 0$).
- Once values $i_0, \ldots, i_n$ of $X_0, \ldots, X_n$ have been found:
  - Generate a *proposed* value $i_{n+1}^* \in E$, from the auxiliary conditional distribution $Y_{n+1} \mid Y_n = i_n$.
  - If $\mu_{i_{n+1}^*}/\mu_{i_n} > u_n$ set $X_{n+1} = i_{n+1}^*$; in this case we say that *the proposal is accepted*. Else set $X_{n+1} = i_n$ and we say that *the proposal is rejected*.

The criterion used to accept or reject the proposal is called *the Metropolis accept/reject rule*. After noting that the acceptance probability is[14]

$$a = 1 \wedge \frac{\mu_{i_{n+1}^*}}{\mu_{i_n}} \qquad (10)$$

---

[13] It is also necessary that the chain constructed be positive recurrent. Also not all positive recurrent chains having the target as equilibrium measure are equally efficient, as the velocity of the convergence to the limit in (7) is of course chain-dependent.

[14] $\wedge$ means min.

(i.e. the proposal is accepted with probability a), it is not difficult to prove the following result:

Theorem 6. *The transitions $X_n \mapsto X_{n+1}$ in the procedure just described satisfy the detailed balance condition (6) with respect to the target distribution μ. Therefore (Theorem 3), the implied chain $\{X_n\}_{n\geq 0}$ is reversible with respect to μ.*

Proof. If $j \neq i$, to reach j at step $n+1$ we require (i) that j is proposed and (ii) that it is accepted. In this way:

$$\mu_i \, p_{ij} = \mu_i \left(p_{ij}^* (1 \wedge \frac{\mu_j}{\mu_i}) \right) = p_{ij}^* (\mu_i \wedge \mu_j),$$

and the last expression is symmetric in i, j in view of (9). □

A few remarks:

- The target distribution μ only enters the algorithm through the ratios in (10) and hence must be known only up to a multiplicative constant. This is an asset: in many applications the normalizing constant of the target distribution is extremely difficult to determine. As an example look at (8) where, for given values of the data y, the denominator is just a real number that normalizes the distribution (i.e. it ensures that, as θ ranges in Θ, the values of $\mu(\theta \mid y)$ add up to 1). In practice, the computation of that denominator may be impossible if the cardinality of Θ is large; when computing the acceptance ratio in the Metropolis algorithm one may substitute the un-normalized values $\phi(y \mid \theta)\mu_0(\theta)$ for the true probabilities $\mu(\theta \mid y)$.
- The rejected values of X_n are part of the chain and must be included to compute the average

$$\frac{1}{N+1} \sum_{k=0}^{N} f(X_k)$$

used to approximate the expectation of f.
- In practice if the starting location i_0 of the chain is far away from the states i for which the target has a significant size the convergence in (7) will be very slow. It is then advisable to run the chain for a *burn in* preliminary period until the states of higher probability are identified. The values of X_n corresponding to the burn in phase are not used to compute the averages.

As we shall see later, it is of interest to consider proposals that do not satisfy the symmetry condition in (9). (For instance we may wish to use proposals $Y_{n+1} = Y_n + Z_n$ where the increments Z_n do not have a symmetric distribution.) As first pointed out by Hastings in 1970 [17], to achieve detailed balance the formula (10) for acceptance probability has then to be changed into

$$a = 1 \wedge \frac{p^*_{i_{n+1} i_n} \mu_{i_{n+1}}}{p^*_{i_n i_{n+1}} \mu_{i_n}}. \tag{11}$$

The proof that this so-called Metropolis-Hastings rule works follows the lines of the proof of Theorem 6. Further possibilities for the acceptance probability recipe exist [17].

5 Metropolis Algorithms for the Continuous Case

For the sake of simplicity, our presentation of the Metropolis-Hastings algorithms has assumed that the target probability is defined on a discrete state space. However the algorithms are equally applicable to sampling from continuous distributions and in fact the next sections will only deal with the continuous case. We begin with a few words on MC with a continuous state space.

5.1 Continuous State Space Markov Chains

We now consider (time-discrete) stochastic processes $\{X_n\}_{n\geq 0}$ where each X_n takes values in $\mathbb{R}^d$. The definition of MC remains the same: $\{X_n\}_{n\geq 0}$ is a MC if the distribution of X_{n+1} conditional on X_n and the distribution of X_{n+1} conditional on $X_n, \ldots, X_0$ coincide. The role played in the discrete state space case by the transition matrix P is now played by a *transition kernel* K (see e.g. [30], Definition 6.2, [10], Chap. VI, Sect. 11). This is a real-valued function $K(\cdot, \cdot)$ of two arguments. For each fixed $x \in \mathbb{R}^d$, $K(x, \cdot)$ is a Borel probability measure in $\mathbb{R}^d$. For each fixed Borel set $A \subseteq \mathbb{R}^d$, $K(\cdot, A)$ is a Borel measurable function. The value $K(x, A)$ represents the probability of jumping from the point $x \in \mathbb{R}^d$ to a set A in one step of the chain. Hence the formula (1) for the evolution of the distributions $\mu^{(n)}$ of the X_n now becomes

$$\mu^{(n+1)}(A) = \mathbb{P}(X_{n+1} \in A) = \int_{\mathbb{R}^d} \mu^{(n)}(dx)\, K(x, A),$$

or in shorthand

$$\mu^{(n+1)}(dy) = \int_{\mathbb{R}^d} \mu^{(n)}(dx)\, K(x, dy). \tag{12}$$

The condition (3) for a stationary or invariant probability distribution is correspondingly

$$\mu(A) = \int_{\mathbb{R}^d} \mu(dx)\, K(x, A)$$

(for each measurable A) or

$$\mu(dy) = \int_{\mathbb{R}^d} \mu(dx)\, K(x, dy), \tag{13}$$

and the detailed balance condition (6) reads

$$\int_A \mu(dx)\, K(x, B) = \int_B \mu(dy)\, K(y, A), \tag{14}$$

or

$$\mu(dx) K(x, dy) = \mu(dy) K(y, dx).$$

While conditions exist that guarantee the existence of a stationary probability distribution and the validity of a corresponding ergodic theorem (see [27, 30], Chap. 6) the technicalities are much more intricate than in the discrete state space case and will not be studied here.

In practice the kernel K often possesses a density $k(x, y)$ (with respect to the standard Lebesgue measure in $\mathbb{R}^d$), i.e. K is expressed in terms of the function k of two variables $x \in \mathbb{R}^d$, $y \in \mathbb{R}^d$ through the formula

$$K(x, A) = \int_{y \in A} k(x, y)\, dy.$$

In that case, if $\mu^{(n)}$ (i.e. X_n) has a density $\pi^{(n)}(x)$, then $\mu^{(n+1)}$ has a density (see (12))

$$\pi^{(n+1)}(y) = \int_{\mathbb{R}^d} \pi^{(n)}(x)\, dx\, k(x, y).$$

The density of a stationary distribution satisfies (see (13))

$$\pi(y) = \int_{\mathbb{R}^d} \pi(x)\, dx\, k(x, y).$$

and the detailed balance condition (14) becomes

$$\pi(x) k(x, y) = \pi(y) k(y, x).$$

5.2 Accept/Reject with Continuous Targets

If the target has a density π, the Metropolis acceptance probability for the discrete case given in (10) has to be replaced by

$$a = 1 \wedge \frac{\pi(x_{n+1}^*)}{\pi(x_n)}, \tag{15}$$

where x_n is the value of X_n (current location of the chain) and x_{n+1}^* is the proposal for the next location. This formula requires that the proposal be based on a symmetric kernel (in the case with densities the density of the proposal kernel must satisfy $k^*(x, y) \equiv k^*(y, x)$). The Hastings formula (11) may be similarly adapted.

5.3 The Random Walk Metropolis Algorithm

As mentioned in the discrete case, a simple way of generating proposals is to use a random walk format: $Y_{n+1} = Y_n + Z_n$ where now the Z_n are independent identically distributed continuous random variables in $\mathbb{R}^d$. A common choice is to take each Z_n to be a *normal (Gaussian) d-variate* distribution $N(m, C)$ with density given by:

$$\frac{1}{(2\pi)^{d/2} \det(C)^{1/2}} \exp\left(-\frac{1}{2}(x-m)^T C^{-1}(x-m)\right). \tag{16}$$

Here $m \in \mathbb{R}^d$ is the expectation $\mathbb{E}(Z_n)$ and C is the $d \times d$ symmetric positive definite matrix of the covariances of the d (scalar) components $Z_{n,i}$ of Z_n, i.e.

$$c_{ij} = \mathbb{E}((Z_{n,i} - m_i)(Z_{n,j} - m_j)).$$

Of course in the scalar ($d = 1$) case, (16) becomes

$$\frac{1}{(2\pi)^{1/2}\sigma} \exp\left(-\frac{1}{2\sigma^2}(x-m)^2\right), \tag{17}$$

where σ^2 is the variance.

For $m = 0$ the normal distribution (16) is symmetric and therefore the proposal satisfies the symmetry condition required to apply the Metropolis accept/reject formula (15). The overall procedure is then called a Metropolis Random Walk (RW) algorithm. In the absence of other information, it is reasonable to use in (16) a scalar covariance matrix $C = h^2 I_d$ (so that the scalar components $Z_{n,i}$ of the random vector Z_n have a common variance h^2 and are uncorrelated). Then the RW proposal is

$$X_{n+1}^* = X_n + hZ_n, \qquad Z_n \sim N(0, I_d). \tag{18}$$

Let us present an example. Assume that the (univariate) target probability density is[15]

$$\propto \exp(-\beta V(x)), \qquad V(x) = x^4. \tag{19}$$

---

[15]The symbol $\propto$ means proportional to. To obtain a probability density it is necessary to divide $\exp(-\beta V(x))$ by the normalizing constant $\int_{\mathbb{R}} V(x)\,dx$. As pointed out before the Metropolis algorithm does not require the knowledge of the normalizing constant.

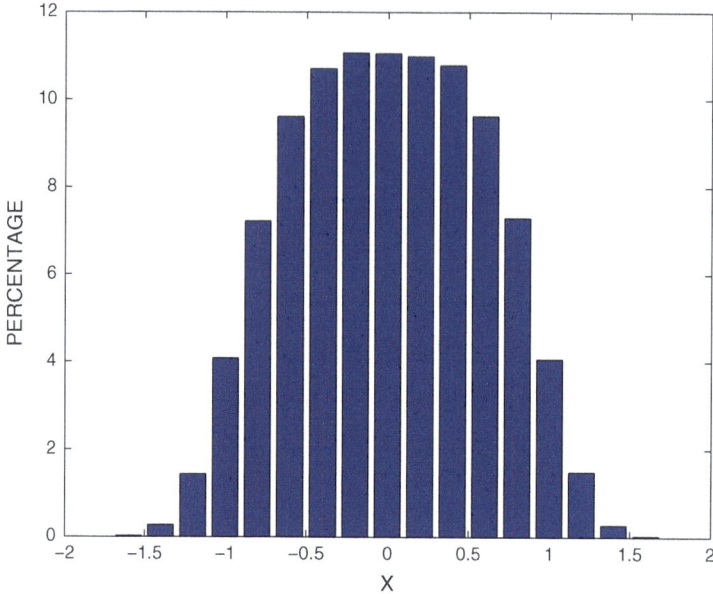

Fig. 6 Histogram of the target density $\propto \exp(-x^4)$ obtained by the RW algorithm with $h = 1$

Anticipating here some elements of the discussion in Sect. 8.3, we mention that (19) would arise in statistical physics as Boltzmann density for a system consisting of a single particle in a potential well with potential $V(x)$ interacting thermally with the environment at absolute temperature $\propto 1/\beta$ (more precisely $\beta = 1/(k_B T_a)$ where k_B is Boltzmann's constant and T_a the absolute temperature). If β is close to ∞ (low temperature) the particle will be at the location $x = 0$ of minimum potential energy. As the temperature increases, the particle is hit e.g. by moving molecules in the environment, and it may leave the minimum $x = 0$. In an ensemble of such systems the value of x will be distributed as in (19).

We have applied to the target (19), the RW algorithm (18). With $\beta = 1, h = 1$ and $N = 1,000,000$ steps, we obtained the histogram in Fig. 6.

Of course the N correlated samples of the target generated by the algorithm contain less information than N independent samples would afford and, therefore, high correlation impairs the usefulness of the samples delivered by the algorithm. In this connection, the choice of the standard deviation h in (18) has a marked effect on the performance of RW. A lower value of h leads to fewer rejections, but the progress of the chain is then slow (i.e. the locations x_{n+1} and x_n at consecutive steps are close to one another). Therefore the correlation between the random variables X_n and X_{n+1} is then high. Large values of h lead to more frequent rejections. Since at a rejected step the chain does not move, $x_{n+1} = x_n$, this also causes an increase of the correlation between X_n and X_{n+1}.

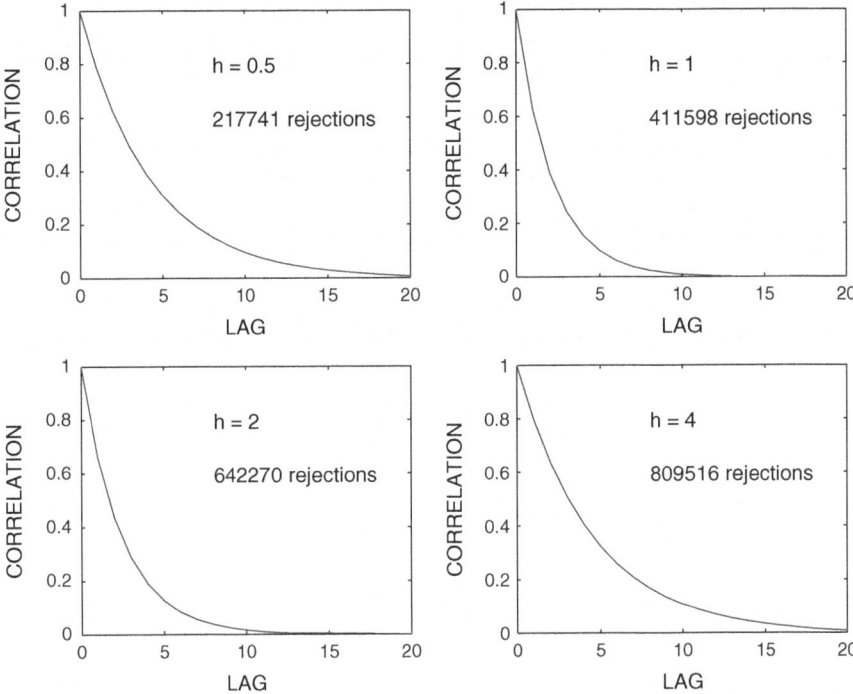

Fig. 7 RW: decay of the empirical correlation coefficient ρ_ν between X_n and $X_{n+\nu}$ as a function of the lag ν

The empirical *auto-covariance* with lag ν of the samples $x_0, \ldots, x_N$ given by the algorithm is

$$\gamma_\nu = \frac{1}{N+1} \sum_{i=0}^{N-\nu} (x_i - \widehat{m})(x_{i+\nu} - \widehat{m}), \qquad \widehat{m} = \frac{1}{N+1} \sum_{i=0}^{N} x_i,$$

and accordingly $\rho_\nu = \gamma_\nu/\gamma_0$ represents the empirical *auto-correlation coefficient*. We would then like that the value ρ_ν approaches zero as quickly as possible as ν increases. Figure 7 illustrates the behavior of ρ_ν as h varies. Note that the number of rejections increases with h and that $h = 1, 2$ are the best choices among those considered.[16]

---

[16] For further details of the statistical analysis of the sequence of samples x_i the reader is referred to [13].

6 Stochastic Differential Equations

In the RW algorithm the choice of proposals is completely independent of the target distribution. It is plausible that, by incorporating into the proposals some knowledge of the target, MCMC algorithms may take large steps from the current position without drastically reducing the chance of having the proposal accepted. Stochastic differential equations (SDEs) provide a means to improve on random walk proposals.

The rigorous study of continuous-time stochastic processes in general and of SDEs in particular is rather demanding mathematically. Here, in the spirit of [18], we present an algorithmic introduction, focused on how to simulate such processes in the computer. This kind of simulation provides much insight and is a very useful first step for those wishing to study these issues.

6.1 The Brownian Motion

It is well known that the Brownian motion of pollen grains in water is named after the Scottish botanist R. Brown who described it 1827. For three quarters of a century the causes of the motion remained unclear, until in 1905 A. Einstein offered a complete explanation in terms of shocks provided by the molecules of water, thus furnishing the definitive proof of the molecular nature of matter. The mathematical Brownian motion (also called the Wiener process or Wiener-Bachelier process) was first studied by Bachelier in 1900 and then by Wiener in the 1920s. The standard or normalized, scalar Brownian motion is a real-value stochastic process B_t ($t \in [0, \infty)$) with the following characteristic features, [12], Chap. 3, [5], Sect. 37:

1. $B_0 = 0$.
2. It has independent increments (i.e. if $0 \leq s_1 < t_1 < s_2 < t_2$, the random variables $B_{t_1} - B_{s_1}$ and $B_{t_2} - B_{s_2}$ are independent).
3. For $t > s$, $B_t - B_s \sim N(0, t-s)$,[17] $0 \leq s < t$ (see (16)).
4. It has continuous paths $t \mapsto B_t$.

A d-dimensional Wiener process takes values in $\mathbb{R}^d$ and its components are independent one-dimensional Wiener processes. The mathematical construction of B may be seen e.g. in [12], Chap. 3, [5], Sect. 37.[18]

After discretizing the variable t on a grid $t = 0, \Delta t, 2\Delta t, \ldots$, the d-dimensional Wiener process may be simulated [18] by the recursion

$$B_{n+1} = B_n + \sqrt{\Delta t}\, Z_n,$$

---

[17]$\sim$ means "has a distribution."

[18]These references also show that (4) is essentially a consequence of (1), (2) and (3).

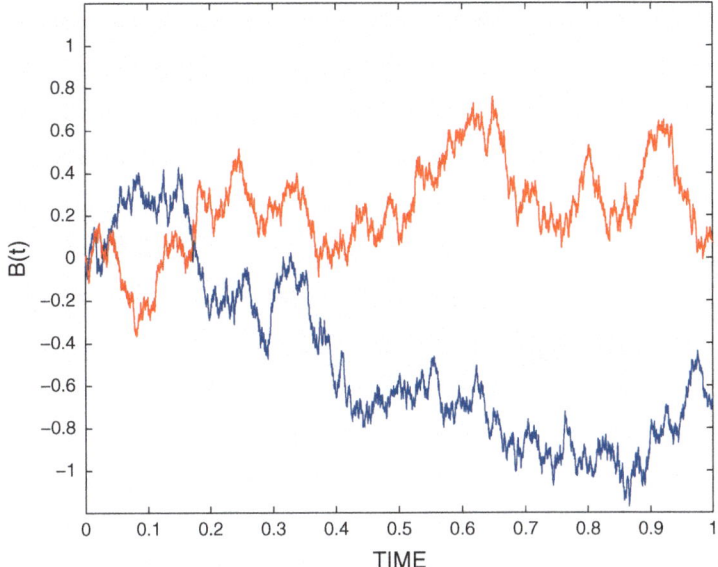

Fig. 8 Two simulated trajectories of the standard Brownian motion

where B_n is the approximation corresponding to $t_n = n\Delta t$ and the Z_n are independent variables with distribution $N(0, I_d)$ (see (16)). Note that the simulated (discrete time) process $\{B_n\}$ has then independent Gaussian increments with the right expectation $\mathbb{E}(B_n) = 0$ and variance $\mathrm{Var}(B_n) = t_n$.

Figure 8 depicts two simulated trajectories for $0 \leq t \leq 1$, $\Delta t = 0.0001$. Since over a time interval of small length Δt the increment $B_{n+1} - B_n$ has the relatively large standard deviation $\sqrt{\Delta t}$, simulated paths have a rugged look. In fact, before discretization, the Wiener paths are almost surely nowhere differentiable, [5], Theorem 37.3, [12], Chap. 3, Theorem 2.2.[19]

6.2 The Euler-Maruyama Method

A stochastic differential equation has the form ([24], Chap. 2, [12], Chap. 5)

$$dX_t = f(X_t, t)dt + \sigma(X_t, t)dB_t, \qquad (20)$$

where f takes values in $\mathbb{R}^d$, σ takes values in the $d \times d'$ real matrices and B_t is a d'-dimensional Wiener process. The first term in the right-hand side provides a

---

[19] The trajectories of the Wiener process are in fact complex objects. For instance, with probability 1, the set $Z(\omega)$ of values t for which a trajectory $B_t(\omega)$ vanishes is closed, unbounded, without isolated points and of Lebesgue measure 0, [5], Theorem 37.4.

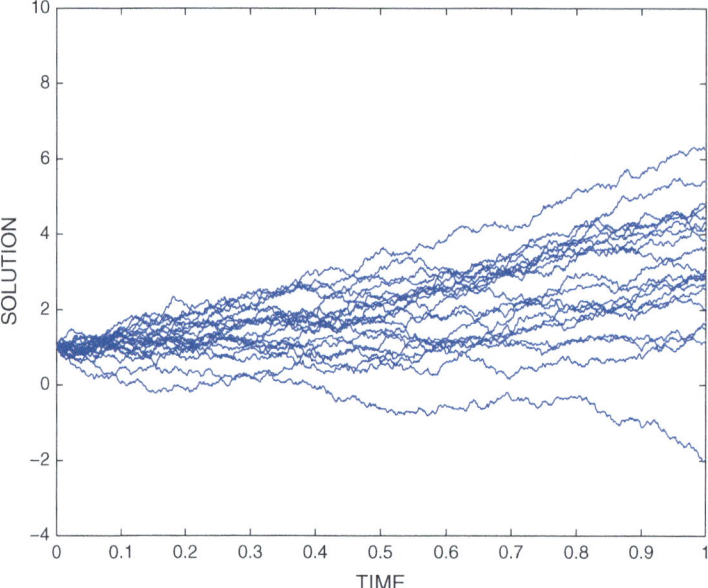

Fig. 9 Ten trajectories of the stochastic differential equation (21)

deterministic *drift;* the second random *noise or diffusion*. The expression dB_t does not make sense, since, as pointed out above, the paths $t \mapsto B_t$ are non-differentiable. In fact the differential equation is shorthand for the integral equation

$$X_t = X_0 + \int_0^t f(X_s, s)ds + \int_0^t \sigma(X_s, s)dB_s,$$

where the last term is an Ito integral ([12], Chap. 4, [24], Chap. 1). Simulations may be carried out by the Euler-Maruyama discretization [18]:

$$X_{n+1} = X_n + \Delta t f(X_n, n\Delta t) + \sqrt{\Delta t}\sigma(X_n, n\Delta t)Z_n,$$

where the Z_n are independent $\sim N(0, I_{d'})$.

As a simple example consider the scalar problem

$$dX_t = X_t dt + dB_t, \quad t > 0, \qquad X_0 = 1 \tag{21}$$

(the initial condition here is deterministic, but cases where X_0 is a random variable often occur). Without noise, the solution would of course be $X_t = \exp(t)$. The Euler-Maruyama discretization is ($Z_n \sim N(0, 1)$):

$$X_{n+1} = X_n + \Delta t X_n + \sqrt{\Delta t} Z_n, \quad n = 0, 1, \ldots, \qquad X_0 = 1.$$

Ten trajectories of the simulated solution X_t, $0 \leq t \leq 1$, may be seen in Fig. 9 ($\Delta t = 0.0001$). Clearly the paths exhibit an upward drift, corresponding to the

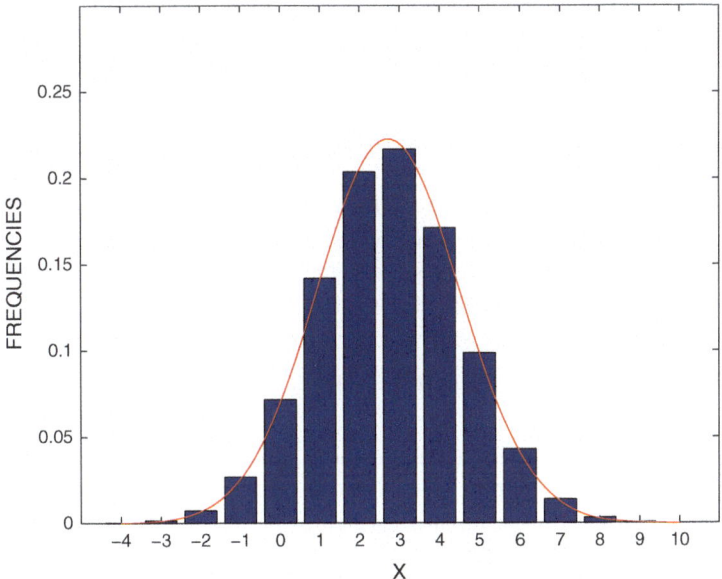

Fig. 10 Histogram of 100,000 samples of $X_{t=1}$, where X_t is the solution of the stochastic differential equation (21). The *line* is the Gaussian density with mean e and variance $(e^2-1)/2$

tern $X_t dt$ in (21), while at the same time showing a diffusion whose variance increases with t. In order to visualize the drift better, we have simulated 100,000 trajectories in $0 \le t \le 1$ with step-size $\Delta t = 0.001$, recorded the value of $X_{t=1}$ for each trajectory and produced an histogram by distributing those 100,000 values into 15 bins centered at $-4, -3, \ldots, 10$. The result may be seen in Fig. 10. A Gaussian density with mean e and variance $(e^2-1)/2$ provides an excellent fit to the distribution of $X_{t=1}$.

6.3 The Fokker-Planck Equation

How did we find the probability distribution of the solution X_t at $t = 1$? The densities $\pi(x,t)$, $x \in \mathbb{R}^d$, of the solution X_t of the SDE (20) obey the *Fokker-Planck* equation [23], Sect. 2.2.1[20]

$$\partial_t \pi(x,t) + \sum_{i=1}^d \partial_i \left(f^i \, \pi(x,t) \right) = \frac{1}{2} \sum_{i,j=1}^d \partial_i \partial_j \left(a^{i,j} \, \pi(x,t) \right), \qquad (22)$$

---

[20] The terminology Fokker-Panck is used in physics; in probability the equation is known as Kolmogorov's forward equation, see e.g. [10], Chap. X, Sect. 5.

where $a = \sigma\sigma^T$, superscripts denote components and f and a are of course evaluated at (x, t). Let us see some examples.

6.3.1 Fokker-Planck Equation: No Drift

Consider first the scalar equation without drift: $dX_t = dB_t$. When $X_0 = 0$ the solution is of course B_t and may be seen as describing the abscissa of a particle that moves due to random shocks from the environment. The Fokker-Planck equation (22) is the familiar heat equation

$$\partial_t \pi(x, t) = \frac{1}{2} \partial_{xx} \pi(x, t);$$

this governs the diffusion of the trajectories of the particle corresponding to different realizations of the process or, in the language of ensembles, the diffusion of an ensemble of particles, initially located at the origin, that evolve randomly independently from one another.

If the initial condition for the partial differential equation is a unit mass located at $x = 0$ (the initial location of the particle is 0 independently of the chance), then the solution, by definition, is the *fundamental solution of the heat equation*, which has the familiar expression:

$$\pi(x, t) = \frac{1}{\sqrt{2\pi t}} \exp\left(-\frac{1}{2}\frac{x^2}{t}\right).$$

Comparing with (17), we conclude that $X_t \sim N(0, t)$; this of course matches the fact that $B_t \sim N(0, t)$ as we know.

6.3.2 Fokker-Planck Equation: No Diffusion

Assume now that the SDE is the standard ordinary differential equation (ODE) $dX_t = f(X_t, t)dt$ so that the dynamics are deterministic. The Fokker-Planck equation (22) reads

$$\partial_t \pi + \nabla \cdot (\pi f) = 0,$$

where we recognize the familiar Liouville equation for the transport of densities by the ODE (see e.g. [21], Sect. 10.1). The trajectories of the ODE are characteristic curves of this first-order linear partial differential equation.

6.3.3 Fokker-Planck Equation: A Linear Example

For the problem (21), the Fokker-Planck equation (22) is given by

$$\partial_t \pi + \partial_x (x\pi) = \frac{1}{2} \partial_{xx} \pi.$$

The solution with initial data given by a unit mass at $x = 1$ is found to be:

$$\frac{1}{\sqrt{2\pi}\sigma(t)} \exp\left(-\frac{1}{2} \frac{(x - m(t))^2}{\sigma^2(t)}\right),$$

with

$$m(t) = \exp(t), \qquad \sigma^2(t) = \frac{\exp(2t) - 1}{2}.$$

Comparison with (17) shows that the solution has, at each t, the Gaussian distribution with variance $\sigma^2(t)$ and expectation $m(t)$ (this average $\exp(t)$ coincides with the solution when the noise is turned off so that the equation becomes $dX_t = X_t dt$).

7 Metropolis Adjusted Langevin Algorithm

The Metropolis adjusted Langevin algorithm (MALA) [32] is an instance of a MCMC where proposals are based on an SDE for which the target π is an invariant density, cf. [30], Sect. 7.8.5. Without loss of generality (densities are positive) the target density is written as $\pi \propto \exp(\mathcal{L})$. Then the proposal is (cf. (18))

$$X^*_{n+1} = X_n + \frac{h^2}{2} \nabla \mathcal{L}(X_n) + h Z_n, \qquad Z_n \sim N(0, I_d).$$

The middle term in the right-hand side (absent in the RW proposal) provides an increment in the direction of steepest ascent in $\mathcal{L}$, thus biasing the exploration of the state space towards high-probability areas. Since the proposal kernel is not symmetric, the Hastings accept/reject mechanism must be used.

More generally, one may use a *preconditioned* version of the proposal

$$X^*_{n+1} = X_n + \frac{h^2}{2} M^{-1} \nabla \mathcal{L}(X_n) + h \sqrt{M^{-1}} Z_n,$$

with M a symmetric positive definite $d \times d$ constant matrix. The idea is that M should be taken "large" in those directions in state space where smaller increments are desirable because the target varies more quickly.

The recipe for the proposal is an Euler-Maruyama step with step-length $\Delta t = \sqrt{h}$ for the SDE

$$dX_t = \frac{1}{2} M^{-1} \nabla \mathscr{L}(X_t) dt + \sqrt{M^{-1}} dB_t$$

whose Fokker-Planck equation has the target $\propto \exp(\mathscr{L}(x))$ as a stationary (time-independent) solution. This implies that, at stationarity in the chain, the proposals will be distributed (except for the Euler discretization error) according to the target: therefore high acceptance rates will be attained. In practice it is of course impossible to start the chain from the stationary distribution, but chains are likely to be asymptotically stationary (Sect. 3.6) and high acceptance rates may be expected in that case.

The paper [33] proves that if the target consists of d independent copies of the same distribution then the RW algorithm requires $h \propto 1/d$ to have $\mathscr{O}(1)$ acceptance probabilities as $d \to \infty$. MALA improves on that because, as shown in [31], it may operate with larger values of h, namely $h \propto (1/d)^{1/3}$. These papers also show that these algorithms perform best when the acceptance probability is approximately 0.234 for the RW case and 0.574 for MALA. These results have recently been extended [25, 29] to situations where the target is not product of equal copies.

As an example of the use of MALA, consider target density

$$\propto \exp(-(1/2)k(r-1)^2), \qquad r = |x|, \qquad x \in \mathbb{R}^d.$$

This is the Boltzmann density (Sect. 8.3) for the motion of a spring in d dimensions. Here we take $d = 100$, $k = 100$; the probability is concentrated in the neighborhood of the unit sphere $|r| = 1$, i.e. essentially on a manifold of dimension 99. We have applied the RW and MALA algorithms. Figures 11 (RW) and 12 (MALA) show the projections of the draws x onto the two-dimensional plane (x_1, x_2) (the corresponding marginal distribution is concentrated in the neighborhood of the origin in the (x_1, x_2)-plane and the correlation in the variable x_1. Clearly MALA is able to take larger steps allowing for a faster exploration of the distribution.

8 Hamiltonian Dynamics

We now turn our attention to Hamiltonian systems, a topic that is essential to formulate the Hybrid Monte Carlo method to be discussed in the next section.

8.1 An Example: Systems of Point Masses

The study of the motion of a conservative system of point masses in three-dimensional space is of much interest in many branches of science. Examples

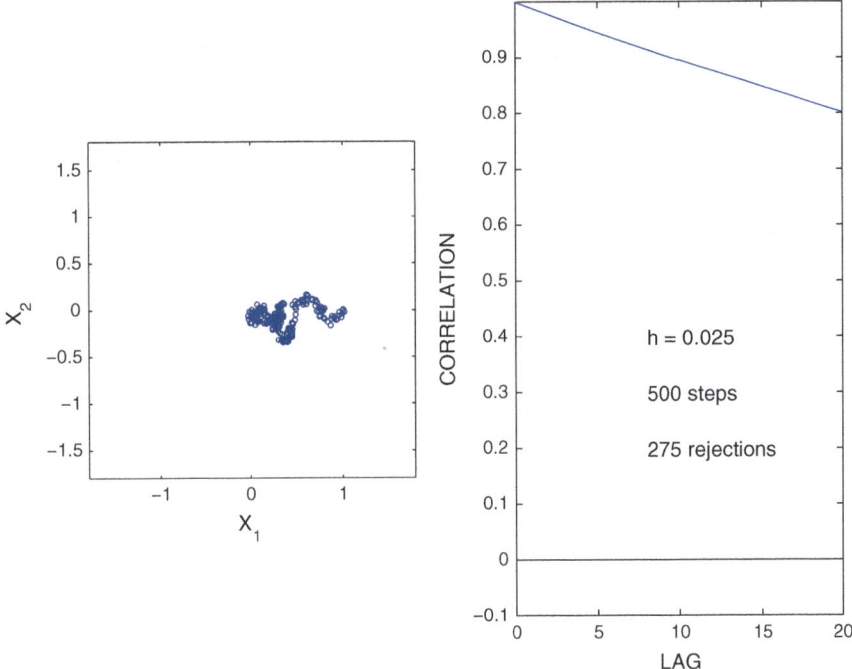

Fig. 11 RW results for a stiff spring in $\mathbb{R}^{100}$. Samples of the coordinates x_1, x_2 and autocorrelation in x_1

range from molecular dynamics, where the particles are molecules or atoms, to astrophysics, where one deals with stars or galaxies. If ν is the number of particles, m_i the mass of the i-th particle, and $\mathbf{r}_i \in \mathbb{R}^3$ its radius vector, Newton's second law reads:

$$m_i \frac{d^2}{dt^2}\mathbf{r}_i = -\nabla_i V(\mathbf{r}_1, \ldots, \mathbf{r}_\nu), \quad i = 1, \ldots, \nu,$$

where the scalar V is the *potential* and $-\nabla_i V$ is the net force on the i-th particle (∇_i means gradient with respect to $\mathbf{r}_i$). This is of course a system of 3ν *second-order* scalar differential equations for the 3ν cartesian components $r_{i,j}$ of the $\mathbf{r}_i$, $j = 1, 2, 3$. After introducing the *momenta*

$$\mathbf{p}_i = m_i \frac{d}{dt}\mathbf{r}_i, \quad i = 1, \ldots, \nu, \qquad (23)$$

the equations may be rewritten in *first-order* form:

$$\frac{d}{dt}\mathbf{p}_i = -\nabla_i V(\mathbf{r}_1, \ldots, \mathbf{r}_\nu), \quad i = 1, \ldots, \nu.$$

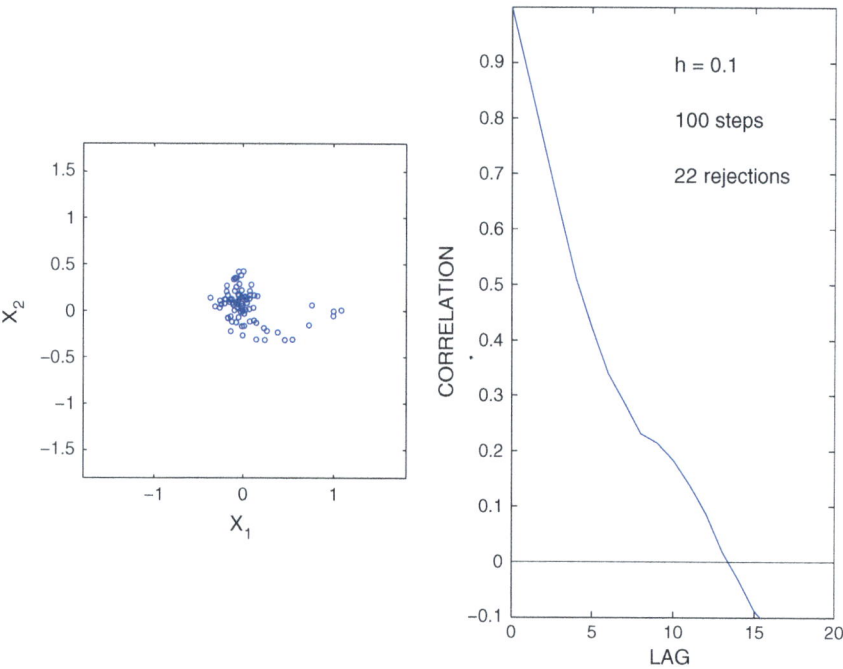

Fig. 12 MALA results for a stiff spring in $\mathbb{R}^{100}$. Samples of the coordinates x_1, x_2 and autocorrelation in x_1

These 3ν scalar equations, together with the 3ν scalar equations in (23) provide a system of $D = 2 \cdot 3 \cdot \nu$ first-order scalar differential equations for the D cartesian components of the vectors $\mathbf{r}_i$ and $\mathbf{p}_i$, $i = 1, \ldots, \nu$. With the introduction of the *Hamiltonian* function

$$H = \sum_i \frac{1}{2m_i} \mathbf{p}_i^2 + V(\mathbf{r}_1, \ldots, \mathbf{r}_\nu) \qquad (24)$$

the first-order system takes the very symmetric *canonical* form:

$$\frac{d}{dt} p_{i,j} = -\frac{\partial H}{\partial r_{i,j}}, \quad \frac{d}{dt} r_{i,j} = +\frac{\partial H}{\partial p_{i,j}}, \quad i = 1, \ldots, \nu, \quad j = 1, 2, 3. \qquad (25)$$

Note that H represents the total mechanical *energy* in the system, composed of a *kinetic* part

$$\sum_i \frac{1}{2m_i} \mathbf{p}_i^2 = \sum_i \frac{1}{2} m_i \left(\frac{d}{dt} \mathbf{r}_i\right)^2$$

and a *potential* part V.

The use of the Hamiltonian format in lieu of Newton's equations is essential in statistical mechanics and quantum mechanics.

8.2 Hamiltonian Systems

In the *phase space* $\mathbb{R}^D$, $D = 2d$, of the points (p, x),

$$p = (p_1, \ldots, p_d) \in \mathbb{R}^d, \qquad x = (x_1, \ldots, x_d) \in \mathbb{R}^d,$$

to each smooth real-valued function $H = H(p, x)$ (the Hamiltonian) there corresponds a first-order differential system of D *canonical* Hamiltonian equations (cf. (25)):

$$\frac{d}{dt}p_j = -\frac{\partial H}{\partial x_j}, \qquad \frac{d}{dt}x_j = +\frac{\partial H}{\partial p_j}, \qquad j = 1, \ldots, d. \qquad (26)$$

In mechanics, as it was the case in (25), the variables $x \in \mathbb{R}^d$ describe the *configuration* of the system, the variables p are the momenta conjugate to x [2] and d is the number of *degrees of freedom*.

Canonical Hamiltonian systems appear very frequently in science; virtually all situations where dissipation is absent or may be neglected may be brought into Hamiltonian form. At the same time, Hamiltonian systems possess properties not frequently found in "general" systems. Before discussing such special properties, it is convenient to introduce the *flow* $\{\Phi_t\}_{t \in \mathbb{R}}$ of the system (26). For each fixed (but arbitrary) t (see [35], Sect. 2.1), Φ_t is a map in phase space, $\Phi_t : \mathbb{R}^D \to \mathbb{R}^D$, defined as follows: for each point (p_0, x_0), $\Phi_t(p_0, x_0)$ is the value at time t of the solution $(p(t), x(t))$ of the canonical equations (26) with value (p_0, x_0) at time 0. The simplest example has $d = 1$ and $H = (1/2)(p^2 + x^2)$, the canonical system is $(d/dt)p = -x$, $(d/dt)x = p$ (the harmonic oscillator). The solution with initial value (p_0, x_0) is

$$p(t) = p_0 \cos t - x_0 \sin t, \qquad x(t) = p_0 \sin t + x_0 \cos t,$$

and therefore Φ_t is the rotation in the plane that moves the point (p_0, x_0) to the point

$$\Phi_t(p_0, x_0) = (p_0 \cos t - x_0 \sin t, \ p_0 \sin t + x_0 \cos t);$$

$\{\Phi_t\}_{t \in \mathbb{R}}$ is the one-parameter family of rotations in the plane. Note that, in general, $\Phi_t(p_0, x_0)$ means:

- If t is varied while keeping (p_0, x_0) fixed: the solution of (26) with initial condition (p_0, x_0).
- If t is fixed and (p_0, x_0) regarded as a variable: a transformation Φ_t in phase space.

- If t is regarded as a parameter: a one-parameter family $\{\Phi_t\}_{t \in \mathbb{R}}$ of transformations in phase-space. This family is a *group*: $\Phi_t \circ \Phi_s = \Phi_{t+s}$, $\Phi_{-t} = \Phi_t^{-1}$.

We now describe the properties of Hamiltonian systems that we shall require when formulating and analyzing the Hybrid Monte Carlo method.

8.2.1 Properties of Hamiltonian Systems: Conservation of Energy

The function H is a *conserved quantity* or *first integral* of (26). In fact, along solutions:

$$\frac{d}{dt} H(p(t), x(t)) = \sum_j \left(\frac{\partial H}{\partial p_j} \frac{d}{dt} p_j + \frac{\partial H}{\partial x_j} \frac{d}{dt} x_j \right)$$

$$= \sum_j \left(-\frac{\partial H}{\partial p_j} \frac{\partial H}{\partial x_j} + \frac{\partial H}{\partial x_j} \frac{\partial H}{\partial p_j} \right) = 0,$$

and therefore

$$H(p(t), x(t)) = H(p(0), x(0)).$$

In terms of the flow, this property simply reads $H \circ \Phi_t = H$ for each t.

For the example (25) with ν point masses, we pointed out that H measures the total energy; therefore the conservation of H corresponds to *conservation of energy*. This is also the case for most Hamiltonian problems.

8.2.2 Properties of Hamiltonian Systems: Conservation of Volume

For each t, Φ_t is a *volume preserving* transformation in phase space ([35], Sect. 2.6): for each (Borel) subset $A \subset \mathbb{R}^D$,

$$\mathrm{Vol}(\Phi_t(A)) = \mathrm{Vol}(A).$$

For the simple example of the harmonic oscillator, this corresponds to the obvious fact that the area of a planar set A does not change when the set is rotated. In general, conservation of volume is a direct consequence of Liouville's theorem: the solution flow of a differential system $\dot{z} = G(z)$ is volume preserving if and only if the corresponding vector field G is divergence-free ([2], Sect. 16). Indeed for a canonical system the divergence is

$$\sum_{j=1}^{d} \left(\frac{\partial}{\partial p_j} \left(-\frac{\partial H}{\partial x_j} \right) + \frac{\partial}{\partial x_j} \frac{\partial H}{\partial p_j} \right) = 0.$$

In terms of the flow, conservation of volume simply reads: $\det(\Phi'_t) \equiv 1$, for each t where Φ'_t denotes the Jacobian of Φ_t.

8.2.3 Properties of Hamiltonian Systems: Reversibility

Consider now the momentum-flip symmetry S in phase-space defined by

$$S(p, x) = (-p, x)$$

and assume that $H \circ S = H$, i.e. the Hamiltonian is an even function of the momenta as in (24) and many other mechanical systems. If $(p(t), x(t))$ is a solution of the canonical equations (26), so is $(\widehat{p}(t), \widehat{x}(t)) := (-p(-t), x(-t))$. The proof is simple:

$$\frac{d}{dt}\widehat{p}_i(t) = \frac{d}{dt}p_i(-t) = -\frac{\partial H}{\partial x_i}(p(-t), x(-t)) = -\frac{\partial H}{\partial x_i}(\widehat{p}(t), \widehat{x}(t))$$

and similarly for x_i. Since $(\widehat{p}(0), \widehat{x}(0)) = S(p, x)$, this fact, called *reversibility* of the flow, may be compactly written as

$$\Phi_t \circ S = S \circ \Phi_{-t};$$

see Fig. 13 that portraits the two-dimensional phase space of the Hamiltonian function

$$H(p, x) = \frac{1}{2}p^2 + V(x), \qquad V(x) = (x^2 - 1)^2. \tag{27}$$

The significance of reversibility is well known: if a movie is made of a motion of a reversible system and projected backwards what we see is also a possible (forwards) motion of the system. In Fig. 13, if the forwards movie shows the sequence of configurations $x(t)$ from $t = 0$ to $t = T$ (top circle to top diamond), when projected backwards will display in reversed order the same configurations and that sequence of configurations corresponds to the solution that at time $t = 0$ starts at the lower diamond and reaches the lower circle at $t = T$.

8.3 The Canonical Density

Consider a mechanical system whose time-evolution, when isolated from rest of the universe, is governed by the canonical equations associated with a Hamiltonian $H(p, x)$ (for instance our earlier system of ν point masses). As discussed above, the value of $H(p(t), x(t))$ (the energy) along the evolution remains constant. Assume

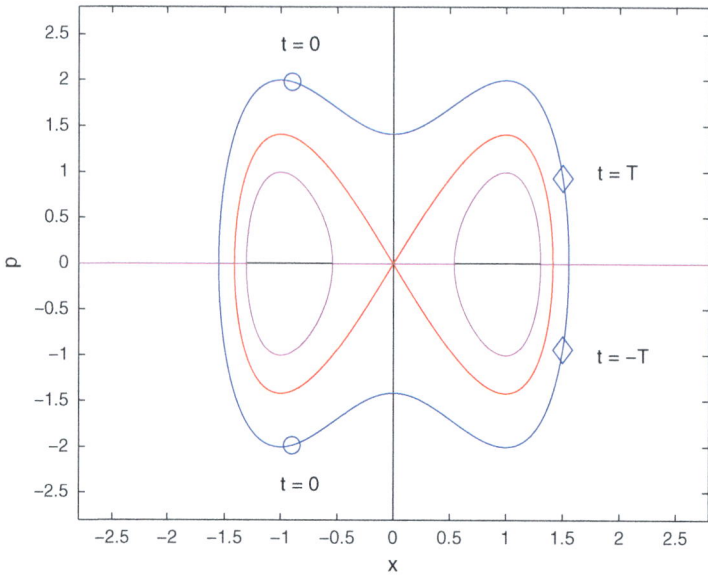

Fig. 13 Reversibility: $\Phi_t \circ S = S \circ \Phi_{-t}$. Begin from the *lower circle*, flip the momentum to get the *upper circle* and use the solution flow to reach the *upper diamond* after T units of time. The final result is the same as one would get by first evolving $-T$ units of time to reach the *lower diamond* and then flipping the momentum

now that this system is not isolated but interacts with an environment at constant temperature (for instance our point masses collide with water molecules of a heat bath that surrounds them). There will be exchanges of energy between the system and the environment, the value of H will not remain constant and the canonical equations (26) will not describe the time evolution. The energy exchanges with the environment must be modeled as random and therefore, for each fixed t, $(p(t), x(t))$ are random variables that need to be described statistically. Once thermal equilibrium with the environment has been reached, the stochastic process $(p(t), x(t))$ possesses a stationary probability distribution: this is the *Maxwell-Boltzmann* distribution with density

$$\propto \exp(-\beta H(p, x)), \qquad \beta = \frac{1}{k_B T_a} \qquad (28)$$

(here T_a is the absolute temperature of the environment and k_B the Boltzmann constant). The corresponding ensemble is called the *canonical ensemble* ([36], Sect. 12.5.3, [21], Sect. 10.2). Thus, at any temperature, a canonical ensemble contains few systems at locations of high energy. However, as the temperature T_a increases locations of high energy become more likely.

8.3.1 The Maxwell-Boltzmann Distribution

As an example consider again the system of point masses with Hamiltonian (24). The canonical density (28) is

$$\propto \exp\left(-\beta\left(\sum_{i=1}^{\nu} \frac{1}{2m_i}\mathbf{p}_i^2 + V(\mathbf{r}_1, \ldots, \mathbf{r}_\nu)\right)\right),$$

and since the exponential may be rewritten as a product, the $\nu + 1$ random vectors $\mathbf{p}_1 \in \mathbb{R}^3, \ldots, \mathbf{p}_\nu \in \mathbb{R}^3$, $(\mathbf{r}_1, \ldots, \mathbf{r}_\nu) \in \mathbb{R}^{3\nu}$ are mutually independent. The density of $\mathbf{p}_i$ is then ([11], Sect. 40.4), as first established by Maxwell in 1859,

$$\propto \exp(-\frac{\beta}{2m_i}\mathbf{p}_i^2),$$

and comparison with (16) shows that the distribution of $\mathbf{p}_i$ is Gaussian with zero mean and covariance matrix $(m_i/\beta)I_3$. In particular there is no correlation among the three cartesian components $p_{i,j}$ of $\mathbf{p}_i$ and each of these components has variance $m_i/\beta = m_i k_B T_a$. It follows that the kinetic energy $p_{i,j}^2/(2m_i)$ of the i-th mass along the j-axis has an ensemble average $(1/2)k_B T_a$; in other words the absolute temperature coincides, up to a normalizing factor, with the kinetic energy in any of the 3ν degrees of freedom of the system (in fact this is the *definition* of absolute temperature [11], Sect. 39.4).

The configuration, specified by $(\mathbf{r}_1, \ldots, \mathbf{r}_\nu) \in \mathbb{R}^{3\nu}$, is, as noted above, independent of the momenta, and possesses the density

$$\propto \exp(-\beta V(\mathbf{r}_1, \ldots, \mathbf{r}_\nu)).$$

In statistical mechanics this is called the *Boltzmann* density for the potential V ([11], Sect. 40.2).

8.3.2 Preservation of the Canonical Density by the Hamiltonian Flow

We shall need later the following result:

Theorem 7. *For each fixed t, the canonical density (28) is preserved by the flow Φ_t of the Hamiltonian system (26):*

$$\int_{\Phi_t(A)} \exp(-\beta H(p,x))\, dp\, dx = \int_A \exp(-\beta H(p,x))\, dp\, dx,$$

for each (Borel) subset A of the phase space $\mathbb{R}^D$.

Proof. Change variables $(p, x) = \Phi_t(\tilde{p}, \tilde{x})$ in the first integral; $H(\Phi_t(\tilde{p}, \tilde{x})) = H(\tilde{x}, \tilde{p})$ by conservation of energy and the required Jacobian determinant is unity by conservation of volume. □

In an image due to Gibbs, one places at $t = 0$ points in the phase space $\mathbb{R}^D$ of H in such a way that they are distributed with probability density $\propto \exp(-\beta H)$. Each point represents a system in the canonical ensemble. As t varies each point will move in the phase space following (26); the theorem implies that the density at any point (p, x) will remain constant.

8.4 Numerical Methods for Hamiltonian Problems

The analytic integration of Hamilton's canonical equations (26) is usually impossible and one has to resort to numerical integrators. In the last 25 years it has become clear that when integrating Hamiltonian problems it is essential in many applications to use numerical methods that possess conservation properties similar to those shared by Hamiltonian systems, like reversibility, conservation of volume, etc. The construction and analysis of such numerical methods is part of the field of *geometric integration,* a term coined in [34]. An introductory early monograph is [35] and a more recent expositions are given in [16, 22]. Here we limit ourselves to the material required later to describe the Hybrid Monte Carlo method.

Each one-step *numerical method* to integrate (26) is specified by a map $\psi_{\Delta t} : \mathbb{R}^D \to \mathbb{R}^D$, where Δt represents the step-length. The approximation (p^{m+1}, x^{m+1}) to the true solution value $(p((m+1)\Delta t), x((m+1)\Delta t))$ is obtained recursively by

$$(p^{m+1}, x^{m+1}) = \psi_{\Delta t}(p^m, x^m).$$

In this way the approximate solution at time T (for simplicity we assume that $T/\Delta t$ is an integer) is obtained by applying $T/\Delta t$ times the mapping $\psi_{\Delta t}$, or, in other words, the true T-flow $\Phi_T = \Phi_{\Delta t}^{T/\Delta t}$ is approximated by $\Psi_T := \psi_{\Delta t}^{T/\Delta t}$, the composition of $\psi_{\Delta t}$ $T/\Delta t$ times with itself.

It turns out that it is impossible to construct a general integrator $\psi_{\Delta t}$ that exactly preserves energy and volume ([35], Sect. 10.3.2). Faced with this impossibility, it is advisable to drop the requirement of exact conservation of energy and demand exact conservation of volume.[21] The best-known volume preserving, reversible algorithm to integrate Hamiltonian systems is the Verlet/Stoermer/leapfrog algorithm applicable when the Hamiltonian has the special (but common) form (cf. (24))

$$H = \frac{1}{2} p^T M^{-1} p + V(x);$$

---

[21]More precisely it is customary to insist in the integrator being *symplectic* [35], Chap. 6; symplecticness implies conservation of volume and satisfactory—but not exact—conservation of energy, [35], Sect. 10.3.3. The Verlet scheme is symplectic.

M a constant positive definite symmetric matrix—the so-called mass matrix. For our purposes we may think of this method as a *splitting* (fractional step) algorithm, see [35], Sect. 12.4. The Hamiltonian H is written as a sum $H = H^{(1)} + H^{(2)}$ of potential and kinetic parts with

$$H^{(1)}(p,x) = V(x), \quad H^{(2)}(p,x) = \frac{1}{2} p^T M^{-1} p.$$

For the Hamiltonian $H^{(1)}$, the equations of motion are $(d/dt)p = -(\partial/\partial x)V(x)$, $(d/dt)x = 0$, with solutions

$$p(t) = p_0 - t \frac{\partial}{\partial x} V(x_0), \quad x(t) = x_0.$$

For the Hamiltonian $H^{(2)}$, the equations of motion are $(d/dt)p = 0$, $(d/dt)x = M^{-1}p$ leading to

$$p(t) = p_0, \quad x(t) = x_0 + tM^{-1}p_0.$$

Then the method is defined by the familiar Strang's splitting recipe:

$$\psi_{\Delta t} := \Phi^{(1)}_{\Delta t/2} \circ \Phi^{(2)}_{\Delta t} \circ \Phi^{(1)}_{\Delta t/2}$$

($\Phi^{(i)}$ is the flow of $H^{(i)}$). In this way, given (p^m, x^m), we compute the approximation $(p^{m+1}, x^{m+1}) = \psi_{\Delta t}(p^m, x^m)$ at the next time level by means of the three fractional steps:

$$p^{m+1/2} = p^m - \frac{\Delta t}{2} \frac{\partial}{\partial x} V(x^m),$$

$$x^{m+1} = x^m + \Delta t M^{-1} p^{m+1/2},$$

$$p^{m+1} = p^{m+1/2} - \frac{\Delta t}{2} \frac{\partial}{\partial x} V(x^{m+1}).$$

Since the individual transformations

$$(p^m, x^m) \mapsto (p^{m+1/2}, x^m),$$
$$(p^{m+1/2}, x^m) \mapsto (p^{m+1/2}, x^{m+1}),$$
$$(p^{m+1/2}, x^{m+1}) \mapsto (p^{m+1}, x^{m+1})$$

are flows of canonical systems they preserve volume. As a result $\psi_{\Delta t}$ (which is the composition of the three) and $\Psi_T = \psi_{\Delta t}^{T/\Delta t}$ preserve volume.

The reversibility of $\psi_{\Delta t}$ (and hence that of Ψ_T i.e. $S \circ \Psi_T = \Psi_T^{-1} \circ S$) is easily checked and is a consequence of the symmetric pattern of the Strang splitting.

More sophisticated reversible, volume preserving splitting algorithms exist, but the Verlet method is commonly used in molecular simulations and other application areas.

9 The Hybrid Monte Carlo Method

The Hybrid Monte Carlo (HMC) algorithm originated in the physics literature [8] and, while it may be used in other application fields such as Bayesian statistics (see e.g. [14]), its description requires to think of the given problem in physical terms. Let us first present the idea that underlies the method.

9.1 The Idea

Without loss of generality, we write the target density $\pi(x)$ in the state space $\mathbb{R}^d$ as $\exp(-V(x))$ and, regardless of the application in mind, think of $x \in \mathbb{R}^d$ as specifying the configuration of a mechanical system and of $V(x)$ as the corresponding potential energy. We choose arbitrarily $T > 0$ and a positive definite symmetric matrix M (M is often diagonal). Next we consider the Hamiltonian function

$$H = \frac{1}{2} p^T M^{-1} p + V(x)$$

and think of p as momenta and of M as a mass matrix. For the canonical probability distribution in the phase space $\mathbb{R}^D$, $D = 2d$, with density (we set $\beta = 1$ for simplicity)

$$\propto \exp(-H) = \exp\left(-\frac{1}{2} p^T M^{-1} p\right) \times \exp\left(-V(x)\right)$$

the random vectors $p \in \mathbb{R}^d$ and $x \in \mathbb{R}^d$ are stochastically independent (we found a similar independence in Sect. 8.3.1). The (marginal) distribution of x is our target $\pi(x) = \exp(-V(x))$ and p has a Gaussian density $\propto \exp\left(-(1/2) p^T M^{-1} p\right)$ so that samples from p are easily available. In this set-up, Theorem 7 suggests a means to construct a Markov chain in $\mathbb{R}^d$ reversible with respect to the target $\pi(x)$:

Theorem 8. *Define the transitions $x_n \mapsto x_{n+1}$ in the state space $\mathbb{R}^d$ by the following procedure:*

- *Draw p_n from the Gaussian density $\propto \exp\left(-(1/2) p^T M^{-1} p\right)$.*
- *Find $(p_{n+1}^*, x_{n+1}) = \Phi_T(p_n, x_n)$, where Φ_T is the T-flow of the canonical system (26) with Hamiltonian function H.*

Then $x_n \mapsto x_{n+1}$ defines a Markov chain in $\mathbb{R}^d$ that has the target $\pi(x) \propto \exp(-V(x))$ as an invariant probability distribution. Furthermore this Markov chain is reversible with respect to $\pi(x)$.

Proof. The Markov property is obvious: the past enters the computation of x_{n+1} only through the knowledge of x_n. If X_n is distributed $\sim \pi(x)$, then, by the choice of p_n, the random vector (P_n, X_n) has the canonical density $\propto \exp(-H)$. By Theorem 7, (P_{n+1}^*, X_{n+1}) also has density $\propto \exp(-H)$; accordingly, the density of X_{n+1} will be the marginal $\pi(x)$. The reversibility of the chain is a simple consequence of the reversibility of the flow Φ_T. □

The main appeal of this procedure is that, unlike the situation in RW or MALA, the transitions $x_n \mapsto x_{n+1}$ are non-local in the state space $\mathbb{R}^d$, in the sense that x_{n+1} may be far away from the previous state x_n. Figure 13, as we know, corresponds to the double well potential V in (27) with minima at $x = \pm 1$, so that the target $\exp(-V(x))$ has modes (locations of maximum probability density) at $x = \pm 1$. If the current location x_n is the abscissa of the circles in that figure and the drawing of p_n leads to the point (p_n, x_n) depicted by the upper circle, then the T-flow of the Hamiltonian system yields the upper diamond and x_{n+1} will be the corresponding abscissa. In this way the procedure has carried out, in a single step of the Markov chain, a transition from the neighborhood of the mode at $x = 1$ to the neighborhood of the mode at $x = -1$.

Note that, once x_{n+1} has been determined, the momentum vector p_{n+1}^* is discarded and a fresh p_{n+1} is drawn. Therefore the next starting location (p_{n+1}, x_{n+1}) will have $H(p_{n+1}, x_{n+1}) \neq H(p_{n+1}^*, x_{n+1}) = H(p_n, x_n)$. This makes it possible to explore the whole phase space in spite of the fact that each point only flows within the corresponding level set of the energy H.

Unfortunately the procedure in Theorem 8 cannot be implemented in practice: the required flow Φ_T is not explicitly known except in simple academic examples!

9.2 The Algorithm

In order to turn the procedure we have studied into a practical algorithm, the exact flow Φ_T is replaced by a numerical approximation Ψ_T as in Sect. 8.4 and an accept/reject mechanism is introduced to assure that the resulting chain still has the target as an invariant distribution. The accep/reject recipe is greatly simplified if integrator Ψ_T is *volume preserving and reversible*, something we assume hereafter (Verlet is integrator of choice).

The transition $x_n \mapsto x_{n+1}$ in HMC is as follows:

- Draw a value p_n from the density $\exp\left(-(1/2)p^T M^{-1} p\right)$.
- Find $(p_{n+1}^*, x_{n+1}^*) = \Psi_T(p_n, x_n)$ (i.e. perform $T/\Delta t$ time-steps of the chosen numerical integrator with step-length Δt). Discard p_{n+1}^* and take x_{n+1}^* as proposal.
- Set $x_{n+1} = x_{n+1}^*$ with probability

$$1 \wedge \exp\left(-\left(H(p_{n+1}^*, x_{n+1}^*) - H(p_n, x_n)\right)\right)$$

(acceptance). If the proposal is rejected set $x_{n+1} = x_n$.

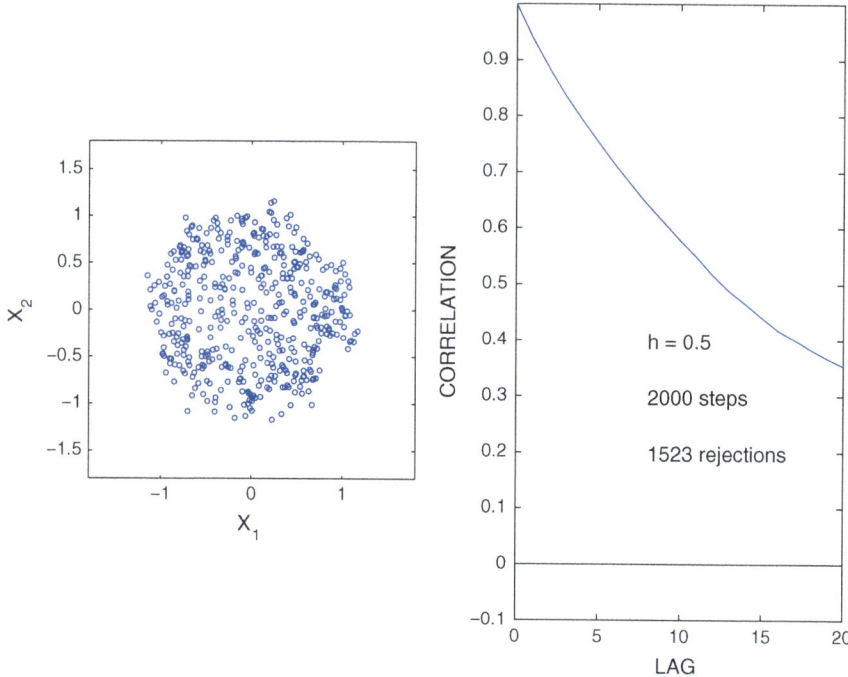

Fig. 14 RW results for a stiff spring in $\mathbb{R}^3$. Samples of the coordinates x_1, x_2 and autocorrelation in x_1

Analyses of HMC are given in [7, 37]. The following result, whose proof is postponed to Sect. 9.3, holds:

Theorem 9. *In the situation just described, the transitions $x_n \mapsto x_{n+1}$ define a Markov chain reversible with respect to the target $\pi(x) \propto \exp(-V(x))$.*

Some comments are in order. If the exact flow Φ_T were known and we used it as "numerical integrator", i.e. $\Psi_T = \Phi_T$, then, by conservation of energy,

$$\exp\left(-\left(H(p^*_{n+1}, x^*_{n+1}) - H(p_n, x_n)\right)\right) = 1$$

and every proposal would be accepted: one is then back in the procedure covered by Theorem 8. In a similar vein, the better the numerical scheme Ψ_T preserves H the higher the probability of acceptance.[22]

---

[22]In this connection it may be worth noting that the proof of Theorem 9 demands that the mapping Ψ_T is time reversible and volume preserving, but would work even if Ψ_T were not an approximation to the true Φ_T. However if Ψ_T is not close to Φ_T, the acceptance probability will be low.

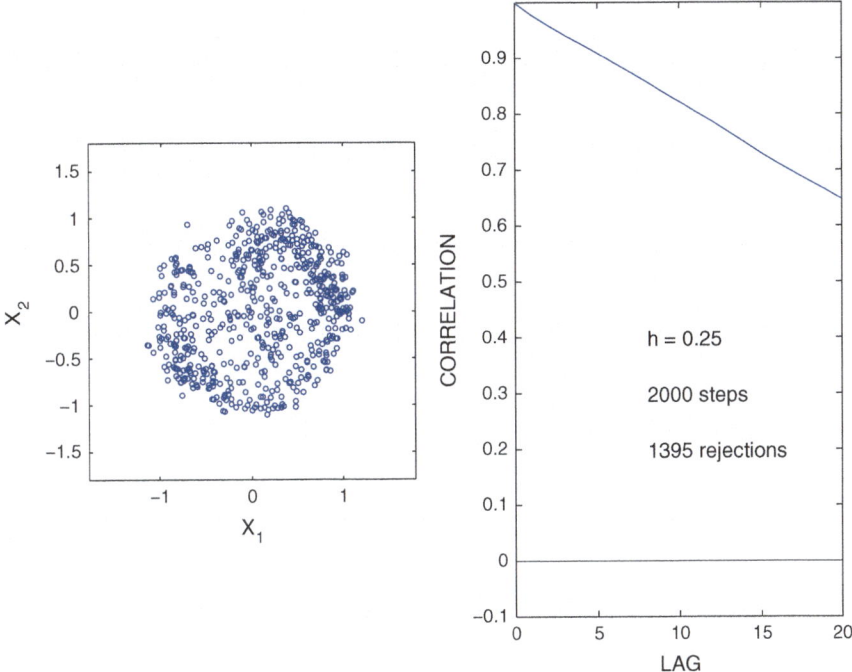

Fig. 15 MALA results for a stiff spring in $\mathbb{R}^3$. Samples of the coordinates x_1, x_2 and autocorrelation in x_1

With HMC, a *Markov chain step* $x_n \mapsto x_{n+1}$ requires $T/\Delta t$ *time-steps* of the numerical integrator. In the particular case where the integrator is the Verlet scheme and $\Delta t = T$, so that there is a single time-step per step of the chain, it is easy to check that HMC is identical to MALA with $h = \Delta t$ (more precisely, given x_n, the proposal x_{n+1}^* and the accept/reject mechanism are the same in both algorithms). This equivalence is somewhat surprising as MALA proposals are motivated by an SDE, whereas HMC proposals are based on deterministic Hamiltonian dynamics.[23] After this equivalence MALA/HMC one may think of HMC as a non-local version of MALA.

The paper [4] shows that if the target consists of d independent copies of the same distribution, then the Verlet time-step Δt should be chosen $\propto (1/d)^{1/4}$ to have $\mathcal{O}(1)$ acceptance probabilities as $d \to \infty$. For reversible, volume preserving integrators of (necessarily even) order 2ν, $\Delta t \propto (1/d)^{1/(4\nu)}$. This compares favorably with the corresponding relations for RW and MALA reported in Sect. 7.

---

[23]Recall that when the MALA proposal is seen as an Euler-Maruyama step for an SDE, the MALA parameter h coincides with the *square* of the time-step Δt. However in the relation of MALA with HMC studied in this section, $h = \Delta t$ as we have just pointed out.

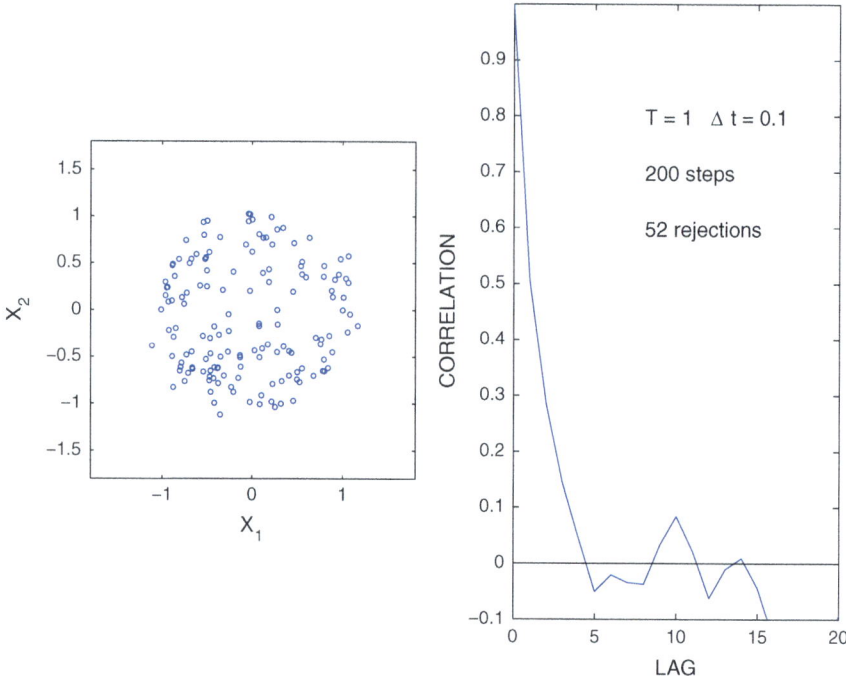

Fig. 16 HMC results for a stiff spring in $\mathbb{R}^3$. Samples of the coordinates x_1, x_2 and autocorrelation in x_1

In our description in Sect. 8.4 we observed that the Verlet algorithm is based on splitting H into its kinetic and potential part. This is not the only possibility of splitting, see [3,38]. Modifications of HMC may be seen in [1,19,20], among others.

We now turn our attention to an example. As in Sect. 7, consider the target $\propto \exp(-(1/2)k(r-1)^2)$, $k = 100$ but now $d = 3$. We show the draws in the two-dimensional plane of the random vector (x_1, x_2) (the corresponding marginal is approximately uniform on the unit disk) and the correlation in the variable x_1. Figures 14–16 show results for RW, MALA and HMC (with Verlet integration) respectively. The superiority of HMC in this example is manifest.

9.3 Proofs

The proof of Theorem 9, borrowed from [23], is based on some lemmas. We shall use repeatedly the fact that the momentum-flip symmetry S, $S(p,x) = (-p,x)$, preserves the canonical probability measure μ: $\mu(S(A)) = \mu(A)$ for each Borel subset A of $\mathbb{R}^D$.

Lemma 1. *Consider a Borel probability measure μ in $\mathbb{R}^D$ that is preserved by the momentum-flip symmetry S and a transition kernel K in the phase space $\mathbb{R}^D$ that satisfies the following analogue of the detailed balance condition (14):*

$$\int_A \mu(d\xi)\, K(\xi, B) = \int_B \mu(d\eta)\, K(S(\eta), S(A)) \tag{29}$$

for each (Borel measurable) A, B. Then (cf. Theorem 3):

- *The measure μ is invariant with respect to K.*
- *At stationarity, the chain $\Xi_0, \ldots, \Xi_N$ generated by K is statistically the same as the chain $S(\Xi_N), \ldots, S(\Xi_0)$.*

Proof. With $A = \mathbb{R}^D$, the hypothesis (29) implies,

$$\int_{\mathbb{R}^D} \mu(d\xi) K(\xi, B) = \int_B \mu(d\eta) K(S(\eta), \mathbb{R}^D) = \int_B \mu(d\eta) = \mu(B);$$

this proves stationarity (see (13)).

By definition of conditional probability,

$$\mathbb{P}\big(S(\Xi_n) \in S(A) \mid S(\Xi_{n+1}) \in S(B)\big)$$

$$= \mathbb{P}\big(\Xi_n \in A \mid \Xi_{n+1} \in B\big) = \frac{\mathbb{P}\big(\Xi_n \in A \wedge \Xi_{n+1} \in B\big)}{\mathbb{P}\big(\Xi_{n+1} \in B\big)}. \tag{30}$$

Let us rewrite the last fraction. At stationarity $\mathbb{P}(\Xi_{n+1} \in B) = \mathbb{P}(\Xi_n \in B)$; furthermore, after the change of variables $\eta = S(\xi)$, (29) becomes (recall that $\mu(S(d\xi)) = \mu(d\xi)$)

$$\int_A \mu(d\xi)\, K(\xi, B) = \int_{S(B)} \mu(d\xi)\, K(\xi, S(A)),$$

which means

$$\mathbb{P}\big(\Xi_n \in A \wedge \Xi_{n+1} \in B\big) = \mathbb{P}\big(\Xi_n \in S(B) \wedge \Xi_{n+1} \in S(A)\big)$$

Taking these results back to (30)

$$\mathbb{P}\big(S(\Xi_n) \in S(A) \mid S(\Xi_{n+1}) \in S(B)\big)$$
$$= \frac{\mathbb{P}\big(\Xi_n \in S(B) \wedge \Xi_{n+1} \in S(A)\big)}{\mathbb{P}\big(\Xi_n \in S(B)\big)} = \mathbb{P}\big(\Xi_{n+1} \in S(A) \mid \Xi_n \in S(B)\big).$$

□

Lemma 2. *As above, let μ be a measure preserved by the momentum-flip map S. Assume that K^* is a (proposal) Markov kernel in $\mathbb{R}^D$, such that the measures*

$$K^*\big(S(\eta), S(d\xi)\big)\,\mu(d\eta), \qquad K^*(\xi, d\eta)\,\mu(d\xi)$$

in $\mathbb{R}^D \times \mathbb{R}^D$ are equivalent (i.e. each has a density with respect to the other), so that there is a function r such that

$$r(\xi, \eta) = \frac{K^*\big(S(\eta), S(d\xi)\big)\,\mu(d\eta)}{K^*(\xi, d\eta)\,\mu(d\xi)}. \tag{31}$$

Define a Markov transition $\xi_n \mapsto \xi_{n+1}$ in $\mathbb{R}^D$ by:

- *Propose ξ_{n+1}^* according to $K^*(\xi_n, \cdot)$.*
- *Accept ($\xi_{n+1} = \xi_{n+1}^*$) with probability $1 \wedge r(\xi_n, \xi_{n+1}^*)$. If rejection occurs set $\xi_{n+1} = S(\xi_n)$.*

The chain defined in this way satisfies the generalized detailed balance condition (29) and, in particular, μ is an invariant measure.

Proof. The kernel of the chain is:

$$K(\xi, d\eta) = \big(1 \wedge r(\xi, \eta)\big)\,K^*(\xi, d\eta) + \big(1 - \alpha(\xi)\big)\,\delta_{S(\xi)}(d\eta),$$

where

$$\alpha(\xi) = \int_{\mathbb{R}^D} \big(1 \wedge r(\xi, \eta)\big)\,K^*(\xi, d\eta)$$

is the probability of acceptance conditioned to $\Xi_n = \xi$ and δ denotes a point unit mass (Dirac's delta).

We have then to show that

$$\big(1 \wedge r(\xi, \eta)\big)\,K^*(\xi, d\eta)\,\mu(d\xi) + \big(1 - \alpha(\xi)\big)\,\delta_{S(\xi)}(d\eta)\,\mu(d\xi) = \tag{32}$$
$$\big(1 \wedge r(S(\eta), S(\xi))\big)\,K^*\big(S(\eta), S(d\xi)\big)\,\mu(d\eta) + \big(1 - \alpha(S(\eta))\big)\,\delta_\eta(S(d\xi))\,\mu(d\eta),$$

a task that we carry out by proving that the first and second term in the left-hand side coincide with the first and second term in the right-hand side respectively. For the second terms, if ϕ is a test function, the change of variables $\xi = S(\xi')$ enables us to write

$$\int_{\mathbb{R}^D \times \mathbb{R}^D} \phi(\xi, \eta)\,\big(1 - \alpha(S(\eta))\big)\,\delta_\eta(S(d\xi))\,\mu(d\eta) =$$
$$\int_{\mathbb{R}^D \times \mathbb{R}^D} \phi(S(\xi'), \eta)\big(1 - \alpha(S(\eta))\big)\,\delta_\eta(d\xi')\,\mu(d\eta)$$

and, by definition of δ, the last integral has the value

$$\int_{\mathbb{R}^D} \phi(S(\eta), \eta)(1 - \alpha(S(\eta))) \, \mu(d\eta).$$

Now the change of variables $\eta = S(\xi)$ and the definition of δ allow us to continue

$$\int_{\mathbb{R}^D} \phi(S(\eta), \eta)(1 - \alpha(S(\eta))) \, \mu(d\eta) = \int_{\mathbb{R}^D} \phi(\xi, S(\xi))(1 - \alpha(\xi)) \, \mu(d\xi)$$

$$= \int_{\mathbb{R}^D \times \mathbb{R}^D} \phi(\xi, \eta)(1 - \alpha(\xi)) \, \delta_{S(\xi)}(d\eta) \, \mu(d\xi).$$

This proves that the second terms in (32) are equal. For the first terms note that $r(\xi, \eta) = 1/r(S(\eta), S(\xi))$. Thus:

$$(1 \wedge r(\xi, \eta)) \, K^*(\xi, d\eta) \, \mu(d\xi) = (r(S(\eta), S(\xi)) \wedge 1) \, r(\xi, \eta) \, K^*(\xi, d\eta) \mu(d\xi)$$

and by definition of r this has the value:

$$(1 \wedge r(S(\eta), S(\xi))) \, K^*(S(\eta), S(d\xi)) \, \mu(d\eta).$$

$\square$

Lemma 3. *Let μ be the measure in $\mathbb{R}^D$ with density $\exp(-H)$, where $H \circ S = H$ and assume that Ψ_T is a reversible and volume preserving transformation in phase space (in particular the numerical solution operator associated with a reversible, volume preserving integrator for the Hamiltonian system associated with H). Define a transition kernel by*

$$K^*(\xi, d\eta) = \delta_{\Psi_T(\xi)}(d\eta).$$

Then μ and K^* satisfy the requirements in Lemma 2 and the Metropolis-Hastings ratio r in (31) has the value $\exp\bigl(-(H(\Psi_T(\xi)) - H(\xi))\bigr)$.

Proof. For the measure in the numerator of (31), the integral of a test function ϕ is

$$I_N = \int_{\mathbb{R}^D \times \mathbb{R}^D} \phi(\xi, \eta) \, \delta_{\Psi_T(S(\eta))}(S(d\xi)) \, \mu(d\eta).$$

Changing $\xi = S(\xi')$ leads to

$$I_N = \int_{\mathbb{R}^D \times \mathbb{R}^D} \phi(S(\xi'), \eta) \, \delta_{\Psi_T(S(\eta))}(d\xi') \, \mu(d\eta)$$

$$= \int_{\mathbb{R}^D} \phi(S(\Psi_T(S(\eta))), \eta) \, \mu(d\eta).$$

Now use the reversibility of Ψ_T and the definition of μ to write

$$I_N = \int_{\mathbb{R}^D} \phi(\Psi_T^{-1}(\eta), \eta) \, \exp(-H(\eta)) \, d\eta$$

and then change $\eta = \Psi_T(\xi)$

$$I_N = \int_{\mathbb{R}^D} \phi(\xi, \Psi_T(\xi)) \, \exp(-H(\Psi_T(\xi))) \, d\xi.$$

(Note we have used here conservation of volume.)

The integral with respect to measure in denominator of (31) is:

$$I_D = \int_{\mathbb{R}^D \times \mathbb{R}^D} \phi(\xi, \eta) \, \delta_{\Psi_T(\xi)}(d\eta) \, \mu(d\xi) = \int_{\mathbb{R}^D} \phi(\xi, \Psi_T(\xi)) \, \exp(-H(\xi)) \, d\xi$$

and a comparison with I_N leads to the sought conclusion. □

After these preparations we may present the proof of Theorem 9.

Proof. Consider the chain $\mathscr{C}$ in the phase space $\mathbb{R}^D$ of the variable (p, x) such that one step $(p_n, x_n) \mapsto (p_{n+1}, x_{n+1})$ of $\mathscr{C}$ is the concatenation of two sub-steps:

1. Discard the value of the momentum p_n and replace it by a fresh sample from the Maxwell distribution for the momentum.
2. Take a step of the chain defined in Lemmas 2 and 3.

Both sub-steps preserve μ (for the second use Lemma 2) and as a consequence so does the chain $\mathscr{C}$. The x-marginal of $\mathscr{C}$ is the chain in the HMC algorithm and will preserve the marginal density $\exp(-V(x))$. □

Acknowledgements This work has been supported by Project MTM2010-18246-C03-01, Ministerio de Ciencia e Innovación, Spain.

References

1. E. Akhmatskaya, S. Reich, GSHMC: An efficient method for molecular simulations. J. Comput. Phys. **227**, 4934–4954 (2008)
2. V.I. Arnold, *Mathematical Methods of Classical Mechanics*, 2nd edn. (Springer, New York, 1989)
3. A. Beskos, F.J. Pinski, J.M. Sanz-Serna, A.M. Stuart, Hybrid Monte-Carlo on Hilbert spaces. Stoch. Process. Appl. **121**, 2201–2230 (2011)
4. A. Beskos, N. Pillai, G.O. Roberts, J.M. Sanz-Serna, A.M. Stuart, Optimal tuning of the Hybrid Monte-Carlo algorithm. Bernoulli (to appear)
5. P. Billingsley, *Probability and Measure*, 3rd edn. (Wiley, New York, 1995)
6. P. Brémaud, *Markov Chains, Gibbs Fields, Monte Carlo Simulation, and Queues* (Springer, Berlin, 1999)

7. E. Cancès, F. Legoll, G. Stoltz, Theoretical and numerical comparison of some sampling methods for molecular dynamics. Esaim Math. Model. Numer. Anal. **41**, 351–389 (2007)
8. S. Duane, A.D. Kennedy, B. Pendleton, R. Roweth, Hybrid Monte Carlo. Phys. Lett. B **195**, 216–222 (1987)
9. W. Feller, *An Introduction to Probability Theory and Its Applications*, vol. 1, 3rd edn. (Wiley, New York, 1968)
10. W. Feller, *An Introduction to Probability Theory and Its Applications*, vol. 1, 2nd edn. (Wiley, New York, 1971)
11. R.P. Feynman, R.B. Leighton, M. Sands, *The Feynman Lectures on Physics*, vol. 1 (Addison-Wesley, Reading, 1963)
12. A. Friedman, *Stochastic Differential Equations and Applications* (Dover, Mineola, 2006)
13. C.J. Geyer, Practical Markov Chain Monte Carlo. Stat. Sci. **7**, 473–483 (1992)
14. M. Girolami, B. Calderhead, Riemann manifold Langevin and Hamiltonian Monte Carlo methods. J. R. Stat. Soc. B **73**, 123–214 (2011)
15. G. Grimmett, D. Stirzaker, *Probability and Random Processes*, 3rd edn. (Oxford University Press, Oxford, 2001)
16. E. Hairer, C. Lubich, G. Wanner, *Geometric Numerical Integration*, 2nd edn. (Springer, Berlin, 2006)
17. W. Hastings, Monte Carlo sampling methods using Markov chains and their application. Biometrika **57**, 97–109 (1970)
18. D. Higham, An algorithmic introduction to numerical simulation of stochastic differential equations. SIAM Rev. **43**, 525–546 (2001)
19. M.D. Hoffman, A. Gelman, The No-U-Turn sampler: Adaptively setting path lengths in Hamiltonian Monte Carlo, preprint
20. J.A. Izaguirre, S.S. Hampton, Shadow hybrid Monte Carlo: An efficient propagator in phase space of macromolecules. J. Comput. Phys. **200**, 581–604 (2004)
21. I.D. Lawrie, *A Unified Grand Tour of Theoretical Physics* (Institute of Physics Publishing, Bristol, 1990)
22. B. Leimkuhler, S. Reich, *Simulating Hamiltonian Dynamics* (Cambridge University Press, Cambridge, 2004)
23. T. Lelievre, M. Rousset, G. Stoltz, *Free Energy Computations: A Mathematical Perspective* (Imperial College Press, London, 2010)
24. X. Mao, *Stochastic Differential Equations and Applications*, 2nd edn. (Horwood, Chichester, 2008)
25. J.C. Mattingly, N.S. Pillai, A.M. Stuart, Diffusion limits of the random walk Metropolis algorithm in high dimensions authors. Ann. Appl. Probab. **22**, 881–930 (2012)
26. N. Metropolis, A. Rosenbluth, M. Rosenbluth, A. Teller, E. Teller, Equations of state calculations by fast computing machines. J. Chem. Phys. **21**, 1087–1092 (1953)
27. S. Meyn, R. Tweedie, *Markov Chains and Stochastic Stability* (Springer, New York, 1993)
28. R. Neal, Probabilistic Inference Using Markov Chain Monte Carlo Methods. Technical Report CRG-TR-93-1, Department of Computer Science, University of Toronto, 1993
29. N.S. Pillai, A.M. Stuart, A.H. Thiery, Optimal scaling and diffusion limits for the Langevin algorithm in high dimensions. Ann. Appl. Probab. **22**, 2320–2356 (2012)
30. C.P. Robert, G. Casella, *Monte Carlo Statistical Methods*, 2nd edn. (Springer, Berlin, 2004)
31. G.O. Roberts, J.S. Rosenthal, Optimal scaling of discrete approximations to Langevin diffusions. J. R. Stat. Soc. B **60**, 255–268 (1998)
32. G.O. Roberts, R.L. Tweedie, Exponential convergence of Langevin diffusions and their discrete approximations. Bernoulli **2**, 341–263 (1996)
33. G.O. Roberts, A. Gelman, W.R. Gilks, Weak convergence and optimal scaling of random walk Metropolis algorithms. Ann. Appl. Probab. **7**, 110–120 (1997)
34. J.M. Sanz-Serna, Geometric integration, in *The State of the Art in Numerical Analysis*, ed. by I.S. Duff, A.G. Watson (Clarendon, Oxford, 1997), pp. 121–143
35. J.M. Sanz-Serna, M.P. Calvo, *Numerical Hamiltonian Problems* (Chapman & Hall, London, 1994)

36. T. Schilick, *Molecular Modeling and Simulation: An Interdisciplinary Guide*, 2nd edn. (Springer, New York, 2010)
37. C. Schütte, Conformational Dynamics: Modelling, Theory, Algorithmm and Application to Biomolecules. Habilitation Thesis, Free University Berlin, 1999
38. B. Shahbaba, S. Lan, W.O. Johnson, R.M. Neal, Split Hamiltonian Monte Carlo, preprint

Stability and Computation of Dynamic Patterns in PDEs

Wolf-Jürgen Beyn, Denny Otten, and Jens Rottmann-Matthes

Abstract Nonlinear waves are a common feature in many applications such as the spread of epidemics, electric signaling in nerve cells, and excitable chemical reactions. Mathematical models of such systems lead to time-dependent PDEs of parabolic, hyperbolic or mixed type. Common types of such waves are fronts and pulses in one, rotating and spiral waves in two, and scroll waves in three space dimensions. These patterns may be viewed as relative equilibria of an equivariant evolution equation where equivariance is caused by the action of a Lie group. Typical examples of such actions are rotations, translations or gauge transformations. The aim of the lectures is to give an overview of problems related to the theoretical and numerical analysis of such dynamic patterns. One major theoretical topic is to prove nonlinear stability and relate it to linearized stability determined by the spectral behavior of linearized operators. The numerical part focusses on the freezing method which uses equivariance to transform the given PDE into a partial differential algebraic equation (PDAE). Solving these PDAEs generates moving coordinate systems in which the above-mentioned patterns become stationary.

1 Dynamics of Patterns and Equivariance: Traveling Waves in One Space Dimension

The first lecture is of introductory character and serves to introduce basic notions and properties. We use the well known topic of traveling wave solutions in order to illustrate the topics of this course such as equivariance, stability with asymptotic phase, spectral properties, and the associated computational problems. There are by

---

W.-J. Beyn (✉) · D. Otten · J. Rottmann-Matthes
Department of Mathematics, Bielefeld University, Bielefeld, Germany
e-mail: beyn@math.uni-bielefeld.de; dotten@math.uni-bielefeld.de; jrottman@math.uni-bielefeld.de

now quite a few monographs and survey articles that treat this topic and we refer to [29, 38, 59, 68].

1.1 Traveling Fronts and Pulses

Consider a parabolic system in one space variable

$$u_t(x,t) = u_{xx}(x,t) + f(u(x,t)), \quad x \in \mathbb{R}, \ t \geqslant 0, \tag{1}$$

where $f : \mathbb{R}^m \to \mathbb{R}^m$ is assumed to be sufficiently smooth and we look for smooth solutions $u(x,t) \in \mathbb{R}^m$, $x \in \mathbb{R}$, $t \geqslant 0$. In the following we omit arguments in (1) and simply write

$$u_t = u_{xx} + f(u), \quad x \in \mathbb{R}, \ t \geq 0. \tag{2}$$

Definition 1. A special solution of (2), which is of the form

$$u(x,t) = \bar{v}(x - \bar{\mu}t), \quad x \in \mathbb{R}, \ t \in \mathbb{R} \tag{3}$$

for some $\bar{\mu} \in \mathbb{R}$ and some $\bar{v} : \mathbb{R} \to \mathbb{R}^m$, is called a **traveling wave** if the limits

$$\lim_{\xi \to \infty} \bar{v}(\xi) = u_+, \quad \lim_{\xi \to -\infty} \bar{v}(\xi) = u_- \tag{4}$$

exist and satisfy $f(u_\pm) = 0$. The function $\bar{v} : \mathbb{R} \to \mathbb{R}^m$ is called the **profile** of the wave and the value $\bar{\mu} \in \mathbb{R}$ is called its velocity. In case $u_+ \neq u_-$ one speaks of a **front solution** and in case $u_+ = u_-$ of a **pulse solution**.

Note that the wave moves to the right if $\bar{\mu} > 0$ and to the left if $\bar{\mu} < 0$. In case $\bar{\mu} = 0$ we have a standing wave.

Example 1 (Nagumo equation). This well known example with $m = 1$ is given by the equation

$$u_t = u_{xx} + u(1-u)(u-\alpha), \tag{5}$$

where $0 < \alpha < 1$ is a parameter. For this equation there is a simple explicit formula of a traveling front due to Huxley

$$\bar{v}(\xi) = \frac{1}{1 + \exp\left(-\frac{\xi}{\sqrt{2}}\right)}, \ \xi \in \mathbb{R}, \quad \bar{\mu} = \sqrt{2}\left(\alpha - \frac{1}{2}\right), \tag{6}$$

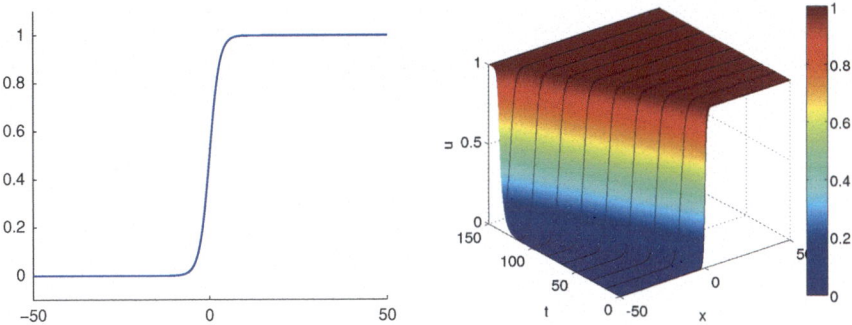

Fig. 1 Profile of the Nagumo front (*left*) and space-time diagram (*right*) for $\alpha = \frac{1}{4}$

with $u_+ = 1$ and $u_- = 0$. The wave travels to the right if $\alpha > \frac{1}{2}$ and to the left if $\alpha < \frac{1}{2}$. The following Fig. 1 shows the profile and the time-dependent solution (3) for the special case $\alpha = \frac{1}{4}$:

1.2 Traveling Waves and ODEs

Usually, both the profile and the velocity of a traveling wave are unknown. Hence the task is to find a function v and a parameter μ such that $u(x,t) = v(x - \mu t)$ solves (2). This leads to solving a second order ordinary differential equation for v with boundary conditions given at infinity

$$0 = v_{xx} + \mu v_x + f(v), \quad \lim_{x \to \pm\infty} v(x) = u_\pm, \quad f(u_\pm) = 0. \tag{7}$$

Introducing $V = \begin{pmatrix} v_1 \\ v_2 \end{pmatrix} = \begin{pmatrix} v \\ v_x \end{pmatrix}$, this can be rewritten as a first order system of dimension $2m$

$$\begin{pmatrix} v_1 \\ v_2 \end{pmatrix}_x = V_x = F(V, \mu) = \begin{pmatrix} v_2 \\ -\mu v_2 - f(v_1) \end{pmatrix}, \quad \lim_{x \to \pm\infty} V(x) = V_\pm = \begin{pmatrix} u_\pm \\ 0 \end{pmatrix}. \tag{8}$$

Traveling pulses and fronts therefore correspond to **homoclinic** and **heteroclinic orbits** that connect two steady states V_- to V_+ of the dynamical system (8) for a specific value of the parameter μ.

Example 2 (Nagumo equation). For an illustration we return to the Nagumo equation (6), where the first order system (8) reads

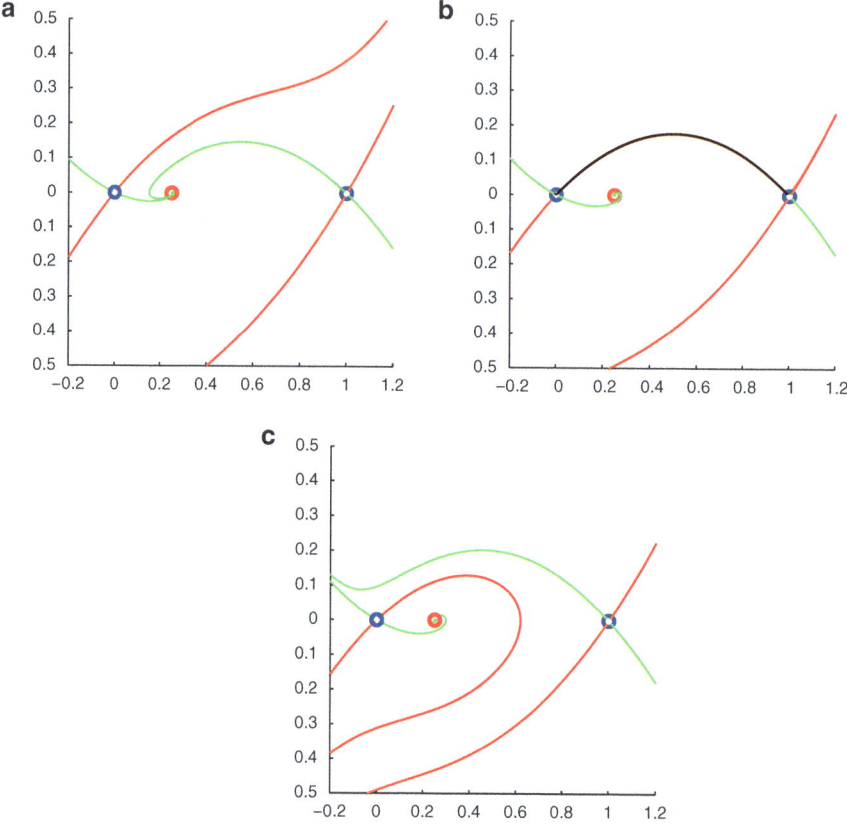

Fig. 2 Phase diagrams of (9) for $\alpha = \frac{1}{4}$ and μ-values close to $\bar{\mu} = -\frac{\sqrt{2}}{4}$. (**a**) $-\frac{2}{4} = \mu < \bar{\mu}$. (**b**) $-\frac{\sqrt{2}}{4} = \mu = \bar{\mu}$. (**c**) $-\frac{1}{4} = \mu > \bar{\mu}$

$$\begin{pmatrix} v_1 \\ v_2 \end{pmatrix}_x = \begin{pmatrix} v_2 \\ -\mu v_2 - v_1(1-v_1)(v_1 - \alpha) \end{pmatrix}, \qquad (9)$$

$$\lim_{x \to -\infty} \begin{pmatrix} v_1(x) \\ v_2(x) \end{pmatrix} = \begin{pmatrix} 0 \\ 0 \end{pmatrix}, \quad \lim_{x \to \infty} \begin{pmatrix} v_1(x) \\ v_2(x) \end{pmatrix} = \begin{pmatrix} 1 \\ 0 \end{pmatrix}. \qquad (10)$$

Figure 2 shows the phase diagrams of the two-dimensional system (9) for values $\mu < \bar{\mu}$, $\mu = \bar{\mu}$, and $\mu > \bar{\mu}$, where $\bar{\mu} = -\frac{\sqrt{2}}{4}$, $\alpha = \frac{1}{4}$. At the value $\mu = \bar{\mu}$ we have a heteroclinic orbit connecting the two saddles $(0, 0)$ and $(1, 0)$.

We briefly discuss how to compute the profile $v(x) \in \mathbb{R}^m$ and $\mu \in \mathbb{R}$ from

$$V_x = F(V, \mu), \quad \lim_{x \to \pm\infty} V(x) = V_{\pm}, \quad F(V_{\pm}, \mu) = 0. \qquad (11)$$

Fig. 3 Linear approximation of the stable manifold by projection boundary conditions

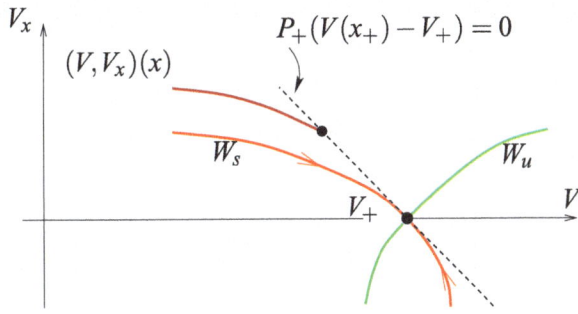

Note that such connecting orbits always come in families. If (V, μ) solves (11) then so does $(V(\cdot - \gamma), \mu)$ for any $\gamma \in \mathbb{R}$. In order to eliminate this ambiguity one introduces a **phase condition** and solves the following boundary value problem for (V, μ)

$$V_x = F(V, \mu), \; x \in \mathbb{R}, \quad \Psi(V) := (\hat{V}_x, V - \hat{V})_{L^2} = 0. \tag{12}$$

Here the phase condition uses an initial approximation or template function $\hat{V}$ which we require to have the correct limits $\lim_{x \to \pm\infty} \hat{V}(x) = V_\pm$ and such that the inner product in (12) exists. In Sect. 2 we will motivate this condition and discuss alternatives.

For numerical computations one chooses a bounded interval $J = [x_-, x_+]$ and then solves the following boundary value problem for $V \in C^1(J, \mathbb{R}^m)$, $\mu \in \mathbb{R}$,

$$V_x = F(V, \mu), \; x \in J, \quad (\hat{V}_x\big|_J, V - \hat{V}\big|_J)_{L^2(J)} = 0 \tag{13}$$

$$P_+(\mu)(V(x_+) - V_+) = 0, \quad P_-(\mu)(V(x_-) - V_-) = 0. \tag{14}$$

The most common choice for the boundary operators $P_\pm$ are **projection boundary conditions** which require the endpoint $V(x_-)$ to lie in the linear approximation of the unstable manifold at V_- and $V(x_+)$ to lie in the linear approximation of the stable manifold at V_+, see Fig. 3. Concretely, one chooses $P_+ = P_+(\mu) \in \mathbb{R}^{m_u \times m}$ of maximal rank such that $P_+(\mu)DF(V_+, \mu) = \Lambda_+ P_+(\mu)$ and such that the spectrum of $\Lambda_+ \in \mathbb{R}^{m_u \times m_u}$ coincides with the spectrum of $DF(V_+, \mu)$ with positive real part. The rows of P_+ then span the left unstable eigenvectors of $DF(V_+, \mu)$ which are orthogonal to the right stable eigenvectors. Similarly, one chooses $P_-(\mu) \in \mathbb{R}^{m_s \times m}$ such that the rows of P_- span the left stable eigenvectors of $DF(V_-, \mu)$. Note that in general the projection matrices depend on μ so that the boundary value problem (13) becomes nonlinear both in V and μ. We refer to [8, 22, 32] for various methods that allow to compute such projection matrices depending smoothly on a parameter. Finally, note that the boundary value problem (13) has the same number of equations and boundary conditions provided $m_s + m_u = m$, which is obviously satisfied in the homoclinic case but an assumption in the heteroclinic case. There is also a well established theory that studies the errors when passing from the infinite problem (12) to the finite problem (13), (14), see [8].

1.3 Dynamics of PDE and Shift Equivariance

Let us return to the time-dependent equation (2) in a slightly more general form

$$u_t = Au_{xx} + f(u, u_x), \ x \in \mathbb{R}, \ t \geq 0, \ u(x,t) \in \mathbb{R}^m, \tag{15}$$

where $A \in \mathbb{R}^{m \times m}$ is assumed to be positive definite and $f : \mathbb{R}^{2m} \to \mathbb{R}^m$ is smooth. For a fixed $\mu \in \mathbb{R}$ we transform into a moving coordinate frame via $u(x,t) = v(x - \mu t, t)$. This leads to

$$v_t = Av_{xx} + \mu v_x + f(v, v_x). \tag{16}$$

This is a parabolic system for which a traveling wave $u(x,t) = \bar{v}(x - \bar{\mu}t)$ now appears as a steady state $(\bar{v}, \bar{\mu})$. In fact, we have a family of steady states $(\bar{v}(\cdot - \gamma), \bar{\mu}), \gamma \in \mathbb{R}$.

In Sect. 3 we will deal with the classical Hodgkin-Huxley system for which $m = 4$. Then the matrix A is only positive semidefinite since there is no diffusion in 3 of 4 variables. The system (15) is then of mixed hyperbolic-parabolic type and this creates extra difficulties, both theoretically as well as numerically, see Sect. 3.

In the following it will be useful to phrase (15) in a more abstract way as

$$u_t = F(u), \quad F(u) = Au_{xx} + f(u, u_x), \tag{17}$$

where we consider F as an operator

$$F : Y = w + H^2(\mathbb{R}, \mathbb{R}^m) \to L^2(\mathbb{R}, \mathbb{R}^m) = X. \tag{18}$$

Here L^2, H^2 are standard Lebesgue and Sobolev spaces, the function $w \in C^2(\mathbb{R}, \mathbb{R}^m)$ satisfies for some $\varepsilon > 0$

$$|w(x) - u_{\pm}| + |w_x(x)| + |w_{xx}(x)| \leq Ce^{-\varepsilon|x|}, \ x \in \mathbb{R},$$

and we assume $f(u_{\pm}, 0) = 0$. We have carefully chosen Y as an affine space in order to incorporate traveling fronts with different limits at $\pm\infty$. Under these assumptions, using Sobolev embedding one can show that F maps Y into X.

Now consider the shift operator as an action of the group $G = \mathbb{R}$ on Y

$$a : G \times Y \to Y, (\gamma, u) \mapsto a(\gamma, u), \quad [a(\gamma, u)](x) = u(x - \gamma), \ x \in \mathbb{R}. \tag{19}$$

Obviously, $a(\gamma, u)$ has the following properties for $u, v \in Y$, $\gamma, \gamma_1, \gamma_2 \in G$, $\lambda \in \mathbb{R}$,

$$a(\gamma_1 + \gamma_2, u) = a(\gamma_1, a(\gamma_2, u)), \qquad \text{homomorphism,}$$
$$a(\gamma, \lambda u + (1-\lambda)v) = \lambda a(\gamma, u) + (1-\lambda) a(\gamma, v) \qquad \text{affine linearity w.r.t. u.}$$

Moreover, the action immediately extends to $X = L^2(\mathbb{R}, \mathbb{R}^m)$ with the same properties. We often write $a(\gamma)u$ instead of $a(\gamma, u)$, in particular when $a(\gamma)$ is a linear operator on X.

The most important property of the operator F is **equivariance under the action of the group**, i.e.

$$a(\gamma) F(u) = F(a(\gamma)u), \quad u \in Y, \ \gamma \in G. \tag{20}$$

This follows from

$$(Fu)(\cdot - \gamma) = Au_{xx}(\cdot - \gamma) + f(u(\cdot - \gamma), u_x(\cdot - \gamma)) = F(u(\cdot - \gamma)), \ \gamma \in \mathbb{R}.$$

Thus, we have recast (15) as an abstract **equivariant evolution equation** (17), (20). Some further notations are useful. For a given element $v \in Y$ the set

$$\mathcal{O}_G(v) = \{a(\gamma)v : \gamma \in G\}$$

is called its **group orbit**. For the shift action (19) the group orbit of a function consists of all its translates. A **relative equilibrium** of (17) is a solution $\bar{u}(t), t \in \mathbb{R}$ that lies in a single group orbit, i.e.

$$\bar{u}(t) = a(\gamma(t))\bar{v}, \quad \text{for some } \bar{v} \in Y, \ \gamma(\cdot) \in C^1(\mathbb{R}, G). \tag{21}$$

In this sense, traveling waves $\bar{u}(x, t) = \bar{v}(x - \bar{\mu}t)$ are relative equilibria w.r.t. shift equivariance where in this special case $\gamma(t) = \bar{\mu}t$.

1.4 Stability with Asymptotic Phase

In the previous section we saw that traveling waves, and relative equilibria in general, always appear in families. In order to take this into account the classical notion of Lyapunov stability is modified as follows.

Definition 2. A traveling wave solution $u(x, t) = \bar{v}(x - \bar{\mu}t)$ of the system (15) is called **asymptotically stable with asymptotic phase** with respect to given norms $\|\cdot\|_1$ and $\|\cdot\|_2$ on Y, if for any $\varepsilon > 0$ there exists a $\delta > 0$ such that for any initial data $u_0 \in Y$ with $\|u_0 - \bar{v}\|_1 \leq \delta$ there exists some $\gamma_\infty \in \mathbb{R}$ with the following property. The Cauchy problem $u_t = Au_{xx} + f(u, u_x), u(\cdot, 0) = u_0$ has a unique solution $u(\cdot, t) \in Y, t \geq 0$ and

$$\|u(\cdot,t)-\bar{v}(\cdot-\bar{\mu}t-\gamma_\infty)\|_2 \begin{cases} \leq \varepsilon & \text{for all } t \geq 0, \\ \to 0 & \text{as } t \to \infty. \end{cases} \quad (22)$$

In general, the value γ_∞ depends on the initial function u_0 and is called the **asymptotic phase**. The definition is not completely rigorous since it leaves open the precise notion of solution and of the associated function spaces. These depend on the particular type of application. Note that our formulation allows an affine space for Y as in (18). For PDEs with hyperbolic parts it is important to use two different norms in the definition, see Sect. 3. Then initial perturbations often must be measured in stronger norms than perturbations of solutions. On the other hand, for parabolic systems it is often possible to use the same Sobolev norm $\|\cdot\|_{H^1}$ for both norms. For various stability theorems we refer to the monographs [38, 68] and to the survey article [59].

An essential feature of all stability results are the spectral properties of the linearized differential operator

$$\Lambda = A\partial_{xx} + (\bar{\mu}I + D_2 f(\bar{v},\bar{v}_x))\partial_x + D_1 f(\bar{v},\bar{v}_x). \quad (23)$$

We introduce coefficient matrices in (23) by writing

$$\Lambda = A\partial_{xx} + B(x)\partial_x + C(x) \quad (24)$$

and note that, due to our assumption, the following limits exist

$$B_\pm = \lim_{x\to\pm\infty} B(x) = \bar{\mu}I + D_2 f(u_\pm,0), \quad C_\pm = \lim_{x\to\pm\infty} C(x) = D_1 f(u_\pm,0). \quad (25)$$

1.5 Spectral Properties of Second Order Operators

In this section we recall basic facts about spectra of second order linear differential operators as they arise from linearizations at traveling waves. Since these operators are defined on the whole line they typically have essential as well as isolated spectrum. The essential spectrum is determined by the limit operators

$$\Lambda_\pm = A\partial_{xx} + B_\pm\partial_x + C_\pm \quad (26)$$

obtained from the coefficients in (25). The spectrum of $\Lambda_\pm$ can be computed by evaluating the so-called **dispersion relation**, see (30) below. On the contrary, it is not so easy to determine the remaining isolated eigenvalues, except for the fact that zero is always an eigenvalue due to shift equivariance.

Fig. 4 Spectrum of
$\Lambda_0 = a\partial_x^2 + b\partial_x + c$

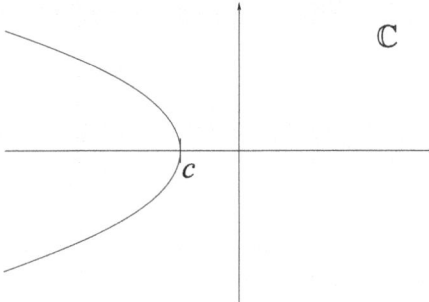

Let us first look at the real scalar case of (26), i.e.

$$\Lambda_0 v = av_{xx} + bv_x + cv, \quad a, b, c \in \mathbb{R}. \tag{27}$$

We look for eigenvalues $s \in \mathbb{C}$ with eigenfunctions of the form

$$v(x) = e^{i\omega x}, \; x \in \mathbb{R}, \; \omega \in \mathbb{R}. \tag{28}$$

This leads us to the dispersion relation

$$s = -a\omega^2 + ib\omega + c, \quad \omega \in \mathbb{R}. \tag{29}$$

Any value $s \in S = \{s = -a\omega^2 + ib\omega + c \mid \omega \in \mathbb{R}\}$ is an eigenvalue of Λ_0 with a bounded eigenfunction $e^{i\omega x}$. Standard function spaces such as L^2 or H^1 will not contain these eigenfunctions, but their presence leads to unbounded resolvents. For the scalar case with $a > 0, b \neq 0$ the algebraic set S is a left open parabola with the vertex at c, see Fig. 4.

For the general operator (26) one has to consider two algebraic sets

$$S_\pm = \{s \in \mathbb{C} : \det(-\omega^2 A + i\omega B_\pm + C_\pm - sI) = 0 \text{ for some } \omega \in \mathbb{R}\}. \tag{30}$$

If, for instance, A is positive definite and $B_\pm$ is the identity then the curves in $S_\pm$ asymptotically attain a parabolic shape $s \sim i\omega - \omega^2 \lambda_j$, where $\lambda_j, j = 1, \ldots, m$ are the eigenvalues of A. Let us first recall some standard definitions from spectral theory.

Definition 3. Let X be a Banach space and let $\Lambda : \mathscr{D}(\Lambda) \subset X \to X$ be a densely defined closed operator. If $sI - \Lambda$ is one-to-one for $\lambda \in \mathbb{C}$ then the operator $R_s(\Lambda) = (sI - \Lambda)^{-1}$ is defined on $\mathscr{D}(R_s(\Lambda)) = \text{Range}(sI - \Lambda)$ and called the **resolvent of** Λ. Then one defines the **resolvent set**

$$\rho(\Lambda) = \{s \in \mathbb{C} : R_s(\Lambda) \text{ exists}, \mathscr{D}(R_s(\Lambda)) \text{ is dense}, R_s(\Lambda) \text{ bounded}\}, \tag{31}$$

the **spectrum** $\sigma(\Lambda) = \mathbb{C} \setminus \rho(\Lambda)$, the **point spectrum**

$$\sigma_{point}(\Lambda) = \{s \in \mathbb{C} \text{ is an isolated eigenvalue of finite multiplicity}\}, \quad (32)$$

and the **essential spectrum** $\sigma_{ess}(\Lambda) = \sigma(\Lambda) \setminus \sigma_{point}(\Lambda)$.

In the following we introduce the crucial

Spectral Condition (SC)

There exist $\beta, \lambda_{min} > 0$ such that $|\text{Re }\lambda| \geq \lambda_{min}$ for all $\lambda \in \mathbb{C}$ which satisfy

$$\det(\lambda^2 A + \lambda B_\pm + C - sI) = 0, \text{ for some } \text{Re } s \geq -\beta. \quad (33)$$

If A is positive definite then a continuation argument shows that the algebraic sets $S_\pm$ lie in the half plane $\{z : \text{Re}(z) < -\beta\}$ and hence are bounded away from the imaginary axis. More generally, the following theorem from [38] shows that the spectral condition is also sufficient to guarantee that the essential spectrum of the variable coefficient operator Λ from (24) lies in this half plane.

Theorem 1 (Essential spectrum of Λ, [38]). *Let the variable coefficient operator Λ from (24) have continuous coefficients such that $B_\pm = \lim_{x \to \pm\infty} B(x)$ and $C_\pm = \lim_{x \to \pm\infty} C(x)$ exist and A is positive definite.*

Then the spectrum of the operator Λ considered in $L^2(\mathbb{R}, \mathbb{R}^m)$ satisfies

$$S_- \cup S_+ \subset \sigma_{ess}(\Lambda) \subset M^c, \quad (34)$$

where the algebraic sets $S_\pm$ are defined in (30) and M is the unique connected component of $\mathbb{C} \setminus (S_- \cup S_+)$ that contains a right half plane $\{z : \text{Re } z \geq \zeta\}$ for some $\zeta \in \mathbb{R}$. Moreover, if the spectral condition SC holds then $\text{Re } \sigma_{ess}(\Lambda) \leq -\beta$.

Since the proof is quite involved, we only describe the main idea, see [38]. Decompose $\Lambda = L + K$, where L has constant coefficients on both $\mathbb{R}_- = (-\infty, 0]$ and $\mathbb{R}_+^* = (0, \infty)$ and K is of lower order and has decaying coefficients

$$L = \begin{cases} \Lambda_- & \text{on } \mathbb{R}_-, \\ \Lambda_+ & \text{on } \mathbb{R}_+^*, \end{cases} \quad K = \begin{cases} (B(x) - B_-)\partial_x + (C(x) - C_-), & x \in \mathbb{R}_-, \\ (B(x) - B_+)\partial_x + (C(x) - C_+), & x \in \mathbb{R}_+^*. \end{cases} \quad (35)$$

Then, one applies the following theorem on invariance of the essential spectrum [35].

Theorem 2. *Let X be a Banach space, $L : \mathscr{D}(L) \subset X \to X$ be a closed linear operator and $K : \mathscr{D}(K) \supset \mathscr{D}(L) \to X$ be a linear operator such that $K(\lambda_0 I - L)^{-1}$ is compact for some $\lambda_0 \in \rho(L)$. Let $U \subset \mathbb{C}$ be open and connected such that*

Fig. 5 Schematic picture of spectrum for a second order linear operator Λ

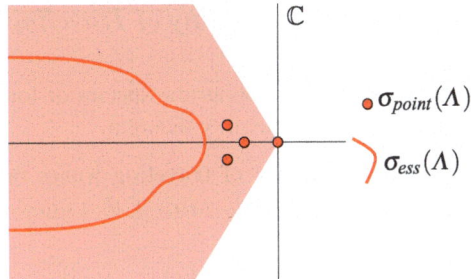

$U \subset \rho(L) \cup \sigma_{\text{point}}(L)$. Then either $U \subset \rho(L+K) \cup \sigma_{\text{point}}(L+K)$ or U contains only eigenvalues of $L+K$.

A perturbation K of an operator L for which $K(\lambda_0 I - L)^{-1}$ is compact, is called **relatively compact**. In the special case (35) one shows that $(\lambda_0 I - L)^{-1}$ is bounded from $L^2(\mathbb{R}, \mathbb{R}^m)$ into $H^1(\mathbb{R}, \mathbb{R}^m)$. Then, using the fact that the coefficients of K vanish as $x \to \pm\infty$ one shows compactness of the operator $K(\lambda_0 I - L)^{-1}$ as an operator in $L^2(\mathbb{R}, \mathbb{R}^m)$ by invoking the Riesz-Fréchet compactness criterion in $L^2(\mathbb{R}, \mathbb{R}^{m \times m})$, see [3].

We close this subsection with some remarks on the point spectrum $\sigma_{\text{point}}(\Lambda)$. Differentiating the equation $0 = \bar{v}_{xx} + \mu \bar{v}_x + f(\bar{v}, \bar{v}_x)$ with respect to x leads to

$$0 = (\bar{v}_x)_{xx} + \mu (\bar{v}_x)_x + D_2 f(\bar{v}, \bar{v}_x)\bar{v}_{xx} + D_1 f(\bar{v}, \bar{v}_x)\bar{v}_x = \Lambda \bar{v}_x. \qquad (36)$$

Hence, we always have $0 \in \sigma_{\text{point}}(\Lambda)$ with eigenfunction $\bar{v}_x = -\frac{d}{d\gamma}v(\cdot - \gamma)\big|_{\gamma=0}$ provided this function is in the appropriate function space. The problem of detecting further eigenvalues in the domain M (see Thm. 2) can be reduced to studying zeros of the so-called **Evans function**, see [2, 51]. Several approaches have been developed for this purpose. However, if the analysis cannot be done explicitly one has to resort to numerical computations for detecting the point spectrum, compare [16, 17, 41, 42, 46, 47, 58, 59]. In Fig. 5 we sketch the typical appearance of the spectrum, where we used that the operator Λ is sectorial in $L^2(\mathbb{R}, \mathbb{R}^m)$.

For the stability results in the next subsection we will need the following

Eigenvalue Condition (EC)

There are no isolated eigenvalues of finite multiplicity for Λ in $\operatorname{Re} s \geq -\beta$ except 0, and the eigenvalue 0 is algebraically simple.

1.6 Nonlinear Stability of Traveling Waves and Applications

With the preparation about the spectra of linear operators we may now formulate the main nonlinear stability theorem.

Theorem 3 (Stability of traveling waves in H^1). *Consider a parabolic system (15) with a smooth nonlinearity f that satisfies*

$$f(v, v_x) = f_1(v)v_x + f_2(v), \tag{37}$$

$$f_1, f_2, f_1', f_2' \in C^1 \text{ globally Lipschitz}. \tag{38}$$

Let $u(x,t) = \bar{v}(x - \bar{\mu}t)$ be a traveling wave of (15) such that the spectral condition (SC) and the eigenvalue condition (EC) are satisfied for the linearized operator Λ in (23). Then the traveling wave $(\bar{v}, \bar{\mu})$ is stable with asymptotic phase in the space $H^1(\mathbb{R}, \mathbb{R}^m)$.

Remark 1. Stability in $H^1(\mathbb{R}, \mathbb{R}^m)$ means that the statement of Definition 2 holds with both norms taken to be $\|\cdot\|_{H^1}$. We refer to [38] for a proof of this result in case f depends only on u, but satisfies weaker assumptions than Lipschitz boundedness. For the version above see [65]. Note that (37) includes the important example of the viscous Burgers equation where $f(v) = v v_x$.

Proof (general idea from [38]). Nonlinear change of coordinates

$$v \to (\gamma, \tilde{v}) \quad \text{where} \quad v = \bar{v}(\cdot - \gamma) + \tilde{v}, \quad (\psi, \tilde{v})_{L^2} = 0, \tag{39}$$

where $\Lambda^* \psi = 0$ (left eigenfunction), $(\psi, \bar{v}_x)_{L^2} = 1$.

The transformed system is

$$\tilde{v}_t = QF(\tilde{v} + \bar{v}(\cdot - \gamma)), \quad v(\cdot, 0) = u_0 \quad \text{(PDE1)}$$

$$\gamma_t = R(\gamma, \tilde{v}), \quad \gamma(0) = 0, \quad \text{(ODE2)}$$

where $Qu = u - \bar{v}_x (\psi, u)_{L^2}$ is the projector onto the orthogonal complement $\psi^\top$. The next steps are:

1. Show that the linearization $Q\Lambda$ of (PDE1) has spectrum Re $\leq -\beta < 0$,
2. prove asymptotic stability of (PDE1) in H^1 uniformly in γ,
3. show $|\gamma_t| \leq C e^{-\frac{\beta t}{2}}$ using (ODE2),
4. determine the asymptotic phase from $\gamma_\infty(\tilde{v}) = \gamma(0, \tilde{v}) + \int_0^\infty \gamma_\tau(\tau, \tilde{v}) d\tau$. □

Example 3 (Nagumo equation). As a first application we study the Nagumo wave from (5), (6). For this equation we have $u_- = 0$, $u_+ = 1$, $f(u_\pm) = 0$, $f'(u_-) = -\alpha$, $f'(u_+) = \alpha - 1$. The dispersion relation (29) leads to the two parabolas

$$S_\pm = \{s = -\omega^2 + \bar{\mu}\omega + f'(u_\pm) : \omega \in \mathbb{R}\}, \tag{40}$$

Fig. 6 Essential spectrum for the Nagumo front

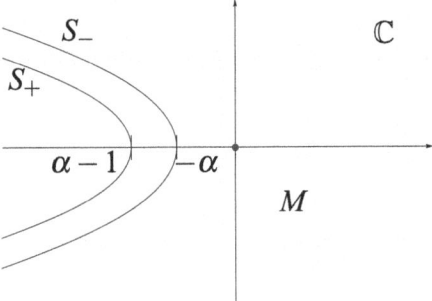

which have their vertices at $-\alpha$ and $\alpha - 1$. Both lie in the negative half plane since $0 < \alpha < 1$, see Fig. 6.

Thus the spectral condition (**SC**) is satisfied. Moreover, one can show that 0 is indeed a simple eigenvalue and there are no further eigenvalues λ with $\operatorname{Re} \lambda > 0$ (see [38]).

Example 4 (FitzHugh-Nagumo system). As another example we mention the well studied FitzHugh-Nagumo system ([30])

$$u_t = \begin{pmatrix} u_1 \\ u_2 \end{pmatrix}_t = \begin{pmatrix} 1 & 0 \\ 0 & \varepsilon \end{pmatrix} u_{xx} + f(u) \tag{41}$$

$$f \begin{pmatrix} u_1 \\ u_2 \end{pmatrix} = \begin{pmatrix} u_1 - \tfrac{1}{3} u_1^3 - u_2 \\ \phi (u_1 + a - b u_2) \end{pmatrix}, \quad \phi, a, b > 0, \; \varepsilon \geqslant 0. \tag{42}$$

For an extensive study of the stability of traveling waves for this system we refer to Evans [24–27]. We first choose parameter values $\varepsilon = 0.1$, $\phi = 0.08$, $a = 0.7$, $b = 3$ for which f has three zeros so that a traveling front occurs. For a parameter setting which leads to the classical FitzHugh Nagumo pulses, we refer to Example 7 below. Figure 7 shows the profile of both components for the traveling front (left) and a space-time plot of the first component of a solution for (41), (42) (right). In this case, we have

$$u_- = \begin{pmatrix} 1.1877 \\ 0.6292 \end{pmatrix}, \quad u_+ = \begin{pmatrix} -1.5644 \\ -0.2881 \end{pmatrix},$$

$$B_- = Df(u_-) = \begin{pmatrix} -0.4106 & -1 \\ 0.08 & -0.24 \end{pmatrix}, \quad B_+ = Df(u_+) = \begin{pmatrix} -1.4474 & -1 \\ 0.08 & -0.24 \end{pmatrix}$$

and the dispersion relation (30) yields the two algebraic sets (cf. [9] for a drawing)

$$S_\pm = \left\{ s \in \mathbb{C} : \det \left(-\omega^2 \begin{pmatrix} 1 & 0 \\ 0 & \varepsilon \end{pmatrix} + i \omega B_\pm - sI \right) = 0 \text{ for some } \omega \in \mathbb{R} \right\}, \quad \varepsilon > 0.$$

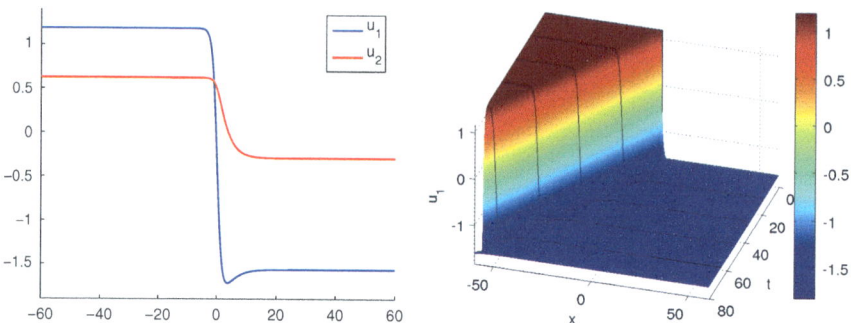

Fig. 7 Profile of the traveling wave for the FitzHugh-Nagumo system (*left*) and space-time diagram of u_1 (*right*) for $\varepsilon = 0.1, \phi = 0.08, a = 0.7, b = 3$

1.7 Equivariant Evolution Equations

In Sect. 1.3 we already mentioned that traveling waves may be viewed as relative equilibria of an abstract evolution equation that has an equivariance property. In this section we will extend this abstract point of view and discuss an application to a wave which is traveling and rotating simultaneously. For some general theory of equivariant evolution equations we refer to [19, 28, 36].

As in (17) we consider the Cauchy problem for a general evolution equation

$$u_t = F(u), \; u(0) = u_0, \tag{43}$$

where we assume $F : Y \subset X \to X$ with X a Banach space and Y a dense subspace. The whole approach can be written in terms of Banach manifolds rather than Banach spaces. But, for the sake of simplicity, we avoid such a generalization. Note, however, that the treatment of traveling fronts already requires to use affine spaces for Y and X, compare Sect. 1.3.

Let G be a **Lie group**, i.e. a finite dimensional manifold with a smooth invertible group operation. By $\mathbb{1}$ we denote the unit element in G. The group operation $\circ$ induces the operators of left and right multiplication via

$$\circ : \begin{cases} G \times G \to G, \\ (\gamma, g) \to \gamma \circ g = L_\gamma g = R_g \gamma. \end{cases} \tag{44}$$

The Lie algebra $\mathscr{A}$ is the tangent space of G at $\mathbb{1}$, i.e. $T_\mathbb{1} G = \mathscr{A}$ and the derivative of the left multiplication $L_\gamma : G \to G$ is denoted by $dL_\gamma(g) : T_g G \to T_{\gamma \circ g} G$.

We further assume that the group G acts on X via

$$a : \begin{array}{l} G \to GL[X], \\ \gamma \to a(\gamma), \end{array} \quad \begin{array}{l} a(\mathbb{1}) = I, \\ a(\gamma_1 \circ \gamma_2) = a(\gamma_1) a(\gamma_2). \end{array} \tag{45}$$

The evolution equation (43) is called **equivariant under the action of the group** if for all $\gamma \in G$,

$$F(a(\gamma)u) = a(\gamma)F(u) \text{ for all } u \in Y, \tag{46}$$

$$a(\gamma)Y \subset Y. \tag{47}$$

It is important to be careful with smoothness assumptions on the action. As our examples will show, it is reasonable to assume that the map $a(\cdot)v : \gamma \to a(\gamma)v$ is continuous for every $v \in X$ and continuously differentiable for every $v \in Y$. We will denote the derivative with respect to $\gamma \in G$ at $\mathbb{1}$ by

$$d\,[a(\mathbb{1})v] = d\,[a(\gamma)v]_{\gamma=\mathbb{1}} : \mathscr{A} = T_{\mathbb{1}}G \to X. \tag{48}$$

Our second example is an equation that is equivariant with respect to a two-dimensional Lie group.

Example 5 (Quintic-cubic Ginzburg Landau equation (QCGL)).

$$\begin{aligned} u_t &= \alpha u_{xx} + f(|u|^2)u, \quad u(x,t) \in \mathbb{C}, \quad \alpha \in \mathbb{C}, \\ f(|u|^2) &= \gamma |u|^4 + \beta |u|^2 + \delta, \quad \beta, \gamma, \delta \in \mathbb{C}. \end{aligned} \tag{49}$$

Note that $u(x,t)$ is complex-valued in this case. But we can rewrite (49) as a real system of dimension 2 which turns out to be parabolic in case $\operatorname{Re}\alpha > 0$. Suitable function spaces for this case are $X = C_{\text{unif}}(\mathbb{R}, \mathbb{C})$, $Y = C^2_{\text{unif}}(\mathbb{R}, \mathbb{C})$. Now the Lie group is $G = \mathbb{R} \times S^1 \ni (\tau, \theta)$ with the action given by

$$\begin{aligned} a(\tau, \theta)v(x) &= e^{-i\theta}v(x - \tau), \quad v \in X, \\ d\,[u(0,0)v](\mu_\tau, \mu_\theta) &= -\mu_\tau v_x - i\mu_\theta v, \quad (\mu_\tau, \mu_\theta) \in \mathscr{A} = \mathbb{R}^2. \end{aligned} \tag{50}$$

Relative equilibria are of the form $u(x,t) = e^{-i\mu_\theta t}\bar{v}(x - \mu_\tau t)$ where μ_τ and μ_θ denote translational and rotational velocities, respectively. If both velocities are different from zero then we have a wave that rotates and travels simultaneously. In fact, for the parameter setting

$$\alpha = \frac{1+i}{2}, \quad \delta = -\frac{1}{2}, \quad \beta = \frac{5}{2} + i, \quad \gamma = -1 - \frac{i}{10},$$

the QCGL exhibits a rotating pulse ($\mu_\theta \neq 0$, $\mu_\tau = 0$) as well as a rotating and traveling wave ($\mu_\theta \neq 0$, $\mu_\tau \neq 0$). The real and imaginary parts of both types of solutions are shown in Figs. 8 and 9. When hitting the boundary with Neumann boundary conditions, the pulse stops traveling but keeps rotating. Finally, recall that a relative equilibrium $\bar{v}$, $\bar{\mu} = (\mu_\tau, \mu_\theta)$ of the QCGL satisfies

$$0 = \alpha v_{xx} + f(|v|^2)v + \mu_\tau v_x + i\mu_\theta v. \tag{51}$$

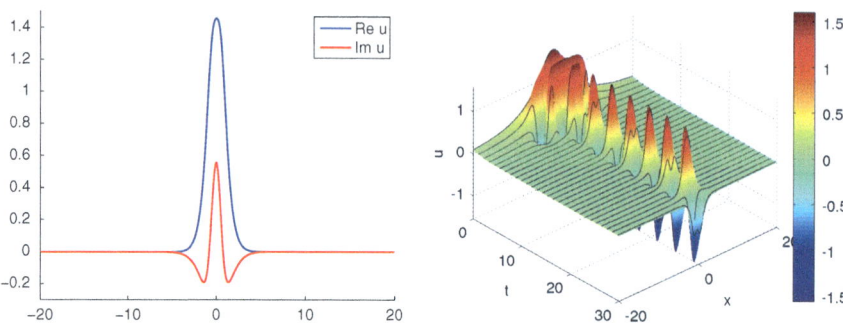

Fig. 8 Profile of the rotating wave of the QCGL (*left*) and space-time diagram of Re u (*right*) for $\alpha = \frac{1+i}{2}, \delta = -\frac{1}{2}, \beta = \frac{5}{2} + i, \gamma = -1 - \frac{i}{10}$

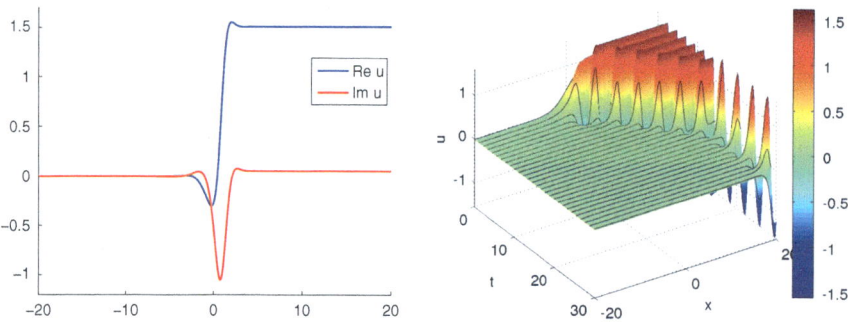

Fig. 9 Profile of the rotating and traveling wave of the QCGL (*left*) and space-time diagram of Re u (*right*) for $\alpha = \frac{1+i}{2}, \delta = -\frac{1}{2}, \beta = \frac{5}{2} + i, \gamma = -1 - \frac{i}{10}$

We also point out that the stability theory outlined in Sect. 1.6 applies only to pulses for which $\lim_{x \to \pm\infty} \bar{v}(x) = 0$.

1.8 Summary

Let us summarize the results of this section:

- Traveling pulses and fronts can be computed from heteroclinic resp. homoclinic orbits of dynamical systems,
- Traveling fronts and pulses may be viewed as relative equilibria with respect to shift equivariance, this is a special case of abstract equivariant evolution equations,
- Nonlinear stability of traveling waves in Sobolev spaces can be derived from linear stability via a nonlinear change of coordinates,
- Linearized differential operators may have essential as well as point spectrum,

- The essential spectrum can be handled theoretically via the dispersion relation, while determining point spectrum often needs numerical computations,
- Nonlinear stability theory does not directly apply to waves that travel and rotate simultaneously.

2 Stability of Traveling Waves and the Freezing Method

In the previous lecture we already studied the stability of traveling waves with asymptotic phase. Our main assumptions are concerned with spectral properties of the differential operator that arises by linearizing about the profile of the wave. The basic idea is to transform into a coordinate system that moves with the velocity of the wave and then to perform a nonlinear transformation which allows to study the exponential decay towards the profile and the dynamics of the phase separately. All these transformations assume the knowledge of the exact wave and hence are not suitable for numerical calculations. In this section we study a numerical method that allows to circumvent this problem: the *freezing method*. The method was independently proposed in [12,57]. Meanwhile, it has been extended to a variety of time-dependent partial differential equations, see [13, 14], and applied to control problems, for example [1]. There is also a parallel development by Cvitanović and co-workers (see [33] for a recent review and an application to a 5-dimensional Lorenz system), where the term 'method of sclices' is used for essentially the same approach.

The method introduces new time-dependent coordinates both in the underlying Lie group and in the function space. The extra degrees of freedom in the group is compensated by a corresponding number of phase conditions that try to keep the current profile as constant as possible. Altogether, one has to solve a partial differential algebraic equation (PDAE). For solutions of Cauchy problems that are close to relative equilibria this allows to adaptively compute moving coordinate systems within which the wave is *frozen*. Simultaneously, the flow on the group provides information about the speed and location of the original profile. The method can be formulated for equivariant evolution equations in general and thus has a wide range of applications. In this section we emphasize stability issues of the freezing method. In particular, we show that stability with asymptotic phase for a traveling wave turns into classical Lyapunov stability for the PDAE formulation.

2.1 Moving Frames: The Freezing Method and Phase Conditions

Consider the Cauchy problem associated with (15),

$$u_t = Au_{xx} + f(u, u_x), \quad u(x,0) = u_0(x), \quad x \in \mathbb{R}, \ t \geq 0. \tag{52}$$

The idea of the freezing method is to introduce new unknowns $v(x,t) \in \mathbb{R}^m$, $\gamma(t) \in \mathbb{R}$, such that the solution of (52) is of the form

$$u(x,t) = v(x - \gamma(t), t), \quad x \in \mathbb{R}, \ t \geq 0. \tag{53}$$

Inserting the ansatz into (52) and introducing $\mu(t) = \gamma_t(t)$ leads to a Cauchy problem for the position $\gamma(t)$ and the profile $v(\cdot, t)$,

$$\begin{aligned} v_t &= Av_{xx} + f(v, v_x) + \mu(t)v_x, & v(\cdot, 0) &= u_0, \\ \gamma_t &= \mu(t), & \gamma(0) &= 0. \end{aligned} \tag{54}$$

We note the similarity to equations (PDE1), (ODE2), but now we have not reduced the function space for v. Therefore, the system is not yet well posed. We compensate the extra variable $\mu(t)$ by an extra condition which is called a **phase condition** as in (12). There, the phase condition was used to remove the ambiguity in the traveling wave profile. Here, we use it to keep the time-dependent solution as constant as possible. We consider two possible choices for the phase condition, both based on a minimization principle.

1. *Fixed phase condition*

    Choose a template function $\hat{v} \in X$ where X is the underlying function space for solutions $u(\cdot, t)$, $v(\cdot, t)$. As an example take the affine space $X = w + H^1(\mathbb{R}, \mathbb{R}^m)$ where $w : \mathbb{R} \to \mathbb{R}^m$ is smooth and bounded and has the desired limit behavior $\lim_{x \to \pm\infty} w(x) = u_\pm$, cf. (7). In this case one may choose $\hat{v} = w$ or $\hat{v} = u_0 \in X$. The phase condition requires $\hat{v}$ to be the closest point to $v(\cdot, t)$ on the group orbit $\{\hat{v}(\cdot - g) : g \in \mathbb{R}\}$, i.e.

    $$\min_{g \in \mathbb{R}} \|v(\cdot, t) - \hat{v}(\cdot - g)\|_{L^2} = \|v(\cdot, t) - \hat{v}(\cdot)\|_{L^2}. \tag{55}$$

    The necessary condition is (cf. (12))

    $$0 = \frac{d}{dg} \|v(\cdot, t) - \hat{v}(\cdot - g)\|_{L^2}^2 \big|_{g=0} = 2(v(\cdot, t) - \hat{v}, \hat{v}_x)_{L^2}. \tag{56}$$

    Thus, instead of (52) the freezing method solved the following partial differential algebraic equation (PDAE)

    $$\begin{aligned} v_t &= Av_{xx} + f(v, v_x) + \mu(t)v_x, & v(\cdot, 0) &= u_0, \\ 0 &= (v - \hat{v}, \hat{v}_x)_{L^2} \\ \gamma_t &= \mu(t), & \gamma(0) &= 0. \end{aligned} \tag{57}$$

    This is a PDAE of index 2. Differentiating the constraint with respect to t and inserting the PDE leads to

$$v_t = Av_{xx} + f(v, v_x) + \mu v_x, \qquad v(\cdot, 0) = u_0,$$
$$0 = \mu(v_x, \hat{v}_x)_{L^2} + (Av_{xx} + f(v, v_x), \hat{v}_x)_{L^2} = \psi_{\text{fix}}(v, \mu) \qquad (58)$$
$$\gamma_t = \mu(t), \qquad \gamma(0) = 0.$$

If $(v_x, \hat{v}_x)_{L^2} \neq 0$ the constraint can be solved for μ and hence (58) is a PDAE of index 1.

2. *Orthogonality phase condition*
   Here we select the phase shift such that $\|v_t(\cdot, t)\|_{L^2}$ is minimal at each time instance t, i.e.

$$0 = \frac{d}{d\mu}\|v_t(\cdot, t)\|^2_{L^2}|_{\mu=\mu(t)} = \frac{d}{d\mu}\|Av_{xx} + f(v, v_x) + \mu v_x\|^2_{L^2}|_{\mu=\mu(t)}$$
$$= 2\left[\mu(t)(v_x, v_x)_{L^2} + (Av_{xx} + f(v, v_x), v_x)_{L^2}\right]. \qquad (59)$$

Therefore, instead of (52) we solve the PDAE

$$v_t = Av_{xx} + f(v, v_x) + \mu v_x, \qquad v(\cdot, 0) = u_0,$$
$$0 = \mu(v_x, v_x)_{L^2} + (Av_{xx} + f(v, v_x), v_x)_{L^2} = \psi_{\text{orth}}(v, \mu), \qquad (60)$$
$$\gamma_t = \mu(t), \qquad \gamma(0) = 0.$$

This PDAE is of index 1 provided $(v_x, v_x)_{L^2} \neq 0$, i.e. if v is nonconstant. Note that ψ_{orth} differs from ψ_{fix} only in replacing the template function $\hat{v}$ by v. Since (60) requires no previous knowledge of a template it is easier to apply far away from any traveling wave. However, close to a traveling wave, the system (58) turns out to be more robust, in particular when fixing $\hat{v} = v(\cdot, T)$ at some later time T and leaving it constant from then on.

To summarize, we replace (52) by a PDAE of the general form

$$v_t = Av_{xx} + f(v, v_x) + \mu v_x, \quad v(\cdot, 0) = u_0,$$
$$0 = \psi(v, \mu) \qquad (61)$$
$$\gamma_t = \mu(t), \qquad \gamma(0) = 0,$$

where $\psi : X \times \mathbb{R} \to \mathbb{R}$. Using a proper notion of solutions, one can show that any solution of (61) leads to a solution of (52) via (53), cf. [12, 52]. Conversely, if $u(\cdot, t)$ solves (52) then we obtain a solution of (61), provided the implicit ODE

$$\psi(u(\cdot + \gamma(t), t), \gamma_t(t)) = 0, \quad \gamma(0) = 0$$

has a unique solution $\gamma(t)$ on the interval under consideration.

2.2 Numerical Experiments with Traveling Fronts and Pulses

For numerical computations we solve the PDAE (61) on a large interval $J = [x_-, x_+]$, and we use two-point boundary conditions given by a map $\mathscr{B} : \mathbb{R}^{4m} \to \mathbb{R}^{2m}$,

$$\begin{aligned} v_t &= A v_{xx} + f(v, v_x) + \mu v_x \text{ in } J \times [0, \infty), & v(\cdot, 0) &= u_0|_J, \\ 0 &= \psi_J(v, \mu), & \mathscr{B}((v, v_x)(x_-), (v, v_x)(x_+)) &= 0, \\ \gamma_t &= \mu(t), & \gamma(0) &= 0. \end{aligned} \quad (62)$$

Examples for $\mathscr{B}$ are Neumann boundary conditions $\mathscr{B}((v, v_x)(x_-), (v, v_x)(x_+)) = (v_x(x_-), v_x(x_+))$ and projection boundary conditions (cf. (14), note that $\mu(t)$ enters into the projection matrices).

Example 6 (Nagumo equation). Our first example is the Nagumo equation (5),

$$\begin{aligned} v_t &= v_{xx} + v(1-v)(v-\alpha) + \mu v_x, & v(\cdot, 0) &= u_0|_J, \\ 0 &= \psi_J(v, \mu) & & (63) \\ \gamma_t &= \mu(t), & \gamma(0) &= 0, \end{aligned}$$

with parameter $\alpha = \frac{1}{4}$, solved on $J = [-50, 50]$, with $\triangle x = 0.1$, $\triangle t = 0.1$. For the nonfrozen system on $J = [x_-, x_+]$, the front forms and travels to the left as we expect, see Fig. 10a. When it reaches the boundary it dies out due to Neumann boundary conditions. On the contrary, the front stabilizes for the frozen system, see Fig. 10b, the variable $\mu(t)$ approaches the final speed $\bar{\mu} = -\frac{\sqrt{2}}{4}$ of the front, see Fig. 10c, while the value of $\gamma(t)$ still indicates the position of the front on the real line.

Our conclusion is that the longtime behavior of the initial boundary value problem on the finite interval (62) can be completely different from the behavior of the original system (52) when truncated to the same interval, although on the infinite line both systems are equivalent. The freezing method aims at a moving coordinate system in which a pattern close to the initial data becomes stationary, and this behavior is stable under truncation to a bounded domain. In the following subsections we will provide theorems which make this observation rigorous.

Example 7 (FitzHugh-Nagumo system). Our second example is the FitzHugh-Nagumo system, which in the frozen form reads

$$\begin{aligned} v_t &= A v_{xx} + f(v) + \mu v_x, & v(\cdot, 0) &= v_0, \\ 0 &= \psi_J(v, \mu), & \mathscr{B}((v, v_x)(x_-), (v, v_x)(x_+)) &= 0, & (64) \\ \gamma_t &= \mu(t), & \gamma(0) &= 0. \end{aligned}$$

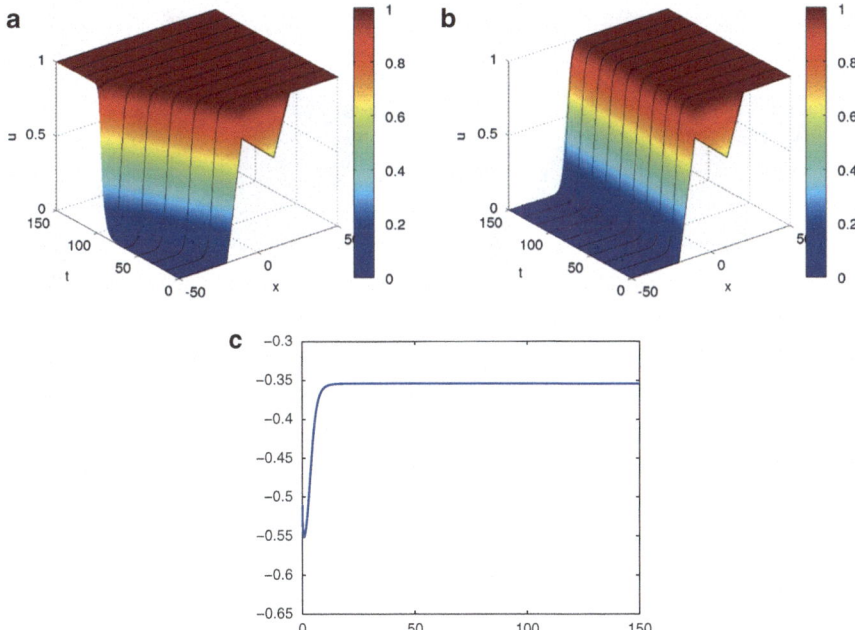

Fig. 10 Traveling front $u(x, t)$ of the original Nagumo equation (**a**), solution $v(x, t)$ of the frozen Nagumo equation (**b**), time dependence of velocity $\mu(t)$ (**c**). Piecewise linear initial function, Neumann boundary conditions, fixed phase condition with reference function $\hat{v} = u_0$ and parameter value $\alpha = \frac{1}{4}$

with $v = (v_1, v_2)^T$,

$$A = \begin{pmatrix} 1 & 0 \\ 0 & \varepsilon \end{pmatrix}, \quad f(v) = \begin{pmatrix} v_1 - \frac{1}{3}v_1^3 - v_2 \\ \phi(v_1 + a - bv_2) \end{pmatrix},$$

parameters $\varepsilon = 0.1$, $a = 0.7$, $b = 0.8$, $\phi = 0.08$, solved on $J = [-60, 60]$, with $\triangle x = 0.1$, $\triangle t = 0.1$. For these parameter values, the function f admits only one zero at $(v_1, v_2)^T = (-1.1994, -0.6243)^T$ and pulses occur. Note that (64) is a parabolic system due to $\varepsilon > 0$. The case $\varepsilon = 0$ leads to a coupled hyperbolic-parabolic system with principal term being $\mu v_{2,x}$ in the second equation. We consider such mixed systems later in Sect. 3. Starting with a ramp-like function for the voltage v_1, the pulse forms as expected and travels to the left until it dies out at the boundary, see Fig. 11a. On the contrary, as above the freezing method captures the shape of the pulse and makes it stationary, see Fig. 11b. Simultaneously, the correct speed of the pulse is attained by the variable $\mu(t)$, which tends to $\bar{\mu} = -0.7892$, see Fig. 11c.

An interesting phenomenon happens when the system is started with a pulse-like initial function. Then two pulses develop, one traveling to the left, the other traveling to the right, see Fig. 12a. Which of these two traveling pulses is captured

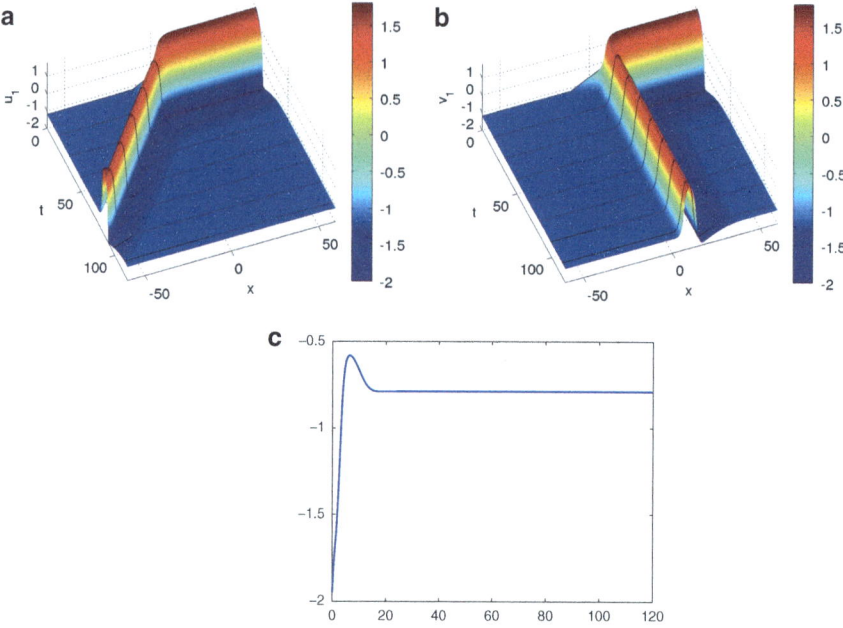

Fig. 11 u_1 component of traveling pulse (**a**) of the nonfrozen and v_1 component of frozen pulse (**b**) with velocity $\mu(t)$ (**c**) of the frozen FitzHugh-Nagumo system with piecewise linear initial function, Neumann boundary conditions, fixed phase condition with reference function $\hat{v} = u_0$ and parameter values $\varepsilon = 0.1, a = 0.7, b = 0.8, \phi = 0.08$

by the freezing system, is unpredictable and can be affected, for instance, by the solver tolerances. In this example, the fixed phase condition happens to freeze the left going pulse while the right going one dies out at the boundary, see Fig. 12b. The velocity $\mu(t)$ of the left frozen pulse nears $\bar{\mu} = -0.7892$, see Fig. 12d. On the contrary, the orthogonal phase condition freezes the right going pulse and the left going one dies out, see Fig. 12c. The velocity $\mu(t)$ of the right frozen pulse tends to $\bar{\mu} = 0.7892$, see Fig. 12e as expected.

The experiments from Example 7 show two new problems that will be dealt with in the following lectures.

1. For parabolic systems coupled to ODEs, the freezing method leads to mixed hyperbolic-parabolic systems with the newly introduced convection term entering the principal part of the equation. The stability theory for such systems is much more subtle than for parabolic systems (cf. Sect. 1.6) and will be topic of Sect. 3.
2. When the solutions exhibit multiple fronts and pulses, there is no longer a common moving frame in which the patterns become stationary. The freezing method then tends to capture one of the patterns and let the others travel towards the boundary. In Sect. 5 we discuss an extension of the freezing method that

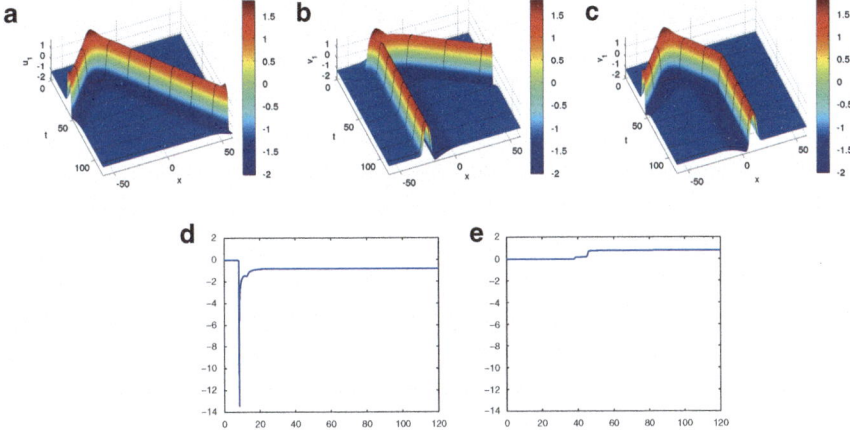

Fig. 12 u_1-component of traveling wave of the nonfrozen system (**a**) for a pulse-like initial function in the FitzHugh-Nagumo system. v_1-component of frozen pulse with fixed phase condition (**b**) and orthogonal phase condition (**c**). Figures (**d**) and (**e**) show the time dependence of the velocity $\mu(t)$ for cases (**b**) and (**c**), respectively. For all plots Neumann boundary conditions and the reference function $\hat{v} = u_0$ are used, parameter values are $\varepsilon = 0.1$, $a = 0.7$, $b = 0.8$, $\phi = 0.08$

allows to handle multiple coordinate frames and to deal with weak and strong interactions of patterns.

2.3 Error Analysis of Numerical Approximations

The experiments of the previous subsection suggest that the PDAE formulation (61) is robust with respect to numerical approximations such as truncation to a bounded interval with two-point boundary conditions and subsequent discretization of time and space. Such expectations have been made rigorous in the work of V. Thümmler [65–67]. It is shown in [66] in which sense relative equilibria of parabolic systems are inherited by numerical discretizations, and in [67] conditions are set up that guarantee asymptotic stability of these discretized equilibria. Moreover, the exponential rate of convergence is proved to be independent of the truncated interval and of the step-size used. Below we will only consider the case of traveling waves and indicate the main results. We also mention the paper [5] where the nonlocal equations obtained by eliminating μ from the phase condition in (61), have been treated directly by truncation to a finite interval.

We consider a finite difference approximation of the PDAE (57) on an equidistant grid $J_h = \{x_n = nh : n = n_-, \ldots, n_+\}$ with step-size $h > 0$. For functions $v_n = v(x_n)$, $x_n = nh$, defined on an extended grid $J_h^e = \{nh : n = n_- - 1, \ldots, n_+ + 1\}$, we use standard difference quotients as follows,

$$\delta_+ v_n = \frac{1}{h}(v_{n+1} - v_n), \quad n = n_- - 1, \ldots, n_+,$$

$$\delta_- v_n = \frac{1}{h}(v_n - v_{n-1}), \quad n = n_-, \ldots, n_+ + 1,$$

$$\delta_+ \delta_- v_n = \frac{1}{h^2}(v_{n-1} - 2v_n + v_{n+1}), \quad n = n_-, \ldots, n_+,$$

$$\delta_0 v_n = \frac{1}{2}(\delta_+ v_n + \delta_- v_n), \quad n = n_-, \ldots, n_+.$$

Leaving time continuous, a spatial discretizaton of (57) leads to the following DAE:

$$\begin{aligned} v_{n,t} &= A\delta_+\delta_- v_n + f(v_n, \delta_0 v_n) + \mu \delta_0 v_n, \quad n = n_-, \ldots, n_+, \\ 0 &= (\delta_0 \hat{v}, v - \hat{v})_{L^2(J_h)} \\ \eta &= P_- v_{n_-} + Q_- \delta_0 v_{n_-} + P_+ v_{n_+} + Q_+ \delta_0 v_{n_+}, \end{aligned} \quad (65)$$

where $P_\pm, Q_\pm \in \mathbb{R}^{2m,m}$ are given boundary matrices and $\eta = P_- u_- + P_+ u_+$ (see Sect. 1.3 for $u_\pm$). Here and in the following we use discrete analogs $L^2(J_h), H^1(J_h), H^2(J_h)$ of the Sobolev spaces $L^2(J, \mathbb{R}^m), H^1(J, \mathbb{R}^m), H^2(J, \mathbb{R}^m)$. The following conditions are imposed on the continuous problem.

Cont1: The function $\bar{v} \in w + H^2(\mathbb{R}, \mathbb{R}^m)$ (cf. (18)) is a traveling wave of speed $\bar{\mu} \in \mathbb{R}$ for (15) such that $u_\pm = \lim_{x \to \pm\infty} \bar{v}(x)$.
Cont2: The nonlinearity f satisfies the condition (37) from Thm. 3.
Cont3: 0 is a simple eigenvalue of the linearized operator Λ from (23), and 0 lies in the set M defined in Thm. 1.

Note that the condition $0 \in M$ is weaker than the spectral condition **SC** (cf. (33)) which requires $\{\text{Re}\, z \geq -\beta\} \subset M$. It implies that the quadratic eigenvalue problems $(\lambda^2 A + \lambda B_\pm + C_\pm)y = 0$ have m eigenvalues with real part positive and m with real part negative. More precisely, there are matrices $Y_\pm^s, \Lambda_\pm^s \in \mathbb{R}^{m,m}$ which solve the quadratic invariant subspace equation,

$$AY_\pm^s (\Lambda_\pm^s)^2 + B_\pm Y_\pm^s \Lambda_\pm^s + C_\pm Y_\pm^s = 0, \quad \text{Re}\,(\sigma(\Lambda_\pm^s)) < 0. \quad (66)$$

such that $\text{rank} \begin{pmatrix} Y_\pm^s \\ Y_\pm^s \Lambda_\pm^s \end{pmatrix} = m$. Similarly, there exist $Y_\pm^u, \Lambda_\pm^u \in \mathbb{R}^{m,m}$ satisfying the same conditions except $\text{Re}\,(\sigma(\Lambda_\pm^u)) > 0$. The next two conditions impose a coupling between boundary matrices and data of the continuous problem.

Discrete 1: $\eta = P_- u_- + P_+ u_+$ (consistency of boundary values)
Discrete 2: $\det \left((P_-\ Q_-) \begin{pmatrix} Y_-^s \\ Y_-^s \Lambda_-^s \end{pmatrix} \Big| (P_+\ Q_+) \begin{pmatrix} Y_+^u \\ Y_+^u \Lambda_+^u \end{pmatrix} \right) \neq 0$

Condition **Discrete 2** ensures that modes increasing at $\pm\infty$ can be controlled by the boundary conditions, see [9]. Note that the columns of $\begin{pmatrix} Y^s_- \\ Y^s_- \Lambda^s_- \end{pmatrix}$ determine growing solutions of (the first order version of) Λ at $-\infty$ while those of $\begin{pmatrix} Y^u_+ \\ Y^u_+ \Lambda^u_+ \end{pmatrix}$ determine growing solutions at $+\infty$.

Theorem 4 ([66]). *Assume **Cont1–3**, let $\hat{v} \in w + H^2(\mathbb{R}, \mathbb{R}^m)$ be a template function such that $(\bar{v}_x, \hat{v}_x)_{L^2} \neq 0$ and let the boundary matrices satisfy **Discrete 1, 2**. Then there exist $C, \rho, T, h_0, \alpha > 0$ such that the DAE (65) has a steady state (v^h, μ^h) for all $0 < h \leq h_0$, $T \leq \min(n_+, -n_-)h$ which is unique in the ball*

$$\|\bar{v}|_{J_h} - v^h\|_{H^2(J_h)} + |\bar{\mu} - \mu^h| \leq \rho.$$

Moreover the following error estimate holds

$$\|\bar{v}|_{J_h} - v^h\|_{H^2(J_h)} + |\bar{\mu} - \mu^h| \leq C\left[h^2 + \exp(-\alpha h \min(n_+, -n_-))\right]. \quad (67)$$

The estimate (67) shows, that the errors due to finite difference approximation and truncation to a bounded interval simply add up. Under the extra conditions that $\hat{v}$ decays exponentially as $x \to \pm\infty$, one can show (see [66, Th.2.6]) that the linearization of the right hand side of (65) at (v^h, μ^h) has an eigenvalue close to 0 and an eigenfunction close to $\bar{v}_x|_{J_h}$ with the same estimate as in (67).

For a stronger statement on the asymptotic stability of the 'discrete traveling wave' (v^h, μ^h), more conditions are needed (see [67]). In particular, the assumptions of the Stability Theorem 3 are assumed to hold. Further, condition **Discrete 2** now is strengthened to

$$\det\left(\begin{pmatrix} P_- & Q_- \end{pmatrix}\begin{pmatrix} Y^s_-(s) \\ Y^s_-(s)\Lambda^s_-(s) \end{pmatrix} \Big| \begin{pmatrix} P_+ & Q_+ \end{pmatrix}\begin{pmatrix} Y^u_+(s) \\ Y^u_+(s)\Lambda^u_+(s) \end{pmatrix}\right) \neq 0, \quad \text{Re}(s) \geq -\beta, \quad (68)$$

where $Y^s_\pm = Y^s_\pm(s)$, $\Lambda^s_\pm = \Lambda^s_\pm(s)$ solve the s-dependent equation

$$AY^s_\pm(\Lambda^s_\pm)^2 + B_\pm Y^s_\pm \Lambda^s_\pm + (C_\pm - sI)Y^s_\pm = 0, \quad \text{Re}(\sigma(\Lambda^s_\pm)) < 0, \quad (69)$$

cf. (66). Under a final condition on the Dirichlet and Neumann parts of the boundary matrices (see [67, Hypothesis 2.6]), the following estimate holds for the solutions $v_n(t), \mu(t)$ of the time-dependent DAE (65),

$$\|v(t) - v^h\|_{H^1(J_h)} + |\mu(t) - \mu^h| \leq Ce^{-\alpha t}, \quad t \geq 0, \ h \leq h_0, \ \pm n_\pm h > T, \quad (70)$$

provided $\|v(0) - v^h\|_{H^1(J_h)} \leq \rho$ and v^0, μ^0 are consistent initial values (cf. [67, Sect.2.1]). It is worth noting that all constants C, ρ, α, T in this result do neither depend on the step-size h nor on the values of n_-, n_+. While the extra condition

[67, Hypothesis 2.6] is satisfied for all standard choices such as Dirichlet, Neumann or periodic boundary conditons, condition (68) is essential for the stability of the discretized traveling wave. As shown in [67, Sec.5.2] by a counterexample, violating (68) at one value of s can destabilize the discrete wave by spurious oscillations while the continuous wave is perfectly stable.

2.4 The Freezing Method in an Abstract Setting

The freezing method of the previous section can be generalized to equivariant evolution systems on Banach manifolds, see [13, 66]. For simplicity, we consider here the setting of Banach spaces as in Sect. 1.7. Generalizing (53), the solution of (43) is written as

$$u(t) = a(\gamma(t))v(t), \quad \gamma(t) \in G, v(t) \in Y. \tag{71}$$

Then, $a(\gamma)F(v) = F(a(\gamma)v) = F(u) = u_t = a(\gamma)v_t + d[a(\gamma)v]\gamma_t$ holds and hence

$$v_t = F(v) - a(\gamma)^{-1}d[a(\gamma)v]\gamma_t. \tag{72}$$

It is convenient to introduce $\mu(t) \in \mathscr{A} = T_{\mathbb{1}}G$ via $\gamma_t(t) = dL_{\gamma(t)}(\mathbb{1})\mu(t)$. Then differentiating $a(\gamma)a(g)v = a(L_\gamma g)v$ with respect to $g \in G$ at $g = \mathbb{1}$ leads to $a(\gamma)d[a(\mathbb{1})v]\mu = d[a(\gamma)v]dL_\gamma(\mathbb{1})\mu$. Therefore, (72) can be rewritten as $v_t = F(v) - d[a(\mathbb{1})v]\mu$. Finally, we add dim G phase conditions $\psi(v, \mu) = 0$ defined by a functional $\psi : Y \times \mathscr{A} \to \mathscr{A}^*$. This leads to the abstract formulation of the freezing method as differential algebraic evolution equation (DAEV)

$$\begin{aligned} v_t &= F(v) - d[a(\mathbb{1})v]\mu, & v(0) &= u_0, \\ 0 &= \psi(v, \mu), & & \\ \gamma_t &= dL_\gamma(\mathbb{1})\mu, & \gamma(0) &= \mathbb{1}. \end{aligned} \tag{73}$$

The last equation of this system is called the **reconstruction equation** in [57]. It is decoupled from the first two equations and can be solved in a post-processing step in order to obtain the orbit within the group. If a continuous inner product $(\cdot, \cdot)_X$ on X is given, the two phase conditions discussed in Sect. 2.1, generalize to

$$(d[a(\mathbb{1})\hat{v}]\lambda, v - \hat{v})_X = 0 \quad \text{for all } \lambda \in \mathscr{A}, \tag{74}$$

$$(d[a(\mathbb{1})v]\lambda, v_t)_X = 0 \quad \text{for all } \lambda \in \mathscr{A}, \tag{75}$$

see Fig. 13. Differentiating (74) with respect to t and inserting the differential equation from (73) into both, leads to the phase fixing operators (cf. (58), (60))

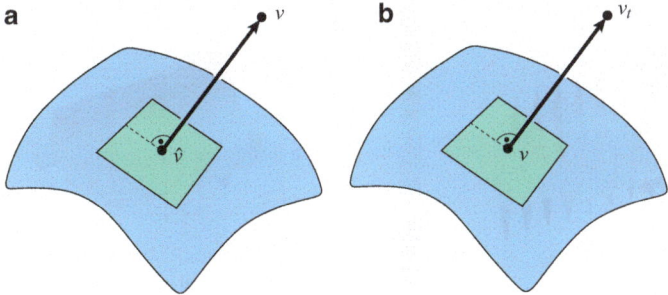

Fig. 13 Illustration of fixed phase condition $v - \hat{v} \perp T_{\hat{v}}\mathcal{O}(\hat{v}) = \mathcal{R}(d[a(\mathbb{1})\hat{v}])$ with group orbit $\mathcal{O}(\hat{v}) = \{a(\gamma)\hat{v} : \gamma \in G\}$ (**a**), illustration of orthogonal phase condition $v_t \perp T_v \mathcal{O}(v) = \mathcal{R}(d[a(\mathbb{1})v])$ with group orbit $\mathcal{O}(v)$ (**b**)

$$\psi_{\text{fix}}(v,\mu)\lambda = (d[a(\mathbb{1})\hat{v}]\lambda, F(v))_X - (d[a(\mathbb{1})\hat{v}]\lambda, d[a(\mathbb{1})v]\mu)_X, \quad \lambda \in \mathscr{A}, \tag{76}$$

$$\psi_{\text{orth}}(v,\mu)\lambda = (d[a(\mathbb{1})v]\lambda, F(v))_X - (d[a(\mathbb{1})v]\lambda, d[a(\mathbb{1})v]\mu)_X, \quad \lambda \in \mathscr{A}. \tag{77}$$

If the map $d[a(\mathbb{1})v] : \mathscr{A} \to X$ is one to one, then the linear system $\psi_{\text{orth}}(v,\mu) = 0$ has a unique solution $\mu \in \mathscr{A}$ and hence (73) is a DAEV of index 1.

2.5 A Numerical Experiment with a Two-Dimensional Group

Example 8 (Quintic-cubic Ginzburg Landau equation (QCGL)). As an example we treat the quintic-cubic Ginzburg Landau equation from Example 2. With the group operations from (50) the DAEV (73) with $\psi = \psi_{\text{fix}}$ yields the following PDAE to be solved

$$\begin{aligned} v_t &= \alpha v_{xx} + f(|v|^2)v + \mu_1 v_x + \mu_2 i v, & v(0) &= u_0, \\ 0 &= (\hat{v}_x, v - \hat{v})_{L^2} = (i\hat{v}, v - \hat{v})_{L^2} = 0, & & \\ \tau_t &= \mu_1, \quad \theta_t = \mu_2, & \tau(0) &= \theta(0) = 0. \end{aligned} \tag{78}$$

For the parameter values

$$\alpha = 1, \quad \beta = 3+i, \quad \gamma = 3+i, \quad \delta = -2.75+i, \tag{79}$$

one finds a rotating pulse with translational velocity $\bar{\mu}_1 = 0$ and rotational velocity $\bar{\mu}_2 =$, see Fig. 14a–c, and a pulse that rotates and travels simultaneously with translational velocity $\bar{\mu}_1 = 1.18$ and $\bar{\mu}_2 = -2.801$, see Fig. 15a–c. In both cases,

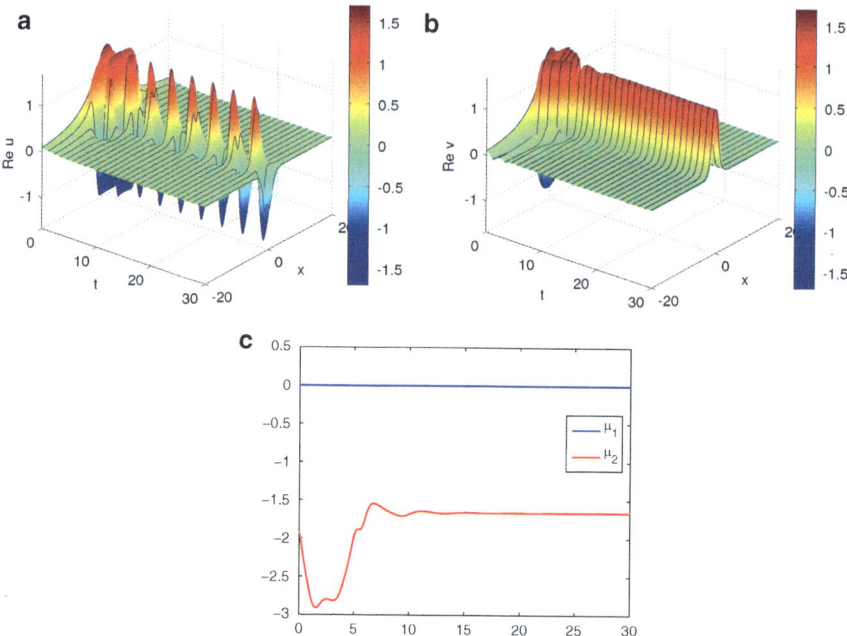

Fig. 14 Real part of the rotating pulse $u(x,t)$ of the nonfrozen system (**a**) and of the frozen pulse $v(x,t)$ (**b**), with velocities (μ_1, μ_2) (**c**) for the frozen QCGL. Solution by COMSOL Multiphysics with piecewise linear finite elements, Neumann boundary conditions, fixed phase condition with template function $\hat{v} = u_0$ and parameter values from (79)

freezing of patterns is successful. Note, however, that the stability analysis of [67] applies only to the first case, since the profile $v(x)$ keeps rotating as $x \to \infty$ and only its absolute value converges.

2.6 Summary

Let us summarize the main results of this section.

- The freezing method allows to automatically generate moving frames in which traveling waves become asymptotically constant.
- For the discretized PDAE formulation, one can prove existence of a "discrete traveling wave" as well as their asymptotic stability with rates independent of the discretization parameters. For such a result one needs the original wave to satisfy the standard stability conditions and the boundary matrices to control unstable modes at $\pm\infty$.
- The freezing method generalizes to abstract equivariant evolution equations posed on a Banach manifold.

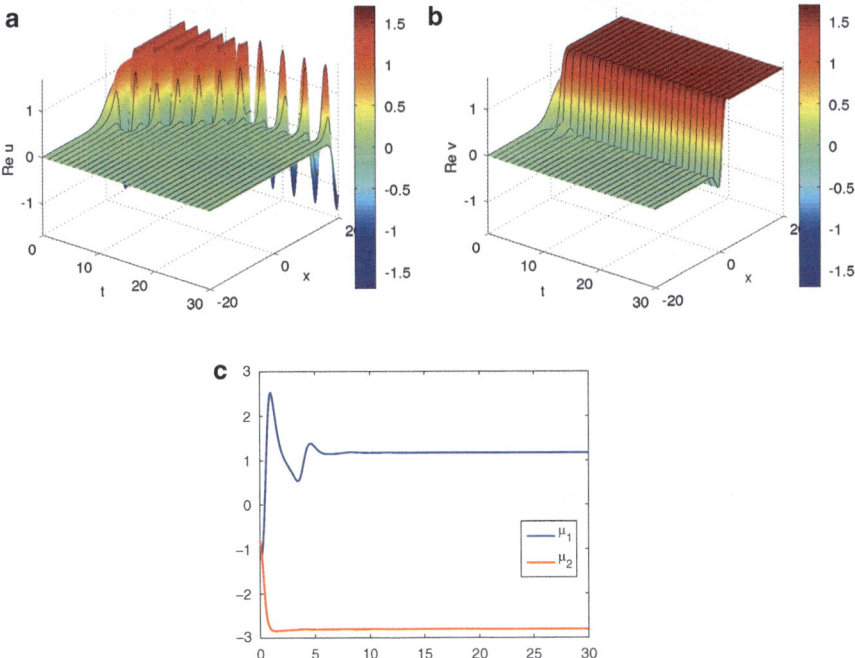

Fig. 15 Real part of the traveling and rotating pulse $u(x,t)$ for the nonfrozen system (**a**) and $v(x,t)$ for the frozen system (**b**), time-dependence of velocities (μ_1, μ_2) (**c**). Further data are as in Fig. 14

- The method successfully freezes waves in the quintic-cubic Ginzburg Landau equation that rotate and travel simultaneously.

3 Patterns in Hyperbolic and Hyperbolic-Parabolic Systems

The freezing method discussed in Lecture 2, leads to challenging problems, both numerically and theoretically, when applied to hyperbolic or hyperbolic-parabolic systems.

A famous example of this type are the Hodgkin-Huxley equations for the propagation of signals in nerve axons. We will use them, both for illustrating the analytical difficulties and for showing numerical applications. In this section we survey results due to J. Rottmann-Matthes [56] on the stability of the freezing method for hyperbolic-parabolic mixed systems of rather general type. The main difficulty arises from the fact that such mixed systems generate only C^0-semigroups so that the techniques for analytic semigroups do no longer apply. Moreover, as in the Hodgkin-Huxley example, nonstrictly hyperbolic parts occur which make the stability analysis even more delicate. The essential tool in resolving these problems is the vector valued Laplace transform applied directly to the PDAE formulation and combined with rather sophisticated resolvent estimates.

We also show an application of the freezing method to Burgers equation for which equivariance includes scalings of the variables. It is then possible to freeze similarity solutions such as N-waves. However the stability of the method for this case, and for conservation laws in general, is largely unsolved.

3.1 Mixed Systems and the Hodgkin-Huxley Example

The spatially extended version of the Hodgkin-Huxley model from 1952 [40] serves as our standard example,

$$\begin{aligned}
V_t &= \frac{a}{2R} V_{xx} - \bar{g}_K n^4 (V - V_K) - \bar{g}_{Na} m^3 h (V - V_{Na}) - \bar{g}_l (V - V_l), \\
n_t &= \alpha_n(V)(1-n) - \beta_n(V) n, \\
m_t &= \alpha_m(V)(1-m) - \beta_m(V) m, \\
h_t &= \alpha_h(V)(1-h) - \beta_h(V) h.
\end{aligned} \quad (80)$$

The system models electric signalling in nerve cells and we refer to [40, 43]) for details of the modelling, in particular for the special form of the nonlinearities $\alpha_n, \alpha_m, \alpha_h, \beta_n, \beta_m, \beta_h$. We note that the system consists of a parabolic PDE that is coupled nonlinearly to a system of three nonlinear ODEs. It is well-known that there exists a traveling wave solution (see for example [37]), a plot of the traveling pulse is given in Fig. 16.

In a co-moving frame, see (16), a term μV_x is added to the first equation, a term μn_x to the second equation and so forth. The resulting system is then **parabolic-hyperbolic**, with a hyperbolic part that is not strictly hyperbolic because of the common factor μ in the convection terms.

In the following we analyze the asymptotic stability of traveling waves for mixed systems of the general form

$$\begin{aligned}
u_t &= A u_{xx} + g(u,v)_x + f_1(u,v), & u(x,0) &= u_0(x), \\
v_t &= B v_x + f_2(u,v), & v(x,0) &= v_0(x),
\end{aligned} \quad (81)$$

where $u(x,t) \in \mathbb{R}^n$, $v(x,t) \in \mathbb{R}^m$. Note that (81) includes the Hodgkin-Huxley system. The following conditions are imposed on (81).

Basic Assumptions

(i) $g, f_1 \in C^3(\mathbb{R}^{n+m}, \mathbb{R}^n)$ $f_2 \in C^3(\mathbb{R}^{n+m}, \mathbb{R}^m)$,
(ii) the u-equation is parabolic, i.e. $A \in \mathbb{R}^{n,n}$ satisfies $A + A^\top \geq \alpha > 0$ in the sense of inner products,
(iii) the v-equation is hyperbolic, i.e. $B \in \mathbb{R}^{m,m}$ is diagonal,

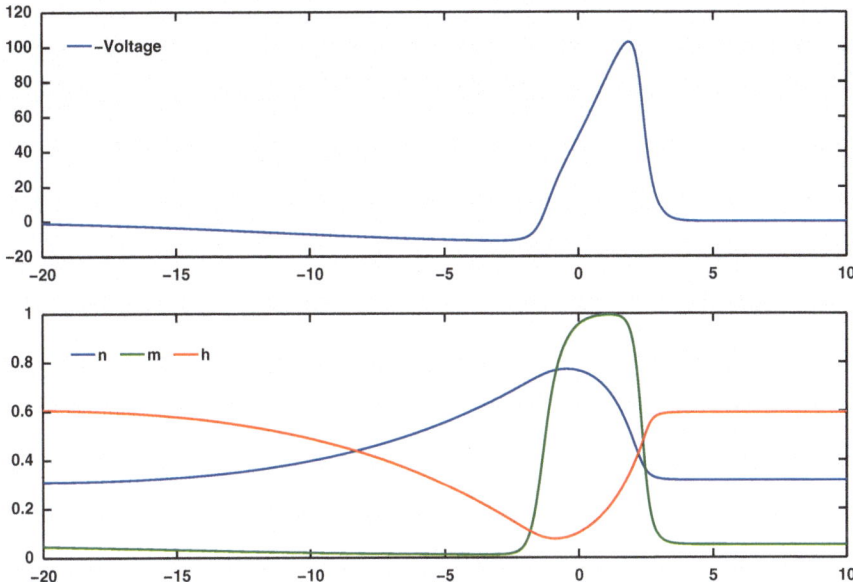

Fig. 16 Traveling pulse of the Hodgkin-Huxley equations

(iv) there exists a traveling wave solution $(u,v)(x,t) = (\bar{u},\bar{v})(x - \bar{\mu}t)$ of the PDE (81) with profile $(\bar{u},\bar{v}) \in C_b^1(\mathbb{R}, \mathbb{R}^{n+m})$ and $\bar{u}_x, \bar{v}_x \in H^2(\mathbb{R}, \mathbb{R}^{n+m})$.

Remark 2. (a) In general, hyperbolicity of the second equation in (81) means that the matrix B is real diagonalizable. By a similarity transformation we can put B into diagonal form and hence assume (iii) without loss of generality.
(b) Note that assumption (iv) allows for pulse as well as front solutions.

Before proving stability of traveling waves for (81), we need existence and uniqueness of solutions to the Cauchy problem, when the initial data belong to a proper neighborhood of the traveling wave.

Theorem 5 ([56, Thm. 2.5]). *Let the **Basic Assumptions** (i)–(iv) hold. Then for all $u_0 \in \bar{u} + H^1, v_0 \in \bar{v} + H^1$ there exists a unique maximal solution of (81). More precisely, there exists $T^* \in (0, \infty]$ and (u^*, v^*) such that $u = u^*|_{[0,T]}$ and $v = v^*|_{[0,T]}$ satisfy (81) in $L^2 \times L^2$ for a.e. $t \in [0,T]$, $0 < T < T^*$, and*

$$u \in C([0,T]; \bar{u} + H^1) \cap L^2(0,T; \bar{u} + H^2) \cap H^1(0,T; \bar{u} + H^1),$$
$$v \in C([0,T]; \bar{v} + H^1) \cap H^1(0,T; \bar{v} + H^1). \tag{82}$$

Conversely, any pair (u,v) for which (82) holds and which solves (81) in $L^2 \times L^2$ for a.e. $t \in [0,T]$, satisfies $u = u^*|_{[0,T]}$ and $v = v^*|_{[0,T]}$. Moreover,

either $T^* = \infty$ or $0 < T^* < \infty$ and $\lim_{t \nearrow T^*} \|u^*(t) - \bar{u}\|_{H^1} + \|v^*(t) - \bar{v}\|_{H^1} = \infty$.

Due to its importance, nonlinear stability of traveling waves in systems of the form (81) has been considered by many authors. We just mention a few: In a series of papers [24–27] J.W. Evans presented a full analysis of Hodgkin-Huxley type equations. By a dynamical systems approach Bates and Jones [7] were able to discuss the stability of systems of the general form (81) without the $g(u, v)_x$ term. But due to a compactness argument, their result does not include the case of fronts. In [45] Kreiss et al. proved stability of traveling waves in systems of the form (81) but they assumed strict hyperbolicity which is not satisfied for the Hodgkin-Huxley model in a co-moving frame. Finally, we mention [34], where parabolic-hyperbolic systems are considered. There the authors allow the spectrum to touch the imaginary axis, but they assume $g(u, v)_x = \tilde{a} u_x$ for a constant matrix $\tilde{a}$ and the v-equation is simply an ODE.

We consider (81) in a moving coordinate frame, see (16). The traveling wave then becomes a steady state of

$$u_t = A u_{xx} + \bigl(g(u, v) + \overline{\mu} u\bigr)_x + f_1(u, v), \qquad (83)$$
$$v_t = (B + \overline{\mu}) v_x + f_2(u, v).$$

As in Sect. 1.4, we expect stability with asymptotic phase. For notational convenience we denote $(u, v)^\top = U$ and $(\overline{u}, \overline{v})^\top = \overline{U}$ and write (83) in the short form

$$U_t = F(U).$$

We aim at a result in the spirit of Thm. 3. The linearization of (83) about the profile reads

$$u_t = A u_{xx} + (\partial_1 \overline{g} + \overline{\mu}) u_x + \partial_2 \overline{g} v_x + (\partial_1 \overline{g}_x + \partial_1 \overline{f}_1) u + (\partial_2 \overline{g}_x + \partial_2 \overline{f}_1) v, \qquad (84)$$
$$v_t = (B + \overline{\mu}) v_x + \partial_1 \overline{f}_2 u + \partial_2 \overline{f}_2 v,$$

where we abbreviate $\overline{g}(x) = g(\overline{u}(x), \overline{v}(x))$, $\partial_1 \overline{g}(x) = g_u(\overline{u}(x), \overline{v}(x))$, etc. The linear operator Λ on the right hand side of (84) has the following block structure

$$\Lambda \begin{pmatrix} u \\ v \end{pmatrix} = \begin{pmatrix} A & 0 \\ 0 & 0 \end{pmatrix} \begin{pmatrix} u \\ v \end{pmatrix}_{xx} + \begin{pmatrix} B_{11} & B_{12} \\ 0 & B_{22} \end{pmatrix} \begin{pmatrix} u \\ v \end{pmatrix}_x + \begin{pmatrix} C_{11} & C_{12} \\ C_{21} & C_{22} \end{pmatrix} \begin{pmatrix} u \\ v \end{pmatrix}. \qquad (85)$$

For simplicity we abbreviate

$$\tilde{A} := \begin{pmatrix} A & 0 \\ 0 & 0 \end{pmatrix}, \quad \tilde{B} := \begin{pmatrix} B_{11} & B_{12} \\ 0 & B_{22} \end{pmatrix}, \quad \tilde{C} := \begin{pmatrix} C_{11} & C_{12} \\ C_{21} & C_{22} \end{pmatrix}.$$

As we already saw in Sect. 1.5 shift equivariance implies $(\overline{u}_x, \overline{v}_x)^\top \in \mathscr{N}(\Lambda)$.

Fig. 17 Sketch of the spectrum of Λ satisfying the **Linear Assumptions**

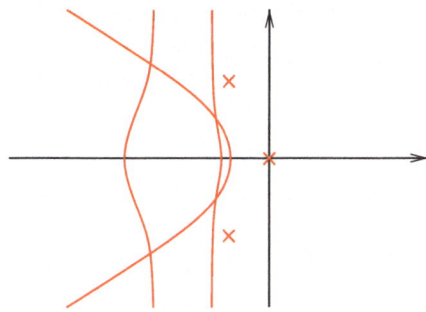

3.2 The Stability Theorem

Similar to Sect. 1 we impose the following conditions on the operator Λ (Fig. 17).

Linear Assumptions

(i) $A \in \mathbb{R}^{n,n}$ satisfies $A + A^\top > 0$,
(ii) $B_{22} = B + \overline{\mu} I \in \mathbb{R}^{m,m}$ is invertible,
(iii) $\tilde{B}, \tilde{C}$ are continuously differentiable with bounded derivatives,
(iv) the limits $\lim_{x \to \pm\infty} \tilde{B} = \tilde{B}_\pm$ and $\lim_{x \to \pm\infty} \tilde{C} = \tilde{C}_\pm$ exist,
(v) if $s \in \mathbb{C}$ satisfies $\det(-\omega^2 \tilde{A} + i\omega \tilde{B}_\pm + \tilde{C}_\pm - sI) = 0$ for some $\omega \in \mathbb{R}$, then $\operatorname{Re} s < -\delta < 0$,
(vi) $\sigma_{\text{point}}(\Lambda) \cap \{\operatorname{Re} s \geq -\delta\} = \{0\}$ and 0 is a simple eigenvalue of Λ.

Note that assumptions (i), (iii), (iv) are already implied by our **Basic Assumptions**, assumption (v) is related to the spectral condition **SC**, and (vi) is just the eigenvalue condition **EC** from Sect. 1.5. The stability of traveling waves in parabolic-hyperbolic systems is the main result of the following theorem.

Theorem 6 (Stability with asymptotic phase [56, Thm. 6.1]). *Assume that the **Basic Assumptions** hold for* (81) *and the linearization Λ about the traveling wave (see* (84), (85)*) satisfies the **Linear Assumptions**.*
Then, for all $0 < \delta_0 < \delta$ there exists $\rho = \rho(\delta_0) > 0$ such that for all initial data with $\|u_0 - \overline{u}\|_{H^2}^2 + \|v_0 - \overline{v}\|_{H^2}^2 \leq \rho$ there is a unique global solution (u^*, v^*) with $T^* = \infty$ for the system (81)*.*
Moreover, there are $C = C(\delta_0), \gamma_\infty = \gamma_\infty(u_0, v_0) \in \mathbb{R}$ with

$$|\gamma_\infty|^2 \leq C \left(\|u_0 - \overline{u}\|_{H^2}^2 + \|v_0 - \overline{v}\|_{H^2}^2 \right)$$

so that for all $t \geq 0$ the following estimate holds:

$$\|u^*(t) - \overline{u}(\cdot - \overline{\mu}t - \gamma_\infty)\|_{H^1}^2 + \|v^*(t) - \overline{v}(\cdot - \overline{\mu}t - \gamma_\infty)\|_{H^1}^2 \\ \leq C \left(\|u_0 - \overline{u}\|_{H^2}^2 + \|v_0 - \overline{v}\|_{H^2}^2 \right) e^{-2\delta_0 t}. \quad (86)$$

As is typical for hyperbolic equations, the initial data are measured in a stronger norm than the solution as $t \to \infty$.

3.3 Central Ideas of the Stability Proof

The proof of Theorem 6 is quite involved and proceeds in four major steps. We only describe the main ideas and refer to [54,56] for the details. Without loss of generality we assume that the problem is posed in a co-moving frame so that the traveling wave is a steady state.

Step 1 [Nonlinear coordinates]: Use a nonlinear change of coordinates, in the spirit of Henry [38]: Choose a linear functional $\psi : H^{-1}(\mathbb{R}, \mathbb{C}^{n+m}) \to \mathbb{R}$ that satisfies the **nondegeneracy assumption** $\psi(\overline{U}_x) \neq 0$ and write the solution in the form

$$U(t) = \tilde{U}(t) + \overline{U}(\cdot - \gamma(t)), \quad \text{where} \quad \tilde{U}(t) \in \mathcal{N}(\psi).$$

This leads to a **partial differential algebraic equation** (PDAE) for $\tilde{U}$:

$$\begin{aligned} \tilde{U}_t &= F(\tilde{U} + \overline{U}(\cdot - \gamma)) + \gamma_t \overline{U}_x(\cdot - \gamma), \\ 0 &= \psi(\tilde{U}). \end{aligned} \quad (87)$$

This change of coordinates is justified because the original PDE problem (83) and the PDAE reformulation (87) are equivalent for U close to $\overline{U}$ and $(\tilde{U}, \gamma)$ close to zero, see [56, Thm. 3.5]. Therefore, it suffices to show that the solution of (87) with transformed initial data $\tilde{U}(0) = \tilde{U}_0$, $\gamma(0) = \gamma_0$ converges exponentially fast.

Step 2 [Linearization]: We introduce $\mu = \gamma_t$ as a new variable. Then the PDAE (87) can be written in the form

$$\begin{aligned} \gamma_t &= \mu, & \gamma(0) &= \gamma_0, \\ \tilde{U}_t &= F(\tilde{U} + \overline{U}(\cdot - \gamma)) + \gamma_t \overline{U}_x(\cdot - \gamma) \\ &= \Lambda \tilde{U} + \overline{U}_x \mu + Q(\tilde{U}, \gamma, \mu), & \tilde{U}(0) &= \tilde{U}_0, \\ 0 &= \psi(\tilde{U}). \end{aligned} \quad (88)$$

Note that in (88) the initial value for μ_0 is given by hidden constraints. The remainder term $Q(\tilde{U}, \gamma, \mu)$ in (88) is of the form

$$Q = \begin{pmatrix} (G_1 + G_2)_x + F_{11} + F_{12} + R_1 \\ F_{21} + F_{21} + R_2 \end{pmatrix},$$

and has estimates which are quadratic in its arguments. For example, we have

$$G_1 = -\int_0^1 D^2 g(\overline{U}(\cdot - s\gamma))[\overline{U}_x(\cdot - s\gamma), \gamma \tilde{U}] ds,$$

$$G_2 = \int_0^1 (1-s) D^2 g(\overline{U}(\cdot - \gamma) + s\tilde{U})[\tilde{U}, \tilde{U}] ds,$$

and similar expressions for F_{ij}. The R_i are quadratic terms in μ and γ.

To prove stability for the nonlinear PDAE problem (88) we treat the higher order terms as inhomogeneities for the linear problem which then reads

$$\gamma_t = \mu,$$
$$\tilde{U}_t = \Lambda \tilde{U} + \overline{U}_x \mu + \begin{pmatrix} F_1 + G_x \\ F_2 \end{pmatrix}, \tag{89}$$
$$0 = \psi(\tilde{U}).$$

In (89) the first equation decouples from the other two equations and can be solved in an additional step. Therefore, we consider only the linear, inhomogeneous PDAE

$$\tilde{U}_t = \Lambda \tilde{U} + \overline{U}_x \mu + \begin{pmatrix} F_1 + G_x \\ F_2 \end{pmatrix} \tag{90}$$
$$0 = \psi(\tilde{U}),$$

subject to consistent initial data $\tilde{U}(0) = (\tilde{u}(0), \tilde{v}(0)) \in H^2 \times H^2$, $\mu(0) \in \mathbb{R}$.

The following linear stability result is the key to nonlinear stability.

Theorem 7 (Linear PDAE stability ([56, Thm. 5.1])). *Let the assumptions be as above and assume $F_1 \in C([0, \infty); L^2)$, $G, F_2 \in C([0, \infty); H^1)$. Then there exists a unique solution (u, v, μ) of (90), and it satisfies*

$$\|u(t)\|_{H^1}^2 + \|v(t)\|_{H^1}^2 + e^{-2\delta_0 t} \int_0^t e^{2\delta_0 \tau} \{\|u\|_{H^1}^2 + \|v\|_{H^1}^2 + |\mu|^2\} d\tau$$
$$\leq C e^{-2\delta_0 t} \left[\|\tilde{u}_0\|_{H^2}^2 + \|\tilde{v}_0\|_{H^2}^2 + \int_0^t e^{2\delta_0 \tau} \{\|G\|_{H^1}^2 + \|F_1\|^2 + \|F_2\|_{H^1}^2\} d\tau \right]. \tag{91}$$

The proof of this theorem will be indicated in **Step 4** below.

Step 3 [From linear to nonlinear stability]: Nonlinear stability of the PDAE (87) can be obtained from Theorem 7 by the following steps:

- Show local existence and uniqueness for (88) for small initial data $\tilde{U}_0, \gamma(0)$.
- Consider $Q(\tilde{U}, \gamma, \mu)$ in (88) as inhomogeneity in (90) and use the linear result, Theorem 7, to obtain a priori estimates for the local solution.

- Use the a priori estimate in a bootstrapping argument to show that the solution can be extended to all positive times and decays exponentially.

Since the PDAE problem (87) and the PDE problem (83) are locally equivalent, this proves Thm. 6.

Step 4 [Proof of linear stability via Laplace-technique]: We indicate the main steps in the proof of Thm. 7. A crucial step is to use the Laplace-technique which, in simple terms, translates resolvent estimates via the Theorem of Plancherel into decay estimates. Using Laplace-transform for stability proofs is standard, but applying the technique in the context of PDAEs such as (90) is a novel approach, see [53].

By homogenizing the initial data, we may assume without loss of generality $\tilde{u}(0) = 0$, $\tilde{v}(0) = 0$ in (90). More precisely, one writes the equations in terms of the new functions $\tilde{u} - e^{-2\delta t}\tilde{u}_0$ and $\tilde{v} - e^{-2\delta t}\tilde{v}_0$. This adds a term of the form $e^{-2\delta t} \Lambda(\tilde{u}_0, \tilde{v}_0)^\top$ to the inhomogeneity (see [53, Thm. 5.3]). The linear inhomogeneous problem is exponentially well-posed so that Laplace transformation of (90) is justified for spectral values s with $\operatorname{Re} s \geq \alpha$ for some sufficiently large α (see [56]). This leads to the following resolvent equation which we write in operator matrix form:

$$\mathscr{A}(s) \begin{pmatrix} \hat{u} \\ \hat{v} \\ \hat{\mu} \end{pmatrix} = \left((sI - \Lambda) - \begin{pmatrix} \overline{u}_x \\ \overline{v}_x \\ \psi & 0 \end{pmatrix} \right) \begin{pmatrix} \hat{u} \\ \hat{v} \\ \hat{\mu} \end{pmatrix} = \begin{pmatrix} \hat{F}_1 + \hat{G}_x \\ \hat{F}_2 \\ 0 \end{pmatrix}. \tag{92}$$

Here $\mathscr{A}(s)$ is an operator on $L^2 \times L^2 \times \mathbb{C}$ with domain $H^2 \times H^1 \times \mathbb{C}$.

We first show how resolvent estimates, i.e. solution estimates for (92) lead to stability. By Plancherel's Theorem we have for $\eta \geq \alpha$,

$$\int_0^\infty e^{-2\eta\tau} \|(\tilde{u}, \tilde{v}, \mu)^\top(\tau)\|_{H^1}^2 d\tau = \frac{1}{2\pi} \int_{-\infty}^\infty \|(\hat{u}, \hat{v}, \hat{\mu})^\top(\eta + i\xi)\|_{H^1}^2 d\xi, \tag{93}$$

where $\|(\hat{u}, \hat{v}, \hat{\mu})^\top\|_{H^1}^2 = \|\hat{u}\|_{H^1}^2 + \|\hat{v}\|_{H^1}^2 + |\mu|^2$. In **Step 6** below, we show estimates for solutions of (92) which hold uniformly in $\{\operatorname{Re} s \geq -\delta_0\}$ for a fixed $\delta_0 < \delta$,

$$\|(\hat{u}(s), \hat{v}(s), \hat{\mu}(s))^\top\|_{H^1}^2 \leq C \|(\hat{F}_1(s), \hat{G}(s), \hat{F}_2(s))^\top\|_{L^2, H^1}^2. \tag{94}$$

Here we used the norm $\|(\hat{F}_1, \hat{G}, \hat{F}_2)^\top\|_{L^2, H^1}^2 = \|\hat{F}_1\|_{L^2}^2 + \|\hat{G}\|_{H^1}^2 + \|\hat{F}_2\|_{H^1}^2$. Note that $\hat{u}, \hat{v}, \hat{\mu}$ in (92) are analytic functions, given as the Laplace transforms of $\tilde{u}, \tilde{v}, \mu$, which is only justified in $\{\operatorname{Re} s \geq \alpha\}$. But the resolvent equation (92) is in fact uniquely solvable in the larger domain $\{\operatorname{Re} s > -\delta\}$ and the solution depends analytically on s in this region. Therefore, $\hat{u}, \hat{v}, \hat{\mu}$ in (94) coincide with the analytic continuations of the Laplace transforms. By [4, Thm. 4.4.13] this implies that the Laplace transforms of $\tilde{u}, \tilde{v}, \mu$ even exist in the larger domain $\{\operatorname{Re} s > -\delta\}$. Then the Payley-Wiener and Plancherel Theorem [4, Sect. 1.8]

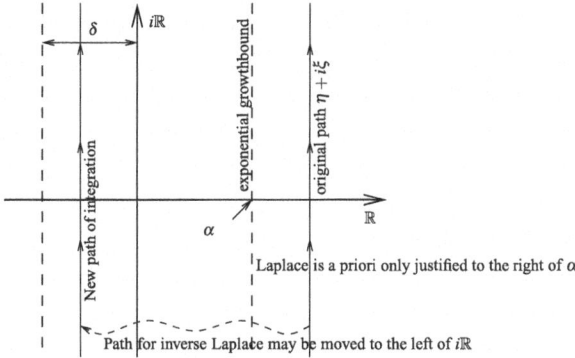

Fig. 18 Initially, equality (93) only holds for a path of integration to the right of α. Then it is justified in a much larger region, and the line of integration is shifted from the right of α to the left of $i\mathbb{R}$

show that (93) even holds for $\eta \geq -\delta_0 > -\delta$. This crucial step of shifting contours in the Laplace transform to the left is illustrated in Fig. 18.

We use (93) for $\eta = -\delta_0$, insert estimate (94) into the right hand side and finally use Plancherel's Theorem for the inhomogeneities to obtain

$$\int_0^\infty e^{-2\eta\tau} \|(\tilde{u}, \tilde{v}, \mu)^\top(\tau)\|_{H^1}^2 \, d\tau \leq \frac{1}{2\pi} \int_{-\infty}^\infty C \|(\hat{F}_1, \hat{G}, \hat{F}_2)^\top(\eta + i\xi)\|_{L^2, H^1}^2 \, d\xi$$
$$= C \int_0^\infty e^{-2\eta\tau} \|(F_1, G, F_2)^\top(\tau)\|_{L^2, H^1}^2 \, d\tau. \tag{95}$$

This estimate proves the linear stability result and leads to the estimate (91).

Step 5 [Fredholm-properties]: It remains to analyze (92) and show (94). We begin with Fredholm properties of $\mathscr{A}(s)$. For this rewrite the second order operator $sI - \Lambda : H^2 \times H^1 \to L^2 \times L^2$ as a first order operator by introducing $z = (u, Au_x, v)^\top$. This leads to an operator $L(s) : H^2 \times H^1 \times H^1 \to H^1 \times L^2 \times L^2$ given by

$$L(s)z = z_x - M(x, s)z,$$

where the matrix valued function $M(x, s) \in \mathbb{C}^{2n+m}$ reads

$$M(x, s) = \begin{pmatrix} 0 & A^{-1} & 0 \\ sI + B_{12}B_{22}^{-1}C_{21} - C_{11} & -B_{11}A^{-1} & -B_{12}B_{22}^{-1}(sI - C_{22}) - C_{12} \\ -B_{22}^{-1}C_{21} & 0 & B_{22}^{-1}(sI - C_{22}) \end{pmatrix}.$$

We employ the following Lemma from [11].

Lemma 1. *The second order operator $sI - \Lambda$ on $L^2 \times L^2$ with domain $H^2 \times H^1$ is Fredholm if and only if the first order operator $L(s)$ on $H^1 \times L^2 \times L^2$ with domain $H^2 \times H^1 \times H^1$ is Fredholm. In this case the Fredholm indices as well as the dimensions of the nullspaces of the two operators coincide.*

It is not difficult to show that the assumptions on the coefficients of the linear operator, in particular **Linear Assumptions** parts (iv) and (v), imply that the limit matrices $\lim_{x \to \pm\infty} M(x,s) = M_\pm(s)$ exist and are hyperbolic matrices. Moreover, the dimensions of the generalized eigenspaces for eigenvalues with real part less than zero is the same for $M_-(s)$ and $M_+(s)$, see [56, App. A]. By classical results of Coppel [20], the linear first order operator $L(s)$ has exponential dichotomies on $\mathbb{R}_\pm$, and a result of Palmer [49] shows the following,

Lemma 2. *For $\operatorname{Re} s > -\delta$ the operator $L(s)$ is Fredholm of index 0 and $\dim \mathcal{N}(L(0)) = \operatorname{codim} \mathcal{R}(L(0)) = 1$.*

Using the bordering lemma, e.g. [8], this proves that the original operator $\mathcal{A}(s)$ is Fredholm of index zero for all $\operatorname{Re} s > -\delta$.

Step 6 [Resolvent estimates]: After these preliminaries, the resolvent estimates are shown separately in the following three regions of the complex plane:

Region I: $\{s \in \mathbb{C} : \operatorname{Re} s \geq -\delta_0, |s| \gg 0\}$, here the parabolic-hyperbolic structure dominates, the **dispersion relation** is crucial for the estimates (this is different from the purely parabolic case!).
Region II: $\Omega \subset \rho(\Lambda) \cap \{\operatorname{Re} s > -\delta_0\}$ compact.
Region III: $\{s \in \mathbb{C} : |s| \ll 1\}$, here the PDAE formulation removes 0 from the spectrum and leads to appropriate estimates.

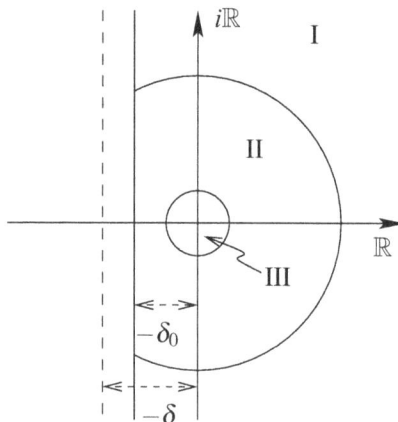

Since Region II is compact in the resolvent sent, the estimates are obvious. In Region I we assume $|s|$ to be large. The **Linear Assumption** (v) then states that

$$s \in \sigma \begin{pmatrix} -\omega^2 A + i\omega B_{11} + C_{11} & i\omega B_{12} + C_{12} \\ C_{21} & i\omega B_{22} + C_{22} \end{pmatrix} \text{ for some } \omega \in \mathbb{R} \text{ implies } \Re s < -\delta.$$

Here and in the following we drop the index $\pm$ for simplicity. By a matrix perturbation result, e.g. from [64], there is $\omega_0 > 0$ so that the matrix is similar to

$$\begin{pmatrix} -\omega^2 A + i\omega B_{11} + C_{11} + \mathcal{O}(|\omega|^{-1}) & i\omega B_{12} + C_{12} \\ 0 & i\omega B_{22} + C_{22} + \mathcal{O}(|\omega|^{-1}) \end{pmatrix}.$$

for all $|\omega| \geq \omega_0$. Together with the **dispersion relation** this shows

Lemma 3. *For every $0 < \delta' < \delta$ there is ω_1 so that for all $\omega \in \mathbb{R}$, $|\omega| > \omega_1$ holds the* hyperbolic dispersion relation

$$s \in \sigma(i\omega B_{22} + C_{22}) \Rightarrow \operatorname{Re} s < -\delta'. \tag{96}$$

For the hyperbolic part we have the following result from [53]:

Proposition 1. *For every $\delta_0 < \delta$ there exist constants $\rho_0, K > 0$ such that the equation*

$$(sI - B_{22}\partial_x - C_{22}(x))v = F \quad \text{in } L^2(\mathbb{R}, \mathbb{C}^m),$$

has a unique solution $v \in H^1$ for all $F \in H^1$ and for all s with $\operatorname{Re} s \geq -\delta_0$, $|s| > \rho_0$. The solution satisfies the estimate

$$\|v\|_{L^2}^2 \leq K \|F\|_{L^2}^2, \quad \|v\|_{H^1}^2 \leq K \|F\|_{H^1}^2.$$

A corresponding result for the parabolic part is proved in [45]:

Proposition 2. *There are constants $c_1, K, \varepsilon > 0$ so that for all $s = re^{2i\theta}$, $r \geq c_1$, $|\theta| \leq \pi/4 + \varepsilon$ there exists a unique solution $u \in H^2$ of*

$$su - Au_{xx} - B_{11}u_x - C_{11}u = f + g_x \quad \text{in } L^2$$

for all $f \in L^2$, $g \in H^1$. The solution satisfies the estimate

$$|s|^2 \|u\|^2 + |s| \|u_x\|^2 \leq K \left(\|f\|^2 + |s| \|g\|^2 \right).$$

By applying Propositions 1 and 2 to the coupled system

$$su - Au_{xx} - B_{11}u_x - C_{11}u = (F_1 + C_{12}v - B_{12,x}v) + (G + B_{12}v)_x,$$
$$sv - B_{22}v_x - C_{22}v = F_2 + C_{21}u,$$

one obtains unique solvability and solution estimates in Region I.

In Region III we benefit from the formulation as a partial differential algebraic equation. First consider $s = 0$ and assume that (W, λ) is in the nullspace of $\mathscr{A}(0)$, i.e.

$$\mathscr{A}(0)\begin{pmatrix} W \\ \lambda \end{pmatrix} = \begin{pmatrix} -\Lambda - \begin{pmatrix} \bar{u}_x \\ \bar{v}_x \end{pmatrix} \\ \psi & 0 \end{pmatrix}\begin{pmatrix} W \\ \lambda \end{pmatrix} = \begin{pmatrix} 0 \\ 0 \end{pmatrix}.$$

By our assumptions, 0 is a simple eigenvalue and, therefore, $(\bar{u}_x, \bar{v}_x)^\top$ is not in the range of Λ, which enforces $\lambda = 0$. Then W must belong to the kernel of Λ which is the one-dimensional space spanned by $(\bar{u}_x, \bar{v}_x)^\top$. The nondegeneracy assumption on ψ then implies that also W vanishes, so that $\mathscr{A}(0) : H^2 \times H^1 \times \mathbb{C} \to L^2 \times L^2 \times \mathbb{C}$ is a linear homeomorphism due to Fredholm index zero. A perturbation argument then allows to deal with small s-values.

Lemma 4. *There exist $c_0, K > 0$ so that for all $s \in \mathbb{C}$ $|s| < c_0$ there is a unique solution $(\hat{u}, \hat{v}, \hat{\lambda}) \in H^2 \times H^1 \times \mathbb{C}$ of*

$$\begin{pmatrix} (sI - \Lambda) - \begin{pmatrix} \bar{u}_x \\ \bar{v}_x \end{pmatrix} \\ \psi & 0 \end{pmatrix}\begin{pmatrix} \hat{u} \\ \hat{v} \\ \hat{\lambda} \end{pmatrix} = \begin{pmatrix} \hat{F}_1 + \hat{G}_x \\ \hat{F}_2 \\ 0 \end{pmatrix},$$

and the solution satisfies

$$\|\hat{u}\|_{H^2} + \|\hat{v}\|_{H^1} + |\hat{\lambda}| \leq K(\|\hat{F}_1\| + \|\hat{G}\|_{H^1} + \|\hat{F}_2\|).$$

This finishes the proof of Theorem 7.

3.4 Freezing Waves in Hyperbolic-Parabolic Systems

In this section we consider the freezing method when solving hyperbolic-parabolic systems in the neighborhood of traveling waves. Recall the general form (61) of the freezing method for a shift-equivariant evolution equation $U_t = F(U)$,

$$V_t = F(V) + \mu V_x, \qquad \text{(Fr}_1\text{)}$$

$$0 = \psi(V, \mu), \qquad \text{(Fr}_2\text{)}$$

where (Fr$_2$) is the phase condition. Further recall the two standard choices for phase conditions from Sect. 2.1:

Fixed phase condition: Given a template function $\hat{V}$, force the solution of (Fr$_1$) to align best under all shifts of $\hat{V}$ with the unshifted template function $\hat{V}$, i.e.

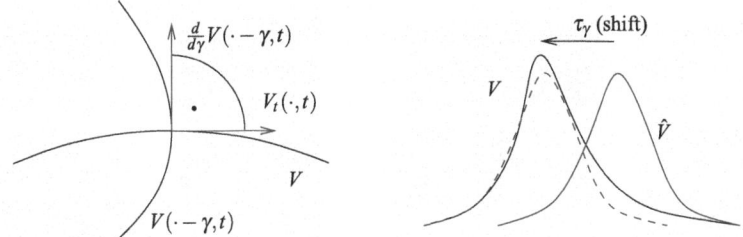

Fig. 19 Geometric interpretation of the orthogonal phase condition (*left*) and of the fixed phase condition (*right*)

$$0 = \mathrm{argmin}_{\gamma \in \mathbb{R}} \|V - \hat{V}(\cdot - \gamma)\|.$$

If the norm is given by some inner product $(\cdot, \cdot)$ and $\hat{V}$ is sufficiently smooth, the condition above implies $0 = (\hat{V}_x, V - \hat{V})$.

Orthogonal phase condition: Force the time evolution of the solution V of (Fr$_1$) to be orthogonal to the orbit of the spatial shifts of V in the Hilbert space L^2:

$$0 = (V_x, V_t) = (V_x, F(V) + \mu V_x).$$

For a sketch of these conditions see Fig. 19.

Example 9 (FitzHugh-Nagumo system). Recall the FitzHugh-Nagumo equation from Example 4 with parameters $a = 0.7$, $b = 3$, $\phi = 0.08$. We set $\varepsilon = 0$, so that we have the hyperbolic-parabolic mixed case as in [55]. The system reads

$$\begin{aligned} u_t &= u_{xx} + u - \tfrac{1}{3}u^3 - v, & u(x,0) &= u_0(x), \\ v_t &= \phi(u + a - bv), & v(x,0) &= v_0(x). \end{aligned} \qquad (97)$$

Figure 20 shows a colorplot of the time-evolution of u and v for the frozen system. The initial data are chosen as a jump function, which equals the rest state at $-\infty$ for $x \leq 0$ and the rest state at $+\infty$ for $x > 0$. A plot of the asymptotic profile, calculated by the solution of a boundary value problem, is given in the left frame of Fig. 21. We indicate the rate at which the solution to the freezing method converges to the asymptotic profile by plotting the L^2-norm of the time-derivative in the right frame of Fig. 21. In [55] this behavior is related to the spectral gap of the linearized operator.

Example 10 (Hodgkin-Huxley system). As a second example we use the freezing method for a long-time simulation of the Hodgkin-Huxley system (80). For suitable initial data, which are chosen as a simple jump in the voltage, the long-time simulation approximates the traveling pulse. The result of one such long-time simulation was used as initial guess to calculate the traveling pulse shown in Fig. 16

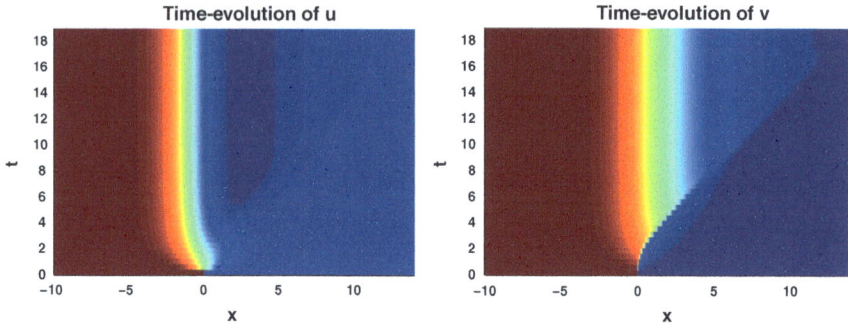

Fig. 20 Plot of frozen FitzHugh-Nagumo solutions $u(x,t)$ (*left*) and $v(x,t)$ (*right*). Note that u appears to be smooth immediately, while v exhibits a discontinuity that decays with time

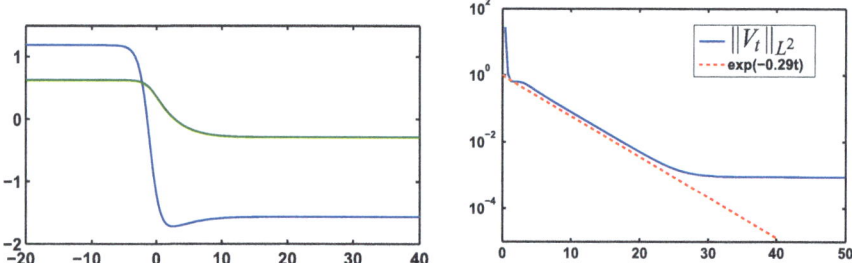

Fig. 21 Space dependence of FitzHugh Nagumo solutions $u(x,t)$ $v(x,t)$ (*left*) and temporal decay of $\|(u_t, v_t)\|_{L^2}$ with time (*right*)

from a boundary value problem (cf. (12)). We sketch the numerical spectrum of the linearization about the traveling pulse in Fig. 22.

3.5 Stability Theorem for the Freezing Method

An obvious question is, whether stability with asymptotic phase for traveling waves in hyperbolic-parabolic problems translates into stability of equilibria for the freezing method. This is in fact true, as the following result will show.

Consider a hyperbolic-parabolic partial differential equation of the general form (81). The freezing method for this system is

$$\begin{aligned}
u_t &= A u_{xx} + g(u,v)_x + f_1(u,v) + \mu u_x, \\
v_t &= B v_x + f_2(u,v) + \mu v_x, \\
0 &= \psi(\hat{u} - u, \hat{v} - v).
\end{aligned} \quad (98)$$

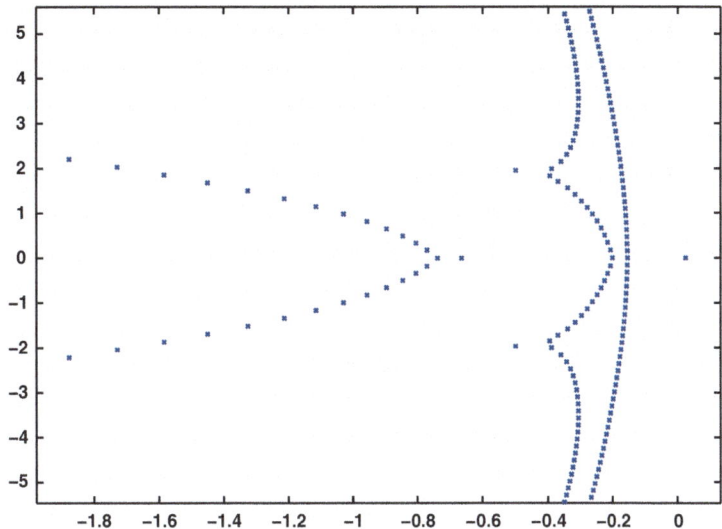

Fig. 22 Numerical approximation of the spectrum near zero for the linearization of the co-moved Hodgkin-Huxley equations about the traveling pulse

We impose the **Basic Assumption** from Sect. 3.1 and the **Linear Assumptions** from Sect. 3.2. In addition, we require for the phase condition:

(i) $\hat{u} - \bar{u}, \hat{v} - \bar{v} \in H^1$,
(ii) $\psi(\hat{u} - \bar{u}, \hat{v} - \bar{v}) = 0$,
(iii) $\psi(\bar{u}_x, \bar{v}_x) \neq 0$.   (non-degeneracy condition)

Condition (ii) implies that $(\bar{u}, \bar{v}, \bar{\mu})$ is a steady state of (98). With these assumptions the following result holds.

Theorem 8 ([54, 55]). *For all $0 < \delta_0 < \delta$ there exists $\rho_0 > 0$ such that for all consistent initial data of (98) with $\|u_0 - \bar{u}\|_{H^2}^2 + \|v_0 - \bar{v}\|_{H^2}^2 \leq \rho_0$ there is a unique global solution (u, v, μ) of the freezing equation (98). The solution satisfies*

$$u - \bar{u} \in C([0, T]; H^1) \cap L^2(0, T; H^2) \cap H^1(0, T; L^2),$$

$$v - \bar{v} \in C([0, T]; H^1) \cap H^1(0, T; L^2), \quad \mu \in C([0, T]; \mathbb{R}),$$

and converges exponentially fast to the asymptotic profile and to the speed of the traveling pulse. More precisely, there is $C = C(\delta_0)$ such that for all $t \geq 0$,

$$\|u(t) - \bar{u}\|_{H^1}^2 + \|v(t) - \bar{v}\|_{H^1}^2 + |\mu(t) - \bar{\mu}|^2$$
$$\leq C\left(\|u_0 - \bar{u}\|_{H^2}^2 + \|v_0 - \bar{v}\|_{H^2}^2\right)e^{-2\delta_0 t}.$$

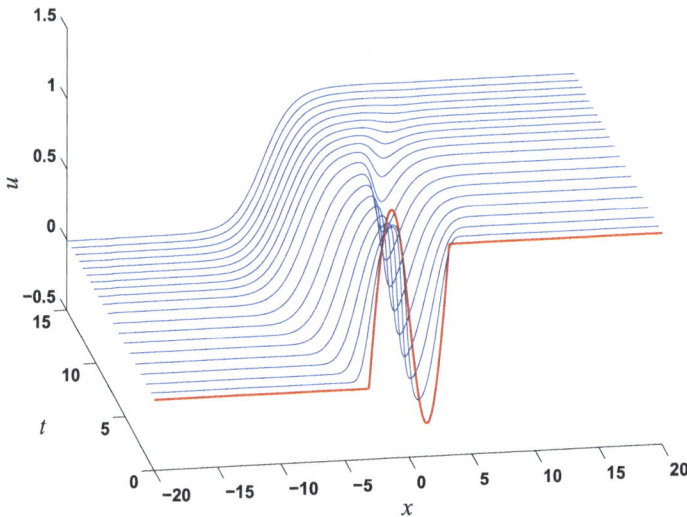

Fig. 23 Numerical simulation of (99) with the freezing method. The *bold line* indicates the initial data (cf. [52])

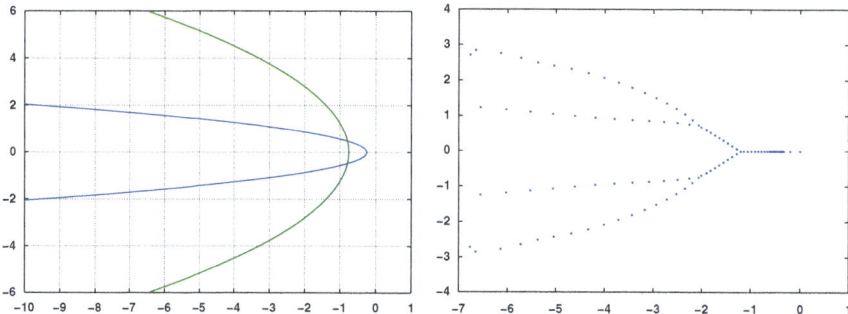

Fig. 24 Dispersion relation for Burgers' equation with a source term (*left*), plot of a numerical approximation for the spectrum of the linearized operator (*right*)

It is worth noting, that the result also applies to viscous conservation laws with a source term. The following numerical experiment shows an example of this type.

Example 11 (Burgers' equation). Consider Burgers' equation with a source term,

$$u_t + \left(\tfrac{1}{2}u^2\right)_x = 0.1 u_{xx} + u(1-u)(u-\tfrac{1}{4}). \tag{99}$$

The following results are taken from [52]. Figure 23 shows the result of a numerical simulation of (99) with the freezing method. The left plot in Fig. 24 shows the dispersion curves and the right plot shows a numerical approximation of the spectrum for the operator linearized about the traveling wave. In numerical experiments one observes an approximate rate of convergence $e^{-0.2t}$ as $t \to \infty$.

Example 12. We present results for hyperbolic systems without a parabolic part. This was first analyzed in [53, 54]. A large number of examples of such systems are obtained by using the so-called Cattaneo-Maxwell flux instead of the usual Fickian law:

Consider a system of reaction diffusion equations

$$v_t + q_x = r(v),$$

where v denotes the concentrations of the substrates, q stands for the fluxes of the substrates, and r contains reaction terms for the substrates. Usually the fluxes are given by the Fickian law ($q = -Dv_x$), but this leads to the unphysical phenomenon of infinite speed of propagation of the substrates. Cattaneo [18] proposed a different flux law for the case of heat transfer. He added a damping term to the Fickian law which reads

$$Tq_t + q = -v_x.$$

This leads to the following semilinear hyperbolic problem

$$\begin{pmatrix} v \\ q \end{pmatrix}_t + \begin{pmatrix} q \\ \frac{1}{T}v \end{pmatrix}_x = \begin{pmatrix} r(v) \\ -\frac{1}{T}q \end{pmatrix}.$$

As is well-known, solutions to hyperbolic problems have a finite speed of propagation. This removes the apparent paradox of infinite speed of propagation for the substrates. Moreover, the Fickian law appears as a singularly perturbed limit if the parameter 'T' becomes large.

In [54] we consider a hyperbolic variant of the Hodgkin-Huxley equations (80) by using the Cattaneo-Maxwell flux. The system then becomes

$$V_t = \frac{a}{2R}q_x - \bar{g}_K n^4(V - V_K) - \bar{g}_{Na} m^3 h(V - V_{Na}) - \bar{g}_l(V - V_l),$$
$$q_t = -V_x - q,$$
$$n_t = \alpha_n(V)(1-n) - \beta_n(V)n, \qquad (100)$$
$$m_t = \alpha_m(V)(1-m) - \beta_m(V)m,$$
$$h_t = \alpha_h(V)(1-h) - \beta_h(V)h,$$

where the nonlinearities α_n, β_n, etc. are the same as in the original equation (80). The freezing method works for this example just as well as for the original Hodgkin-Huxley system. After a first long-time simulation, one can again use the final state of the initial value problem as initial guess for the boundary value problem and then

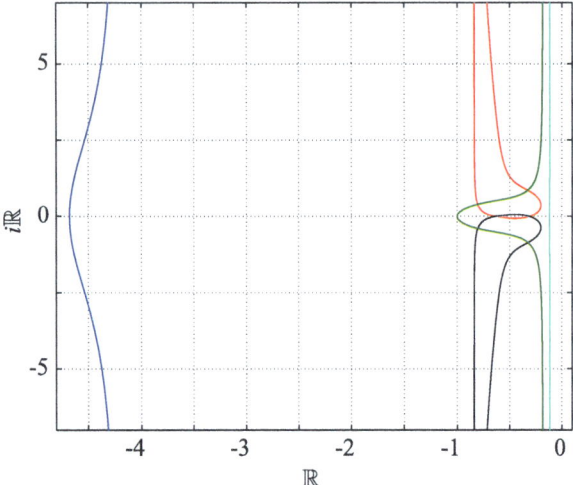

Fig. 25 Dispersion relation for the steady state of the hyperbolic version of the Hodgkin-Huxley equation (cf. [54])

do the standard subsequent analysis for the traveling wave such as computing the spectrum, performing parameter continuation, etc.

We plot in Fig. 25 the dispersion relation of the steady state problem, i.e. of the problem in a co-moving frame with the correct velocity $\overline{\mu}$ of the traveling wave. Due to the hyperbolic character of the equations, all curves approach vertical lines, for more details we refer to [54, Sect. 8] and [52].

3.6 Numerical Experiments for More General Symmetries

Note that equivariance of an evolution equation (43) completely avoids transformation of the time variable. So far we just used equivariance in the form (46) and kept the time variable in the ansatz (71). But it is possible to include more general symmetries that involve the time variable, and this has already been proposed in [57].

As an example we consider the viscous Burgers' equation

$$u_t = -(\tfrac{1}{2}u^2)_x + \nu u_{xx} =: F(u), \quad x \in \mathbb{R}, t \geq 0, \tag{101}$$

and take the Lie group

$$G = \mathbb{R}_+^* \ltimes \mathbb{R},$$

with multiplication for $\gamma, \eta \in G$ given by:

$$(\gamma_1, \gamma_2) \circ (\eta_1, \eta_2) = (\gamma_1\eta_1, \gamma_2 + \gamma_1\eta_2), \quad \mathbb{1} = (1, 0).$$

We use the group actions a and m, given by

$$[a(\gamma)u](x) = \frac{1}{\gamma_1}u(\frac{x-\gamma_2}{\gamma_1}), \quad m(\gamma) = \frac{1}{\gamma_1^2}.$$

A simple computation shows that for sufficiently smooth functions u,

$$F(a(\gamma)u) = m(\gamma)a(\gamma)F(u). \tag{102}$$

Note that, instead of commuting F and $a(\gamma)$ as in (46), we have an additional factor $m(\gamma)$. This factor leads to a time scaling as we will see in the following discussion.

We replace (71) by the following **ansatz:**

$$u(t) = a(\gamma(\tau))v(\tau), \tag{103}$$

where γ is a smooth curve in G, v is a smooth curve in the domain of F and $\tau = \tau(t)$ is a smooth real valued function of t.

Using the symmetry property (102) in the evolution equation (101), we arrive at the following:

$$\begin{aligned} u_t &= F(u) = F\big(a(\gamma(\tau))v(\tau)\big) = m(\gamma)a(\gamma)F(v), \text{ and also} \\ u_t &= a(\gamma)d[a(\mathbb{1})v]dL_{\gamma^{-1}}(\gamma)\gamma_\tau \tau_t + a(\gamma)v_\tau \tau_t. \end{aligned} \tag{104}$$

If we choose τ to satisfy the ordinary differential equation $\tau_t = m(\gamma(\tau))$, then (104) leads to the system

$$v_\tau = F(v) - d[a(\mathbb{1})v]\mu, \tag{105}$$

$$\tau_t = m(\gamma(\tau)), \tag{106}$$

$$\gamma_\tau = dL_\gamma(\mathbb{1})\mu. \tag{107}$$

Note that it is not necessary to solve (105)–(107) simultaneously. In fact, (106) and (107) are only needed for reconstruction, i.e. to obtain the solution in the original coordinates. The relative equilibrium is completely described by a steady solution of (105). For the simulation of (105) the same ideas as in Sect. 2 apply. We only present the results of one simulation in Fig. 26. Looking at the scales in Fig. 26 it is nice to observe that the solution decays to zero in the original variables and spreads to infinity. In the transformed variables however, the solution becomes a steady state. Moreover, using the freezing method, one is able to directly observe a transient behavior: For a very long time (in the original time variable) the solution is close to what is called an N-wave in the theory of hyperbolic conservation laws, before it finally approaches the correct viscosity wave. This behavior has been observed already in [44], but there the authors used that the asymptotic values for $\bar{\mu}$ and, therefore, a correct asymptotic scaling of space and time are known. Our method applies also to systems where the asymptotic scalings are not easily found and, moreover, the method yields these asymptotic scalings as part of its solution.

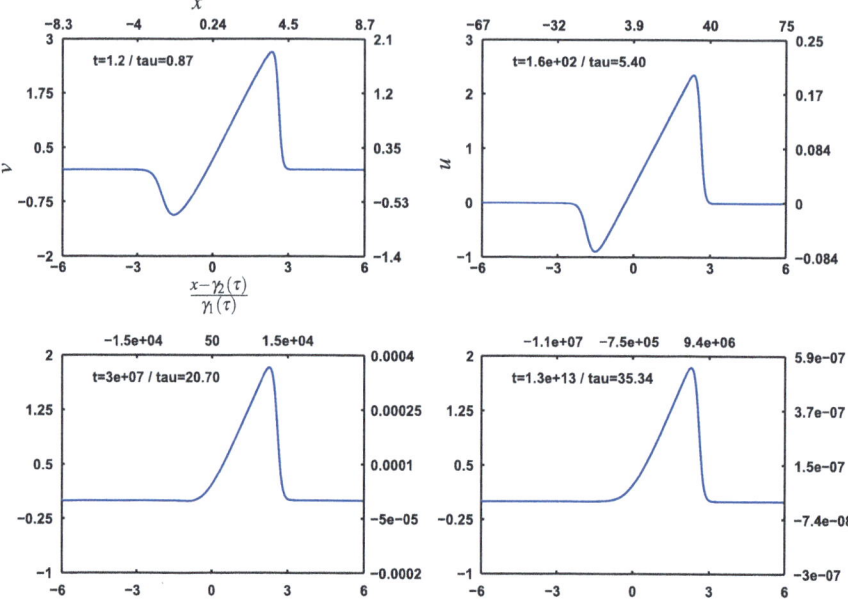

Fig. 26 Result of the freezing method for the viscous Burgers' equation at four different times. In each of the four plots the scale on top is for the original space variable x, the bottom scale is in the transformed variable $\xi = \frac{x-\gamma_2(\tau)}{\gamma_1(\tau)}$, the scale on the *left* is for the transformed dependent variable $v(\xi, \tau)$ and the scale on the *right* is for the original dependent variable $u(x, t)$. In each plot we also give the actual original time t and the transformed time $\tau(t)$ (see [52])

3.7 Summary

A summary of the results of this section is the following:

- Stability with asymptotic phase for general coupled parabolic-hyperbolic systems can be proved under the assumption that the spectrum of the linearization lies strictly to the left of the imaginary axis except for a simple zero eigenvalue,
- The idea of the proof is to use Laplace-transform and derive uniform resolvent estimates for the transformed equation,
- For large spectral values, resolvent estimates are obtained from the parabolic-hyperbolic structure and the dispersion relation,
- For small spectral values, the zero-eigenvalue is removed from the spectrum by the phase condition which appears in the PDAE formulation,
- Numerical experiments confirm the predicted exponential rates,
- The freezing method applies to the computation of similarity solutions.

4 Rotating Patterns in Two and Three Space Dimensions

While the previous lectures were restricted to patterns in one space dimension, we will progress in this section to nonlinear waves in two and three space dimensions. Since the freezing method is formulated in abstract terms it is rather straightforward to make it work for equivariant parabolic systems in $d \geq 2$ dimensions, where the Lie group is now the d-dimensional Euclidean group or consists of products or subgroups thereof. It is much harder, however, to prove nonlinear stability of such patterns under reasonable assumptions on the associated spectra. In this section we discuss the stability result from [10] which applies to two-dimensional rotating patterns that are localized, i.e. which decay at infinity. Again, as in Sect. 3 a major difficulty results from the fact that, due to the angular modes, the linearized operators only lead to C^0-semigroups. Moreover, due to equivariance with respect to the two-dimensional Euclidean group, three eigenvalues now appear on the imaginary axis. We show how to obtain exponential decay of the semigroup in a subspace complementary to the eigenvectors that belong to these three eigenvalues. Then stability with asymptotic phase follows in a suitable Sobolev space. We also show some simulations of the freezing method for cases where the theory does currently not apply: two-dimensional rotating spirals for Barkley's excitable system (see [6]) and three dimensional spinning solitons for the quintic-cubic Ginzburg-Landau equation.

4.1 Reaction Diffusion Systems in $\mathbb{R}^2$ and the Freezing Method

We apply the abstract freezing approach from Sect. 2.4 to reaction diffusion systems in two dimensions

$$u_t = A\Delta u + f(u), \quad t \geq 0, \quad x \in \mathbb{R}^2, \quad u(\cdot, 0) = u_0, \tag{108}$$

where $u(x,t) \in \mathbb{R}^m$ and $A \in \mathbb{R}^{m,m}$ is positive definite. The system (108) is equivariant with respect to the Euclidean group $G = SE(2)$ under the action $a(\gamma)$, given for $\gamma = (\theta, \tau) \in S^1 \times \mathbb{R}^2$ by

$$a(\gamma)v(x) = v(R_{-\theta}(x - \tau)), \quad x \in \mathbb{R}^2, \quad R_\theta = \begin{pmatrix} \cos(\theta) & -\sin(\theta) \\ \sin(\theta) & \cos(\theta) \end{pmatrix}. \tag{109}$$

Here we used the representation of $SE(2)$ as a semi-direct product of S^1 and $\mathbb{R}^2$ with the group operation defined through

$$(\theta_1, \tau_1) \circ (\theta_2, \tau_2) = (\theta_1 + \theta_2, R_{\theta_2}\tau_2 + \tau_1). \tag{110}$$

The derivative of $a(\gamma)v$ with respect to $\gamma \in G$ turns out to be

$$d[a(1)v]\mu = -\mu_1 D_\theta v - \mu_2 D_1 v - \mu_3 D_2 v, \quad \mu = (\mu_1, \mu_2, \mu_3) \in \mathscr{A} = se(2), \tag{111}$$

where $D_\theta v(x) = x_2 D_1 v(x) - x_1 D_2 v(x)$ is the angular derivative and $v : \mathbb{R}^2 \to \mathbb{R}^m$ is assumed to be sufficiently smooth.

Therefore, the freezing system (73) associated with (108) reads as follows

$$\begin{aligned} v_t &= A\Delta v + f(v) + \mu_1 D_\theta v + \mu_2 D_1 v + \mu_3 D_2 v, \quad v(\cdot, 0) = u_0 \\ 0 &= (D_\theta \hat{v}, v - \hat{v})_{L^2} = (D_1 \hat{v}, v - \hat{v})_{L^2} = (D_2 \hat{v}, v - \hat{v})_{L^2}, \\ \gamma_t &= \begin{pmatrix} \theta \\ \tau \end{pmatrix}_t = \begin{pmatrix} 1 & 0 \\ 0 & R_\theta \end{pmatrix} \mu, \quad \gamma(0) = \begin{pmatrix} 0 \\ 0 \end{pmatrix}. \end{aligned} \tag{112}$$

A **rotating wave** solution of (108) is of the form

$$u(x, t) = u_*(R_{-ct} x), \quad x \in \mathbb{R}^2, \quad t \in \mathbb{R}, \tag{113}$$

where u_* denotes the profile and c denotes the angular velocity of the wave. In terms of the group action we may write such a solution as $u(t) = a(ct, 0) u_*$, $t \in \mathbb{R}$ which is a relative equilibrium of (108). For any $\theta \in S^1, \tau \in \mathbb{R}^2$ the function $a((\theta, \tau) \circ (ct, 0)) u_*(x) = u_*(R_{-ct-\theta}(x - \tau))$ is then also a rotating wave, but with phase shift θ and with center of rotation at τ. These solutions are equilibria of the first equation in (112) and solve the reconstruction equation in (112) with initial data $\gamma(0) = (\theta, \tau)$.

Example 13 (Quintic-cubic Ginzburg-Landau equation (QCGL)). Consider the QCGL in two space dimensions (compare Example 5):

$$u_t = \alpha \Delta u + (\delta + \beta |u|^2 + \gamma |u|^4) u, \quad x \in \mathbb{R}^2, \quad u(x, t) \in \mathbb{C}, \tag{114}$$

with parameters $\alpha, \beta, \gamma, \delta \in \mathbb{C}$, Re $\alpha > 0$. In real coordinates $u = u_1 + i u_2$ this leads to a parabolic system of the type (108). According to [21] the system (114) has so called spinning soliton solutions for parameter values

$$\alpha = \frac{1}{2}(1+i), \quad \beta = \frac{5}{2} + i, \quad \gamma = -1 - \frac{i}{10}, \quad \delta = -\frac{1}{2}. \tag{115}$$

Figure 27 shows the result of the numerical computation for these parameter values, both for the given system (108) and for the frozen system (112). The computations were done with COMSOL Multiphysics^{TM} on a ball of radius 20 and with Neumann boundary conditions. Figure 27c shows the final profile of the spinning soliton. In Fig. 27 we plot the time dependence of Re $u(\cdot, 0, t)$ at the cross-section $x_2 = 0$ both for the nonfrozen system (a) and the frozen system (b). The time evolution of the velocities $\mu(t) = (\mu_1(t), \mu_2(t), \mu_3(t))$ are shown in Fig. 27d.

The system (114) is in fact equivariant with respect to the 4-dimensional group $G = S^1 \times SE(2)$ where (109) is replaced by the action $a(\gamma) v(x) = e^{i\varphi}$

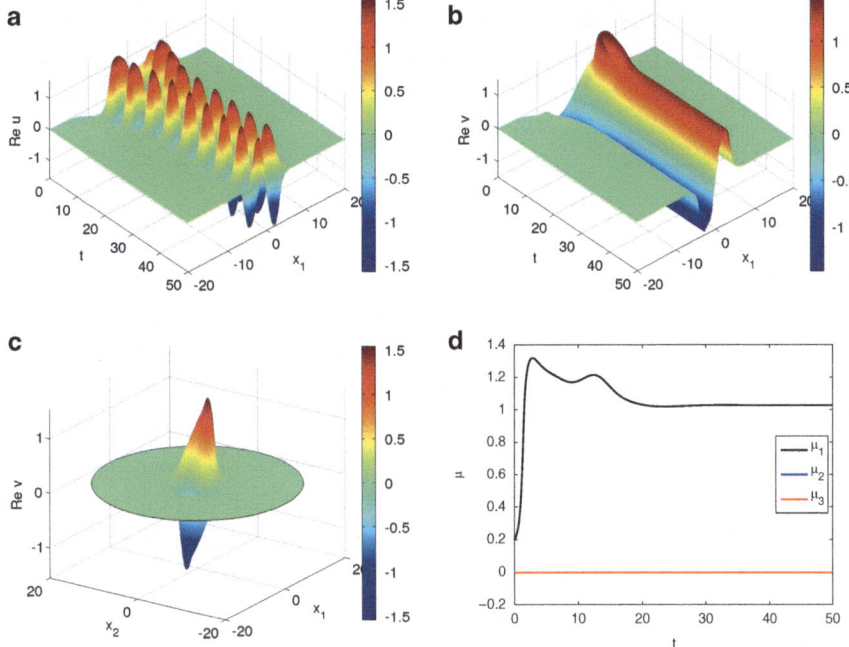

Fig. 27 Spinning solitons in the quintic-cubic Ginzburg-Landau equation. Cross-section at $x_2 = 0$ for the real part of solutions to the nonfrozen system (114) (**a**) and to the frozen system (112) (**b**). Profile $\bar{v}$ of the spinning soliton (**c**) and time-dependence of velocities (μ_1, μ_2, μ_3) (**d**). All solutions obtained by Comsol Multiphysics with piecewise linear finite elements and Neumann boundary conditions. A fixed phase condition as in (112) was used with template function $\hat{v}$ taken from the solution of the nonfrozen system at time $t = 50$. Parameter-values are given by (115)

$v(R_{-\theta}(x - \tau))$, $x \in \mathbb{R}^2$ for $\gamma = (\varphi, \theta, \tau) \in G$. It turns out that the spinning solitons $u(x, t)$ considered here are symmetric in the following sense: $e^{i\varphi}u_*(x) = u_*(R_\varphi x)$. Then there is a nontrivial isotropy subgroup $G(u_*) = \{g \in G : a(g)u_* = u_*\}$ and the linear map $d[a(\mathbb{1})u_*] : \mathscr{A} \to Y$ is no longer one-to-one. This causes problems with the phase conditions (76), (77) which become ill-posed with respect to the parameter μ. In the current example we avoided such complications by considering equivariance only with respect to the three-dimensional group $SE(2)$. In this smaller group the isotropy subgroup becomes trivial.

4.2 Spectra of 2D Rotating Waves: Essential and Point Spectrum

In the following we discuss the behavior of the spectrum of the linear differential operator obtained by linearizing about a rotating wave of the parabolic system (108).

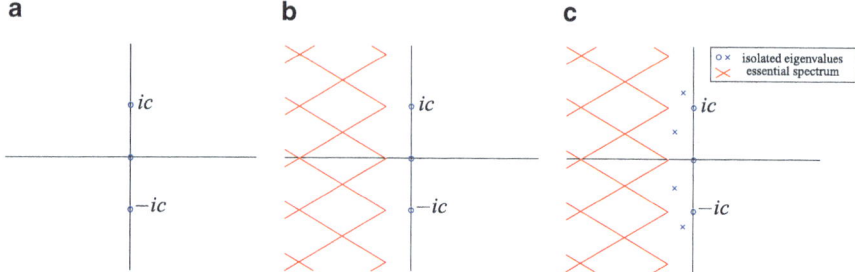

Fig. 28 Critical eigenvalues $0, \pm ic$ for a linearized rotating wave (**a**), essential spectrum for QCGL from Example 13: $s = inc + \delta - \kappa^2(\alpha_1 \pm i\alpha_2)$, $\kappa \in \mathbb{R}$, $n \in \mathbb{Z}$ with parameter values from (115) (**b**), Schematic picture of essential spectrum, critical eigenvalues, and further isolated eigenvalues for the QCGL from Example 13 (**c**)

We consider a localized rotating wave (113). By this we mean that the profile converges to a zero v_∞ of f as $|x| \to \infty$ and that all derivatives up to order 2 converge to zero. By shifting v_∞ into the origin we may assume $f(0) = 0$ and hence

$$\sup_{|x| \geq r} |D^\alpha u_*(x)| \to 0 \quad \text{as} \quad r \to \infty \quad \text{for} \quad |\alpha| \leq 2. \tag{116}$$

We transform (108) into rotating coordinates via $u(x,t) = v(R_{-ct}x, t)$ and obtain

$$v_t = A\triangle v + cD_\phi v + f(v), \text{ with } D_\phi v = -x_2 D_1 v + x_1 D_2 v. \tag{117}$$

Linearizing at the steady state $v = u_*$ of (117) yields the operator

$$\mathscr{L}u = A\triangle u + cD_\phi u + B(x)u, \quad B(x) = f'(u_*(x)), \; x \in \mathbb{R}^2. \tag{118}$$

Applying D_1, D_2, D_ϕ to $A\triangle u_* + cD_\phi u_* + f(u_*) = 0$ and using the commutator relations $[D_1, D_\phi] = D_2, [D_2, D_\phi] = -D_1, [D_\phi, \triangle] = 0$ leads to the equations

$$0 = \mathscr{L}D_\phi u_* = \mathscr{L}(D_1 u_*) + cD_2 u_* = \mathscr{L}(D_2 u_*) - cD_1 u_*, \tag{119}$$

in particular, $\mathscr{L}(D_1 u_* \pm i D_2 u_*) = \pm ic(D_1 u_* \pm i D_2 u_*)$. Therefore, the operator $\mathscr{L}$ has at least the three eigenvalues $0, \pm ic$ in its spectrum (see Fig. 28a) provided the functions $D_1 u_*, D_2 u_*, D_\theta$ lie in the function space under consideration.

Next we discuss the essential spectrum of $\mathscr{L}$. In polar coordinates the operator reads

$$\mathscr{L} = A\left(D_r^2 + \frac{1}{r}D_r + \frac{1}{r^2}D_\theta^2\right) + cD_\theta + f'(u_*(r,\theta)). \tag{120}$$

As $r \to \infty$ we find the constant coefficient operator

$$\mathscr{L}_\infty = AD_r^2 + cD_\theta + f'(0) \tag{121}$$

With $u(r,\theta) = e^{in\theta}e^{i\kappa r}u_\infty$ we obtain $s \in \sigma(\mathscr{L}_\infty)$, if s satisfies for some $\kappa \in \mathbb{R}$ and $n \in \mathbb{Z}$ the **dispersion relation**

$$\det(-\kappa^2 A + inc + f'(0) - s) = 0. \tag{122}$$

For the quintic-cubic Ginzburg-Landau equation from Example 13 the curves from (122) turn out to be infinitely many copies of two half lines shifted along the imaginary axis, see Fig. 28b,

$$s = -\kappa^2\alpha + inc + \delta, \quad s = -\kappa^2\bar{\alpha} + inc + \bar{\delta}, \quad \kappa \in \mathbb{R}, \quad n \in \mathbb{Z}. \tag{123}$$

We indicate why these curves belong to the essential spectrum of the variable coefficient operator $\mathscr{L}$, i.e. $\sigma(\mathscr{L}_\infty) \subset \sigma_{\text{ess}}(\mathscr{L})$. With the eigenfunctions above let

$$u_R(r,\theta) = \psi_R(r)\left(e^{in\theta}e^{i\kappa r}u_\infty\right)$$

where ψ_R is a smooth cut-off function such that

$$\psi_R(r) = \begin{cases} 1, & R \leq r \leq 2R, \\ 0, & 0 \leq r \leq R-1,\ 2R+1 \leq r. \end{cases}$$

By a straightforward computation one shows

$$\|u_R\|_{L^2}^2 \geq CR^2, \quad \|(\mathscr{L}-s)u_R\|_{L^2}^2 \leq C\left(R + R^2\varepsilon_R^2\right),$$

where $\varepsilon_R = \sup_{r \geq R,\theta}|f'(u^*(r,\theta)) - f'(0)| \to 0$ as $R \to \infty$. This contradicts the continuity of $(\mathscr{L}-s)^{-1}$ with respect to $\|\cdot\|_{L^2}$, i.e. $\|u_R\|_{L^2} \leq C\|(\mathscr{L}-s)u_R\|_{L^2}$. For the QCGL from Example 13 we expect further isolated eigenvalues to the right of the essential spectrum, see Fig. 28c for a schematic drawing.

Clearly, since the spectrum of $\mathscr{L}$ is not contained in a sector, we expect the semigroup $e^{t\mathscr{L}}$ to be continuous but not analytic. This has serious implications for the nonlinear stability theory to be discussed in the next subsection.

Figure 29 shows details of the numerical spectrum that is found for a numerical discretization of $\mathscr{L}$ of size 10^4. The detail shows about 400 eigenvalues lying in a ball centered at 3. It turns out that, in addition to the three eigenvalues $0, \pm ic$ on the imaginary axis, there are clusters of eigenvalues which approximate the essential spectrum from Fig. 28b, and there is a total of 8 pairs of complex conjugate eigenvalues (indicated by crosses in Fig. 29) between the imaginary axis and the essential spectrum. The contour plots of the associated eigenfunctions (see [10, Fig. 3]) show that these eigenfunctions are actually localized, i.e. we assume their continuous counterparts to lie in $L^2(\mathbb{R}^2, \mathbb{C})$. On the contrary, the numerical eigenfunctions found for eigenvalues within the clusters, are easily recognized as being non-localized (cf. [10, Fig. 3]).

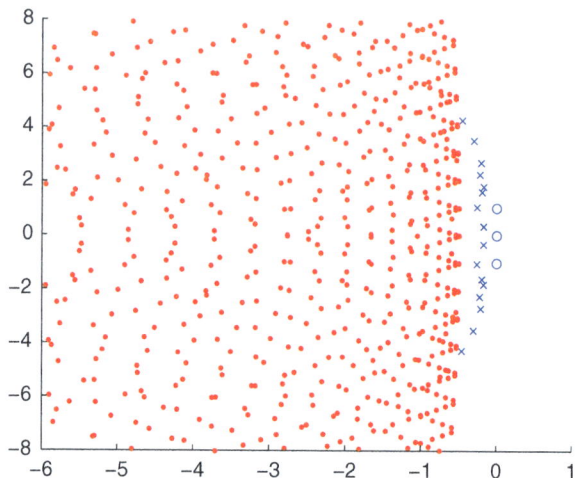

Fig. 29 Plot of numerical spectrum for the QCGL from Example 13 with parameter values from (115). In addition to 3 eigenvalues on the imaginary axis and the clusters approximating the essential spectrum from Fig. 28b, one finds additional pairs of isolated eigenvalues indicated by the *blue crosses*

4.3 A Nonlinear Stability Theorem

In this section we outline the nonlinear stability theory for rotating patterns following [10]. We mention the alternative approach of [61] which uses center manifold reductions. Recall the Sobolev spaces $H^j = H^j(\mathbb{R}^2, \mathbb{R}^m), j = 0, 1, 2$ and the Sobolev embedding $H^2(\mathbb{R}^2) \subset L^\infty(\mathbb{R}^2) \cap C(\mathbb{R}^2)$ and introduce the subspace

$$H^2_{\text{Eucl}} = H^2_{\text{Eucl}}(\mathbb{R}^2, \mathbb{R}^m) = \{u \in H^2 : D_\theta u \in L^2(\mathbb{R}^2, \mathbb{R}^m)\}.$$

As above we assume the existence of a rotating wave (113) for the system (108) with nonvanishing velocity $c \neq 0$, and we impose the following

Wave Conditions

(i) $f \in C^4(\mathbb{R}^m, \mathbb{R}^m)$ and $f(0) = 0$,
(ii) $\sup_{|x| \geq r, |\alpha| \leq 2} |D^\alpha u_*(x)| \to 0$ as $r \to \infty$.
(iii) $f'(0) \leq -2\beta I$ for some $\beta > 0$.
(iv) The eigenvalues $0, \pm ic$ have eigenfunctions $D_\theta u_*, D_1 u_* \pm i D_2 u_*$ in H^2_{Eucl}, and they are algebraically simple for the operator $\mathscr{L} = A\Delta + cD_\theta + f'(u_*)$ in H^2_{Eucl}.
(v) There are no further eigenvalues $s \in \mathbb{C}$ for $\mathscr{L}$ with $\text{Re}(s) \geq -2\beta$.

Theorem 9 ([10]). *Under the wave conditions (i)–(v) above, there exists an $\varepsilon > 0$ such that for any solution of (108) satisfying $\|u(0) - u_*\|_{H^2} \leq \varepsilon$ there is a C^1-function $\gamma(t) = (\theta(t), \tau(t)) \in SE(2), t \geq 0$ and some $(\theta_\infty, \tau_\infty) \in SE(2)$ such that for $t \geq 0$,*

Fig. 30 Decomposition of dynamics near a two-dimensional group orbit

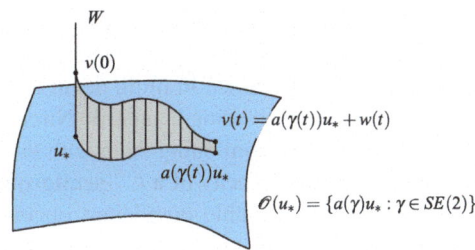

$$\|u(\cdot,t) - a(\gamma(t))u_*\|_{H^2} \leq Ce^{-\beta t} \|u(0) - u_*\|_{H^2},$$
$$|\theta(t) + ct - \theta_\infty| + |\tau(t) - \tau_\infty| \leq Ce^{-\beta t} \|u(0) - u_*\|_{H^2}. \tag{124}$$

Note that this theorem states **stability with asymptotic phase**, as we know it for traveling waves from Theorems 3 and 6.

In the following we provide some ingredients from the proof. First we transform into rotating coordinates (117).

Step 1 [Nonlinear coordinates]: Decompose the solution of (117) and the initial value $v(0) = u_0$ in a way analogous to (87) (see Fig. 30 for an illustration):

$$v(t) = a(\gamma(t))u_* + w(t), \quad \gamma(t) = (\theta(t), \tau(t)) \in S^1 \ltimes \mathbb{R}^2,$$
$$u_0 = a(\gamma_0)u_* + w_0, \quad \gamma(0) = (\theta_0, \tau_0), \tag{125}$$

where $w(t), w_0$ lie in the subspace $W = \{\psi_1, \psi_2, \psi_3\}^\perp$ of H^2. Here, orthogonality holds with respect to $(\cdot,\cdot)_{L^2}$, and the functions $\psi_1 \pm i\psi_2, \psi_3 \in H^2_{\text{Eucl}}$ are eigenfunctions of the adjoint operator $\mathscr{L}^*$ corresponding to the eigenvalues $\pm ic, 0$ (cf. wave condition (iv)).

Step 2 [The decomposed system]: Inserting (125) into (117), expanding the nonlinearities and inverting the linear parts leads to the following system of coupled integral equations for the new variables $w(t), \gamma(t)$:

$$w(t) = e^{t\mathscr{L}} w_0 + \int_0^t e^{(t-\tau)\mathscr{L}} \rho^{[w]}(w(\tau), \gamma(\tau)) d\tau,$$
$$\gamma(t) = e^{tE_c} \gamma_0 + \int_0^t e^{(t-\tau)E_c} \rho^{[\gamma]}(w(\tau), \gamma(\tau)) d\tau, \tag{126}$$

where E_c has the matrix representation $E_c = \begin{pmatrix} 0 & c & 0 \\ -c & 0 & 0 \\ 0 & 0 & 0 \end{pmatrix}$ and $\rho^{[\gamma]}, \rho^{[w]}$ are quadratic remainder terms.

Step 3 [From linear to nonlinear decay estimates]: The crucial step in the proof is the linear decay estimate

$$\|e^{t\mathscr{L}}w\|_{H^2} \leqslant Ce^{-\beta t}\|w\|_{H^2} \quad \text{for} \quad w \in W, \tag{127}$$

which will be discussed in more detail in Step 4 below. The nonlinear estimate (124) is obtained by using Gagliardo Nirenberg type estimates for the remainders $\rho^{[w]}$ resp. $\rho^{[y]}$ and combining them with the linear estimate (127).

Step 4 [Exponential decay of a C^0-semigroup]: We collect the available information for the variable coefficient operator $\mathscr{L}$ from (119) and the constant coefficient operator (121). The wave condition (ii) guarantees $\mathrm{Re}\,(\sigma(\mathscr{L}_\infty)) \leqslant -\beta$. Using condition (iii) one can also prove that $\mathscr{L} = \mathscr{L}_\infty + (f'(u_*) - f'(0)) : H^2_{Eucl} \to L^2$ is a relatively compact perturbation of $\mathscr{L}_\infty$, which by Theorem 2 shows $\mathrm{Re}\,\sigma_{ess}(\mathscr{L}) \leqslant -\frac{\beta}{2} < 0$. But now the problem arises that the spectral mapping theorem for C^0-semigroups holds for the point spectrum, but in general not for the essential spectrum, see [23, 50]. That is, $\exp(\sigma(\mathscr{L})) = \sigma(\exp(\mathscr{L}))$ holds for $\sigma = \sigma_{\mathrm{point}}$ (up to the number 0) but not for σ_{ess}.

However, it turns out that, instead of Theorem 2 one can use the following Theorem on relatively compact perturbation of the semigroup itself.

Theorem 10. *Let $A : D(A) \subset X \to X$ denote the generator of a C^0-semigroup e^{tA} of type*

$$\omega(A) = \inf_{t>0} t^{-1}\log\|e^{tA}\| = \lim_{t\to\infty} t^{-1}\log\|e^{tA}\|,$$

and let $B \in L[X]$ be linear, bounded such that

$$Be^{tA} \quad \text{is compact for all} \quad t > 0.$$

Then $A + B : D(A) \to X$ generates a C^0-semigroup $e^{t(A+B)}$ with

$$\left|\sigma_{\mathrm{ess}}(e^{A+B})\right| \leqslant e^{\omega(A)}. \tag{128}$$

Moreover, $\mathrm{Re}\left[\sigma_{\mathrm{point}}(A+B)\right] \leqslant \omega_+$ implies $\omega(A+B) \leqslant \max\{\omega(A), \omega_+\}$.

For a proof of the theorem we refer to [10, Appendix], and we note that it can also be derived by combining several results from [23]. In our situation we can apply the theorem to the operators $A = \mathscr{L}_\infty$, $A + B = \mathscr{L}$, $B = f'(u_*) - f'(0)$ since $(f'(u_*) - f'(0))e^{t\mathscr{L}_\infty}$ is compact in H^2. Equation (128) then leads to an exponential estimate for $\sigma_{\mathrm{ess}}(e^{t\mathscr{L}})$, in particular $|\sigma_{\mathrm{ess}}(e^{\mathscr{L}})| < 1$. Now one restricts $\mathscr{L}$ to the subspace W which is invariant under $\exp \mathscr{L}$ (but not under $e^{\mathscr{L}_\infty}$!) and applies the spectral mapping theorem to find $|\sigma_{\mathrm{point}}(e^{L_{|W}})| < 1$ from wave condition (v). Combining both results, finally proves the estimate (127).

4.4 Further Experiments with Waves in 2D and 3D

We finish this section with numerical experiments in two and three space dimensions. We note that for these non-localized waves, there is currently no rigorous nonlinear stability analysis available.

Example 14 (Barkley model). The frozen version (112) of the well-known Barkley spiral system [6] reads

$$u_t = \Delta u + \frac{1}{\varepsilon} u(1-u)\left(u - \frac{v+b}{a}\right) + \mu_1(x_2 u_{x_1} - x_1 u_{x_2}) + \mu_2 u_{x_1} + \mu_3 u_{x_2},$$

$$v_t = u - v + \mu_1(x_2 v_{x_1} - x_1 v_{x_2}) + \mu_2 v_{x_1} + \mu_3 v_{x_2},$$

$$0 = (x_2 u_{0,x_1} - x_1 u_{0,x_2}, u - u_0)_{L^2} + \left(x_2 v^0_{x_1} - x_1 v^0_{x_2}, v - v^0\right)_{L^2},$$

$$0 = (u_{0,x_1}, u - u_0)_{L^2} + \left(v^0_{x_1}, v - v^0\right)_{L^2} = (u_{0,x_2}, u - u_0)_{L^2} + \left(v^0_{x_2}, v - v^0\right)_{L^2}.$$

For parameter values

$$\varepsilon = \frac{1}{50}, \quad a = 0.75, \quad b = 0.01, \tag{129}$$

Figure 31d shows the behavior of the 3 group velocities of a frozen spiral.

Let us compute the motion $\gamma(t) = (\theta(t), \tau(t))$ in the group when the solution has reached its relative equilibrium, i.e. we determine $\gamma(t) = \exp(t\bar{\mu}) \in G$ for a given $\bar{\mu} \in \mathscr{A}$ from the reconstruction equation

$$\dot{\gamma} - dL_\gamma(\mathbb{1})\bar{\mu} = \begin{pmatrix} 1 & 0 \\ 0 & R_\theta \end{pmatrix} \bar{\mu}, \quad \gamma(0) = 0. \tag{130}$$

The solution is

$$\bar{\gamma}(t) = \begin{pmatrix} \theta(t) \\ \tau(t) \end{pmatrix} = \begin{pmatrix} \bar{\mu}_1 t \\ (I - R_{\bar{\mu}_1 t}) x_c \end{pmatrix}, \quad \text{where} \quad x_c = \frac{1}{\bar{\mu}_1} \begin{pmatrix} -\bar{\mu}_3 \\ \bar{\mu}_2 \end{pmatrix}. \tag{131}$$

Note that $\tau(t)$ moves on a circle of radius $\|x_c\|$ centered at x_c. Inserting this into the profile u_* we obtain the solution

$$\bar{u}(x,t) = u_*(R_{-\bar{\mu}_1 t}(x + (R_{\bar{\mu}_1 t} - I)x_c)) = u_*(R_{-\bar{\mu}_1 t}(x - x_c) + x_c). \tag{132}$$

If a specific point $\bar{x}$ of the profile u_* is of interest, e.g. the tip of a spiral, then this point will be visible at position $x(t)$ with $\bar{x} = R_{-\bar{\mu}_1 t}(x(t) - x_c) + x_c$, i.e. on the circle given by $x(t) = R_{\bar{\mu}_1 t}(\bar{x} - x_c) + x_c$. Our conclusion is that the freezing method gives the information about the center x_c and the speed of rotation $\bar{\mu}_1$ for free. There is no need to use ad-hoc definitions for locating the tip of a spiral,

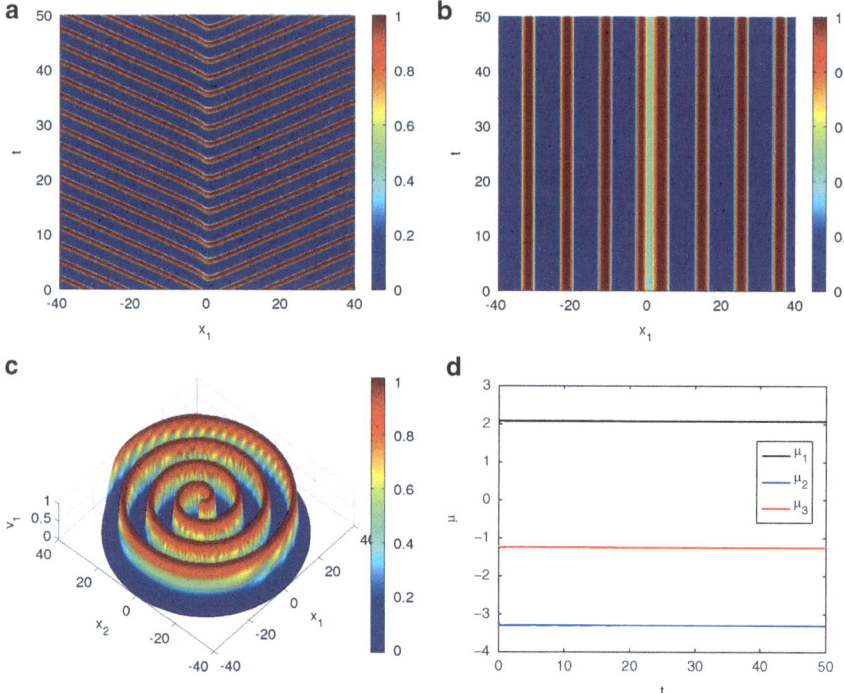

Fig. 31 Cross-section at $x_2 = 0$ for the first component of the Barkley spiral $u(x,t)$ of the nonfrozen system (**a**) and of the frozen system (**b**). First component of the profile $\bar{v}$ of the Barkley spiral (**c**) and time-dependence of velocities (μ_1, μ_2, μ_3) (**d**). Solution by Comsol Multiphysics with piecewise linear finite elements, Neumann boundary conditions, fixed phase condition with template function $\hat{v}$ taken from the solution of the nonfrozen system at time $t = 150$ and parameter values from (129)

for example. A comparison of this method with traditional ways of following the tip of a spiral from a direct simulation of the given system is provided in [13]. However, we note that it can be useful to impose such spiral tip conditions if one aims at phase conditions that lead to global sections. In [31] such an approach is used for freezing not only rigidly rotating spirals (relative equilibria) but also to recognize meandering spirals (relative periodic orbits). The work [39] contains another interesting application of the freezing methodology, namely to follow the large core limit of spiral waves, i.e. to observe the behavior $\mu_1 \to 0$ or $x_c \to \infty$ under parametric perturbations, without solving the equations on extremely large domains.

Example 15 (Quintic-cubic Ginzburg-Landau equation). We continue the QCGL equations from Example 13 in three space dimensions

$$u_t = \alpha \Delta u + (\delta + \beta |u|^2 + \gamma |u|^4)u, \quad x \in \mathbb{R}^3, \quad u(x,t) \in \mathbb{C}, \qquad (133)$$

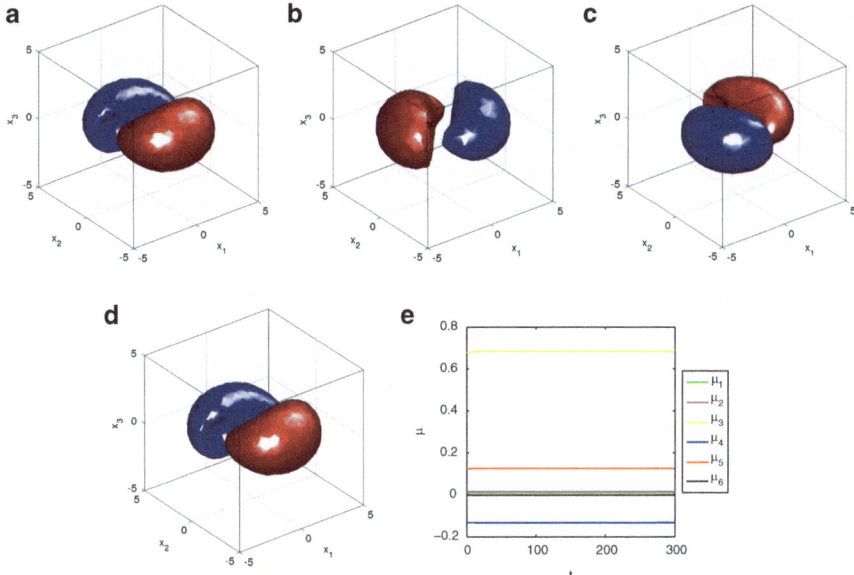

Fig. 32 Isosurfaces $\operatorname{Re} u(x_1, x_2, x_3, t) = \pm 0.5$ at times $t = 0, 3.2, 6.5$ (**a**)–(**c**), isosurfaces of the real part of the profile $\bar{v}$ and time-dependence of velocities $\mu(t)$ for the three-dimensional QCGL (133)

and look for $3D$ spinning solitons. The system is equivariant with respect to the action of the 6-dimensional Euclidean group $G = SE(3) = SO(3) \ltimes \mathbb{R}^3$, given by

$$[a(\gamma)v](x) = v(R^{-1}(x - \tau)), \quad \gamma = (R, \tau) \in SE(3). \tag{134}$$

Recall that the group operation in this representation is $\gamma \circ \tilde{\gamma} = (R\tilde{R}, \tau + R\tilde{\tau})$. The freezing method leads to the PDE

$$\begin{aligned} v_t &= \alpha \Delta v + (\delta + \beta |u|^2 + \gamma |u|^4)u + \mu_4 v_{x_1} + \mu_5 v_{x_2} + \mu_6 v_{x_3} \\ &+ \mu_1(v_{x_2}x_3 - v_{x_3}x_2) + \mu_2(v_{x_3}x_1 - v_{x_1}x_3) + \mu_3(v_{x_1}x_2 - v_{x_2}x_1), \end{aligned} \tag{135}$$

complemented by 6 phase conditions. In Fig. 32 we show the results of a simulation of this system for the same parameter values as in (115). Figure 32a–c shows the spinning solitons for 3 different time instances of the original equation, while Fig. 32d displays the profile of the frozen solution by showing two iso-surfaces of the real part. The behavior of the six algebraic variables $\mu_1, \ldots, \mu_6$ is shown in Fig. 32e. The resulting relative equilibria seem to be localized and stable with asymptotic phase, but we are not aware of any rigorous result in this direction comparable to Theorem 9.

4.5 Summary

Let us summarize the results of this section:

- The freezing method applies to 2D and 3D rotating patterns and automatically generates information about angular velocities and centers of rotation.
- For rotating localized 2D waves one can prove nonlinear stability with asymptotic phase in the Sobolev space H^2 from linear stability.
- Differential operators obtained by linearizing about rotating two-dimensional patterns generate only C^0-semigroups.
- Numerical approximations and convergence of the freezing method are not yet analyzed theoretically. There are also no rigorous theorems on nonlinear stability of nonlocalized rotating patterns such as spiral waves.

5 Decomposition and Freezing of Multi-structures

Many excitable systems discussed in the first sections admit special solutions that are composed of several waves and thus cannot be frozen in a single coordinate frame. Often such patterns travel at different speeds and either move towards each other (the case of *strong interaction*) or repel each other (the case of *weak interaction*). As long as the patterns do not interact strongly they seem to behave like linear superpositions, though this cannot be true in the strict sense for a nonlinear system. In this section we discuss an extension of the freezing method to handle multiple coordinate frames in which the single profiles can stabilize independently while still capturing their nonlinear interaction. The basic idea is to use dynamic partitions of unity in order to decompose the system into a larger system of PDAEs, the dimension of which is determined by the maximal number of patterns. The basic idea is taken from [15] while we follow here the improvement from [63]. In particular, we explain a highly sophisticated stability result from the thesis [63] which applies to weakly interacting fronts and pulses. We also mention that this numerical approach is closely related to an analytical method developed in [62, 69] where so-called exit and shooting manifolds are constructed which are followed by the multi-structures for a certain time.

5.1 Multi-pulses and Multi-fronts

Consider the Cauchy problem for a parabolic (or mixed hyperbolic-parabolic) system in one space variable

$$\begin{aligned} u_t &= Au_{xx} + f(u), \; x \in \mathbb{R}, \, t \geq 0, \\ u(\cdot, 0) &= u_0, \end{aligned} \quad (136)$$

for a function $u(x,t) \in \mathbb{R}^m$ on the real line, where $A \in \mathbb{R}^{m,m}$ is assumed to be positive semidefinite and $f : \mathbb{R}^m \to \mathbb{R}^m$ is assumed to be sufficiently smooth. Multi-pulses and multi-fronts generically appear in a large variety of systems of the form (136) and we mention two standard examples:

Example 16 (FitzHugh-Nagumo system). Recall the FitzHugh-Nagumo system

$$u_t = \begin{pmatrix} u_1 \\ u_2 \end{pmatrix}_t = \begin{pmatrix} 1 & 0 \\ 0 & \varepsilon \end{pmatrix} u_{xx} + f(u)$$

$$f\begin{pmatrix} u_1 \\ u_2 \end{pmatrix} = \begin{pmatrix} u_1 - \frac{1}{3}u_1^3 - u_2 \\ \phi(u_1 + a - bu_2) \end{pmatrix}, \quad \phi, a, b > 0, \ \varepsilon \geq 0.$$

In Example 7 we observed for pulse like initial data the generation of a **double pulse** solution, see Fig. 12. More precisely, the solution to the Cauchy problem develops two pulses, traveling in opposite directions. See Example 7 for the details of the numerical simulation.

We have seen that the numerical method of freezing captures one of the two evolving pulses, while the other leaves the computational domain. In fact, in the numerical experiments the phase condition determined which of the two traveling pulses is captured and which is lost.

Example 17 (Nagumo equation). As an example that generates **double fronts** we consider again the Nagumo equation, compare Example 6,

$$u_t = u_{xx} + u(1-u)(u-a), \quad x \in \mathbb{R}, \ t \geq 0, \quad u(\cdot, 0) = u_0,$$

with parameter $a = \frac{1}{4}$. As initial condition for the Cauchy problem we choose the piecewise linear function

$$u_0(x) = \mathbf{1}_{(-50,0]}(x) \cdot \frac{x+50}{50} + \mathbf{1}_{(0,50)}(x) \cdot \frac{50-x}{50}, \tag{137}$$

where $\mathbf{1}_M(x) = 1$ for $x \in M$ and $\mathbf{1}_M(x) = 0$ for $x \notin M$ is the indicator function of a set M. The solution to this problem consists of two fronts traveling with the same speed in opposite directions. A numerical solution is shown in Fig. 33.

The above examples show that it is important to be able to capture multi-pulses and multi-fronts. Obviously, this cannot be done by using a single moving frame. For patterns sufficiently far apart, the individual parts of the pattern seem not to influence each other and the multi-structures look like linear superpositions, but due to nonlinearities they cannot be linear superpositions. Nevertheless, in the case of **weak interaction**, i.e. when the patterns are far apart for large times, linear superposition is a good model. Currently, there is no theory available in the case of **strong interaction**, i.e. when the individual parts of the multi-structure get close to each other.

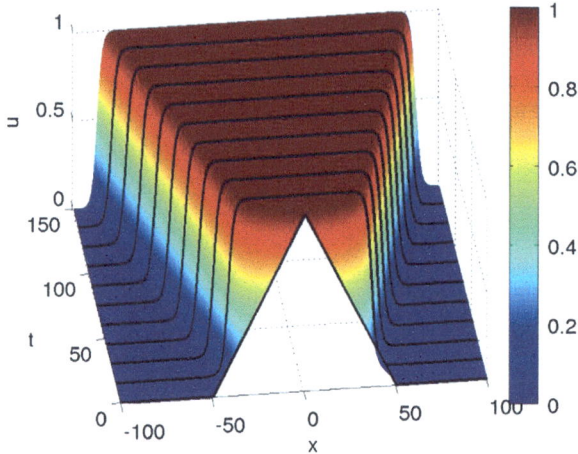

Fig. 33 Space-time diagram of a double front solution to the Nagumo equation for parameter value $\alpha = \frac{1}{4}$ on the domain $\Omega = [-100, 100]$. Solution by Comsol Multiphysics with piecewise linear finite elements, Neumann boundary conditions, $\Delta x = 0.1$, $\Delta t = 0.1$, BDF of order 2 and initial data u_0 from (137)

5.2 Decompose and Freeze Multi-structures

Now consider the Cauchy problem (136) and assume that the solution u consists of N single profiles. To generalize the freezing ansatz to this situation we write the solution as the superposition of N profiles in the following form

$$u(x,t) = \sum_{j=1}^{N} v_j(x - g_j(t), t). \quad (138)$$

Here the function $g_j : [0, \infty[\to \mathbb{R}$ denotes the time dependent position of the j-th profile $v_j : \mathbb{R} \times [0, \infty[\to \mathbb{R}^m$, $(x,t) \mapsto v_j(x,t)$. Of course, due to nonlinearity, the solution u is not just the superposition of N separate profiles as pretended in (138).

To overcome this difficulty and to make use of the fact that well separated profiles basically behave like linear superpositions, we use the idea of partition of unity: Let $\varphi \in C^\infty(\mathbb{R}, \mathbb{R})$ be a positive bump function such that the main mass is located near zero and $0 < \varphi(x) \leq 1$ for every $x \in \mathbb{R}$. A suitable choice for φ is $\varphi(x) = \text{sech}(\beta x) = \frac{1}{\cosh(\beta x)}$ with $\beta > 0$. Then for $g = (g_1, \ldots, g_N) : [0, \infty) \to \mathbb{R}^N$ and $x \in \mathbb{R}$ the functions

$$Q_j(g(t), x) = \frac{\varphi(x - g_j(t))}{\sum_{k=1}^{N} \varphi(x - g_k(t))}, \quad j = 1, \ldots, N \quad (139)$$

have non-vanishing denominators and form a time-dependent partition of unity, i.e.

$$1 = \frac{\sum_{j=1}^{N} \varphi(x - g_j)}{\sum_{k=1}^{N} \varphi(x - g_k)} = \sum_{j=1}^{N} Q_j(g, x).$$

We are interested in solutions of (136) of the form (138). In order to investigate such solutions we insert the ansatz (138) into (136) and use the partition of unity (139). Abbreviating $v_k(*) = v_k(\cdot - g_k(t), t)$ this leads to

$$\sum_{j=1}^{N} \left[v_{j,t}(*) - v_{j,x}(*) g_{j,t} \right] = u_t = A u_{xx} + f(u)$$

$$= \sum_{j=1}^{N} \left[A v_{j,xx}(*) + Q_j(g, \cdot) f \left(\sum_{k=1}^{N} v_k(*) \right) \right] \qquad (140)$$

$$= \sum_{j=1}^{N} \left[A v_{j,xx}(*) + f(v_j(*)) + Q_j(g, \cdot) \left\{ f \left(\sum_{k=1}^{N} v_k(*) \right) - \sum_{k=1}^{N} f(v_k(*)) \right\} \right].$$

Now, we require that the summands on the left and on the right hand side of (140) coincide for every $j = 1, \ldots, N$. The idea is to consider each of the summands in its own co-moving frame and apply the freezing ansatz: We substitute $\xi = x - g_j(t)$, $\mu_j = g_{j,t}$ and $*_{kj} = \xi - g_k(t) + g_j(t)$, add initial and phase condition for each v_j, $j = 1, \ldots, N$ and obtain the following coupled system for $j = 1, \ldots, N$, $\xi \in \mathbb{R}, t \geq 0$

$$v_{j,t}(\xi, t) = A v_{j,\xi\xi}(\xi, t) + v_{j,\xi}(\xi, t) \mu_j(t) + f(v_j(\xi, t))$$

$$+ \frac{\varphi(\xi)}{\sum_{k=1}^{N} \varphi(*_{kj})} \left[f \left(\sum_{k=1}^{N} v_k(*_{kj}, t) \right) - \sum_{k=1}^{N} f(v_k(*_{kj}, t)) \right], \qquad (141)$$

$$0 = \left(v_j(\cdot, t) - \hat{v}_j, \hat{v}_{j,x} \right)_{L^2}, \quad v_j(\cdot, 0) = v_j^0,$$

$$g_{j,t} = \mu_j, \qquad g_j(0) = g_j^0.$$

To enforce that also the initial condition in (136) is satisfied, we additionally require $u_0 = \sum_{j=1}^{N} v_j^0(\cdot - g_j^0)$. It is easy to see that if (v_j, g_j) solves (141) and satisfies the assumption on the initial condition, then $u(\cdot, t) = \sum_{j=1}^{N} v_j(\cdot - g_j(t), t)$ solves the Cauchy problem (136). Note that the decomposition is not unique.

In the case of multi-fronts one has different limits at $\pm \infty$ and it is, even in the linear case, not possible to simply add the single profiles as we did in (138). In order to employ the above procedure also in this case, we define

$$u_j^- = \begin{cases} 0 & , j = 1, \\ \lim_{x \to -\infty} w_j(x) & , j \geq 2, \end{cases} \qquad (142)$$

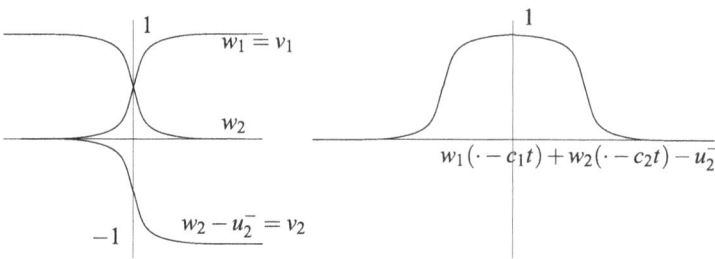

Fig. 34 The sum of two fronts forming a multi-front

where w_j is the expected j-th wave. The situation is depicted in Fig. 34, where a double front is considered. Writing the j-th wave as $v_j(\xi,t) + u_j^-$ and following the recipe from (140), we obtain the coupled PDAE system for $j = 1, \ldots, N$:

$$v_{j,t}(\xi,t) = Av_{j,\xi\xi}(\xi,t) + v_{j,\xi}(\xi,t)\mu_j(t) + f(v_j(\xi,t) + u_j^-)$$

$$+ \frac{\varphi(\xi)}{\sum_{k=1}^N \varphi(*_{kj})} \left[f\left(\sum_{k=1}^N v_k(*_{kj},t)\right) - \sum_{k=1}^N f(v_k(*_{kj},t) + u_k^-) \right],$$

$$0 = \left(v_j(\cdot,t) - \hat{v}_j, \hat{v}_{j,x}\right)_{L^2}, \quad v_j(\cdot,0) = v_j^0,$$

$$g_{j,t} = \mu_j, \qquad g_j(0) = g_j^0.$$

(143)

Again we require $u_0 = \sum_{j=1}^N v_j^0(\cdot - g_j^0)$, so that a solution (v_j, g_j) to (143) yields a solution of (136) via (138). Note that allowing $u_j^- = 0$, (143) includes the case of pulses (141) and also the cases of solutions that consist of both, pulse and front solutions.

We just mention, that the PDAE systems (141) and (143) contain nonlinear and nonlocal coupling terms. For solving the PDAE on a bounded domain $J = [x_-, x_+]$, we have to interpolate by the left and the right limit, respectively, whenever $*_{kj} = \xi - g_k(t) + g_j(t) \notin [x_-, x_+]$. Namely, we extend the function v_j to be constant equal to its boundary values.

Example 18 (Nagumo equation). Consider the Nagumo equation from Example 6 with parameter $a = \frac{1}{4}$,

$$u_t = u_{xx} + u(1-u)(u-a), \quad x \in \mathbb{R}, \ t \geq 0, \quad u(\cdot,0) = u_0.$$

(a) First consider the case of two repelling fronts. This situation occurs, for example, when the initial data form a hat function as in (137). We use the PDAE system (143) with $N = 2$ and $u_2^- = 1$. As initial data we choose

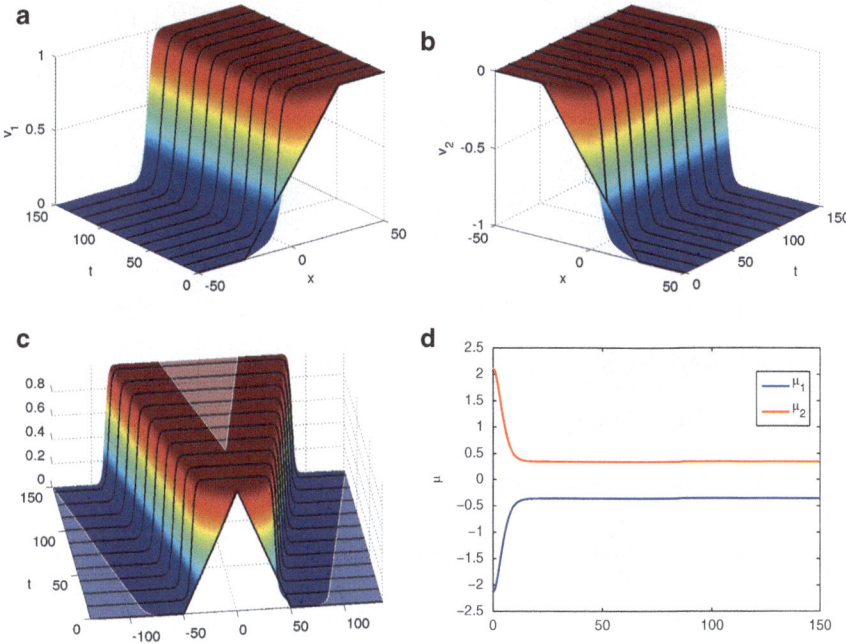

Fig. 35 The decompose and freeze method for a repelling double front in the Nagumo equation. The v_1 component (**a**) and the v_2 component (**b**), plot of the superposition $u(\cdot, t) = v_1(*, t) + v_2(*, t)$ with supports of single fronts indicated by *dark shading* (**c**), evolution of the individual speeds converging to $\bar{\mu}_1 = -\bar{\mu}_2 = 0.3536$ (**d**). Solution of (143) by Comsol Multiphysics with piecewise linear finite elements, Neumann boundary conditions, $\Delta x = 0.1$, $\Delta t = 0.1$, BDF of order 2

$$v_1^0(x) = \mathbf{1}_{(-25,25)}(x) \cdot \frac{x+25}{50} + \mathbf{1}_{[25,\infty)}(x), \quad x \in \mathbb{R},$$

$$v_2^0(x) = -\mathbf{1}_{(-25,25)}(x) \cdot \frac{x+25}{50} - \mathbf{1}_{[25,\infty)}(x), \quad x \in \mathbb{R}$$

for the two profiles and $g_1^0 = -25$, $g_2^0 = 25$ for the initial shifts. The simulation is performed on the finite interval $J = [-50, 50]$ with $\varphi(x) = \operatorname{sech}\left(\frac{x}{2}\right)$ and the solutions v_1 and v_2 are assumed to equal their asymptotic values outside the computational domain. The results of a simulation are plotted in Fig. 35. The performance of the decompose and freeze method can be demonstrated by plotting the difference of the superposition (138) from the result obtained by a direct simulation of the full system (see [15] for such a comparison).

(b) As a second example we consider the case of two colliding fronts. For this we take initial conditions

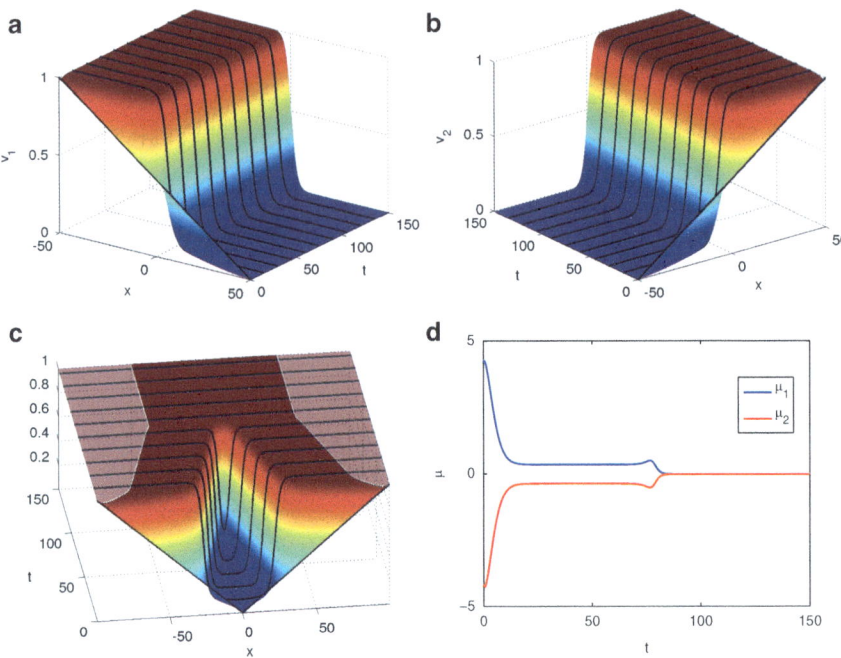

Fig. 36 Freezing two colliding fronts in the Nagumo equation. Plot of v_1 component (**a**), v_2 component (**b**), superposition $u(\cdot,t) = v_1(*,t) + v_2(*,t)$ with supports of single fronts indicated by *dark shading* (**c**), evolution of speeds μ_1, μ_2 of the components v_1, v_2, both converging ultimately to $\bar{\mu}_1 = \bar{\mu}_2 = 0$ (**d**). Solution of (143) by Comsol Multiphysics with piecewise linear finite elements, Neumann boundary conditions, $\Delta x = 0.1$, $\Delta t = 0.1$, BDF of order 2

$$u_0(x) = 1 - \mathbf{1}_{(-100,0]}(x) \cdot \frac{x+100}{100} + \mathbf{1}_{(0,100)}(x) \cdot \frac{x-100}{100}, \quad x \in \mathbb{R},$$

which we split as follows $u_0 = v_1^0(\cdot - g_1^0) + v_2^0(\cdot - g_2^0)$, with

$$v_1^0(x) = \mathbf{1}_{(-\infty,50)}(x) - \mathbf{1}_{(-50,50)}(x) \cdot \frac{x+50}{100}, \quad x \in \mathbb{R},$$

$$v_2^0(x) = \mathbf{1}_{(-50,50)}(x) \cdot \frac{x+50}{100} + \mathbf{1}_{[50,\infty)}(x), \quad x \in \mathbb{R}.$$

In this case we have $u_2^- = 0$ and choose $g_1^0 = -50, g_2^0 = 50$. The other data are as in Example (a). A result of the method is shown in Fig. 36. Note that the decompose and freeze method successfully handles the strong interaction. The single waves assume a common velocity and asymptotically converge to two steady profiles which sum up to the final profile (which is identically 1 in this case). So far, we have no theory which proves this behavior for the case of strong interaction.

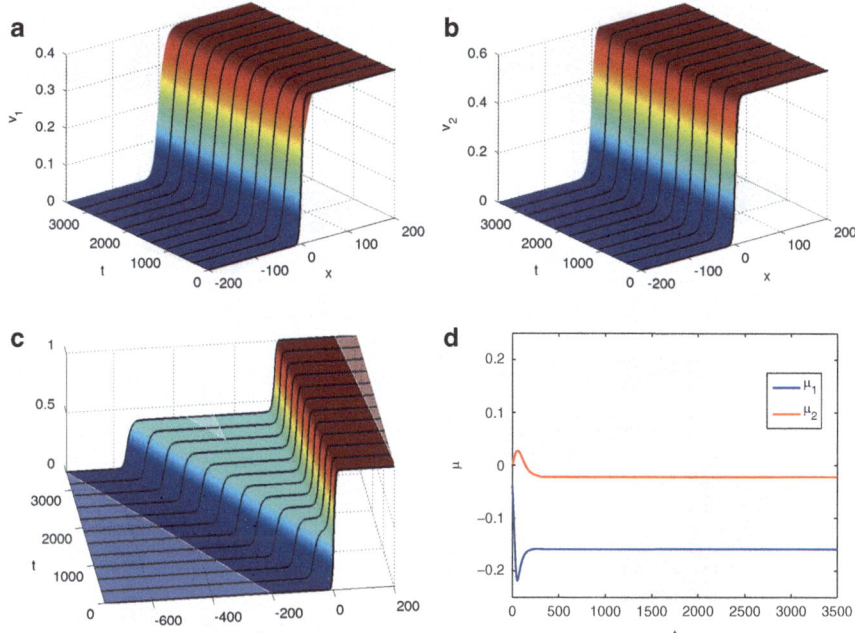

Fig. 37 Two fronts of different speed developing out of a single front in the quintic Nagumo equation (144): convergence of the decompose and freeze method for single fronts v_1 (**a**) and v_2 (**b**), plot of superposition (138) with supports of v_1, v_2 indicated by *dark shading* (**c**), time-dependence of single speeds (**d**)

Example 19 (Quintic Nagumo equation). As an example supporting multi-structures with more than two patterns, we consider the quintic Nagumo equation

$$u_t = u_{xx} - \prod_{i=1}^{5}(u - a_i), \quad x \in \mathbb{R}, \, t \geq 0, \tag{144}$$

with parameters $0 = a_1 < a_2 < a_3 < a_4 < a_5 = 1$. Depending on the choices of a_2, a_3, a_4 one observes different patterns which can be captured by the decompose and freeze method. We present the results for a selection of parameter values. In all cases we solve (143) with Neumann boundary conditions, choose the bump function $\varphi(x) = \text{sech}\left(\frac{x}{20}\right)$ and use spatial step-size $\Delta x = 0.4$.

(a) Parameters: $a_2 = 0.03125$, $a_3 = 0.4$, $a_4 = 0.73$, $\Delta t = 0.8$, initial data: $g_1(0) = g_2(0) = 0$, , $u_2^- = a_3$, $v_1^0(x) = \frac{u_2^-}{2}\left(\tanh\left(\frac{x}{5}\right) + 1\right)$, $v_2^0(x) = \frac{(1-u_2^-)}{2}\left(\tanh\left(\frac{x}{5}\right) + 1\right)$. The solutions are shown in Fig. 37. We start with the superposition of two front-like functions located at the same position. Then two fronts develop, a fast one traveling at speed $\bar{\mu}_1 = -0.1590$ and a slow one with speed $\bar{\mu}_2 = -0.02131$. The single fronts v_1 in (a) and v_2 in (b) converge, Fig. 37 shows their superposition according to (138) with the supports of v_1, v_2 indicated by dark shading. This is a case of weak interaction.

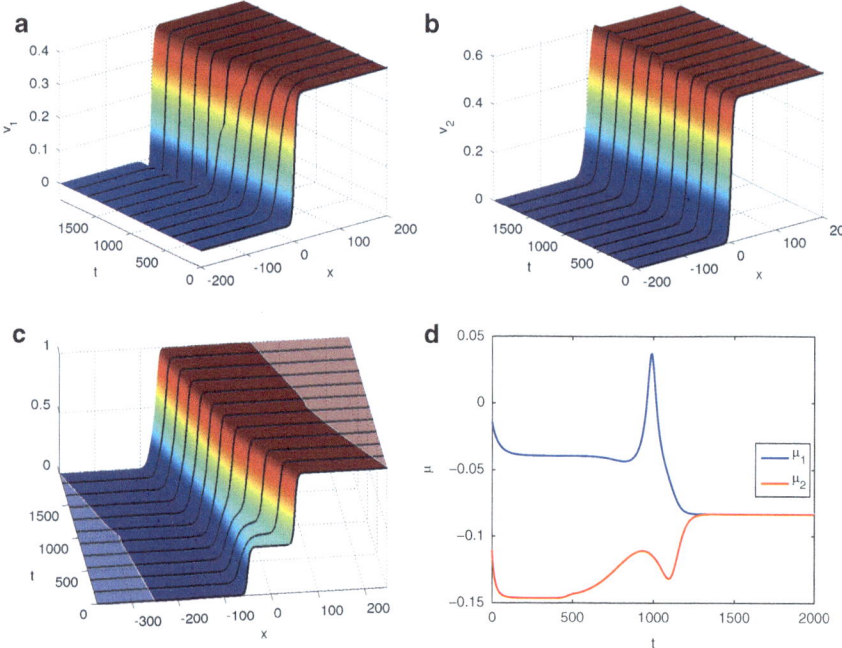

Fig. 38 Strong interaction in the quintic Nagumo equation (144): a fast wave overtaking a slow wave and merging into a single travelling front. Single fronts v_1 (**a**), v_2 (**b**), superposition (138) of both functions with their supports indicated by *dark shading* (**c**), time-dependence of speeds μ_1, μ_2 during strong interaction (**d**)

(b) Parameters: $a_2 = 0.125$, $a_3 = 0.4$, $a_4 = 0.58$, $\Delta t = 0.3$, initial data: $g_1(0) = -50$, $g_2(0) = 50$, $u_2^- = a_3$, $v_1^0(x) = \frac{u_2^-}{2}\left(\tanh\left(\frac{x}{5}\right) + 1\right)$, $v_2^0(x) = \frac{(1-u_2^-)}{2}\left(\tanh\left(\frac{x}{5}\right) + 1\right)$. The results are shown in Fig. 38. Starting with a staircase function, two fronts of different speed develop, with the faster one overtaking the slower one. Then strong interaction takes place and both fronts merge to a single front of speed $\bar{\mu}_1 = \bar{\mu}_2 = -0.08312$, cf. Fig. 38d. The components v_1, v_2 stabilize at profiles with little kinks that add up to the merged travelling front. The decompose and freeze method is able to handle this case of strong interaction.

(c) Parameters: $a_2 = 0.0625$, $a_3 = 0.4$, $a_4 = 0.7$, $\Delta t = 0.8$, initial data: $g_1(0) = -50$, $g_2(0) = 0$, $g_3(0) = 50$, $u_2^- = a_3$, $u_3^- = a_5$, $v_1^0(x) = \frac{u_2^-}{2}\left(\tanh\left(\frac{x}{5}\right) + 1\right)$, $v_2^0(x) = \frac{(1-u_2^-)}{2}\left(\tanh\left(\frac{x}{5}\right) + 1\right)$, $v_3^0(x) = \frac{(u_2^- - u_3^-)}{2}\left(\tanh\left(\frac{x}{5}\right) + 1\right)$. Results are shown in Fig. 39. We start with a multi-front consisting of three stairs. Three fronts develop, one traveling to the right with speed $\bar{\mu}_3 = 0.05088$ and two traveling to the left with speeds $\bar{\mu}_1 = -0.1172$, $\bar{\mu}_2 = -0.05088$. This is a case of weak interaction since the initial locations of fronts are in the same order as the corresponding velocities. The system (143) is now solved with $N = 3$.

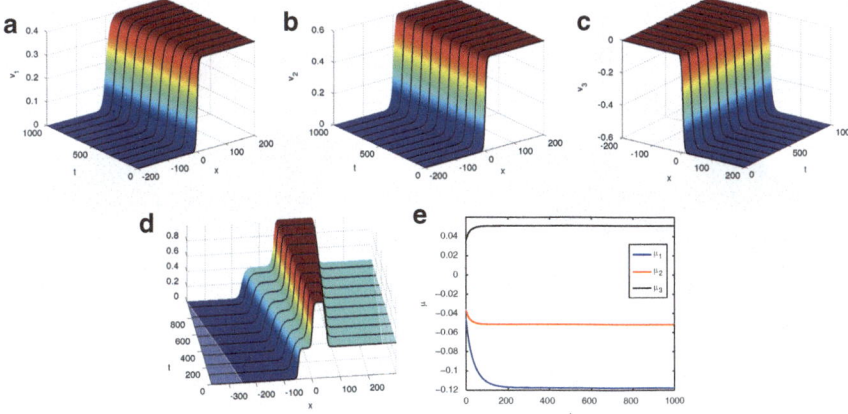

Fig. 39 Weak interaction of three traveling fronts in the quintic Nagumo equation (144). The initial function develops into the superposition of three single fronts (**d**), the functions v_1, v_2, v_3 converge to their limiting profiles (**a**)–(**c**), and the speeds μ_1, μ_2, μ_3 attain their limiting values (**e**)

Example 20 (FitzHugh-Nagumo system). Our final example are repelling and colliding pulses in the FitzHugh-Nagumo system from Example 7. We take the parameter values

$$\varepsilon = 0.1, \quad \phi = 0.08, \quad a = 0.7, \quad b = 0.8,$$

for which we know traveling pulses to exist (see also [48]). The spatial domain is $J = [x_-, x_+] = [-100, 100]$ and we impose Neumann boundary conditions.

(a) In our simulation for two repelling pulses we use the following initial data

$$v_1^0(x) = u_2^- + \begin{pmatrix} \frac{2.5}{1+(\frac{x}{3})^2} \cdot \text{flc2hs}(-x, 5) \\ 0 \end{pmatrix}, \quad v_2^0(x) = \begin{pmatrix} \frac{2.5}{1+(\frac{x}{3})^2} \cdot \text{flc2hs}(x, 5) \\ 0 \end{pmatrix},$$

with initial positions $g_1^0 = g_2^0 = 0$, where u_2^- denotes the unique zero of f from Example 7 and $\text{flc2hs}(x, \text{scale})$ is a smoothed Heaviside function provided by Comsol Multiphysics. Note that the superposition of v_1^0 and v_2^0 coincides with the initial value u_0 from Example 7, see also Fig. 12. For the computation we choose the fixed phase condition in both frames with template functions $\hat{v}_j(x) = v_j^0(x)$ for $j = 1, 2$ and the bump function $\varphi(x) = \text{sech}(bx)$ with $b = 0.5$. We discretize with continuous piecewise linear finite elements in space with stepsize $\Delta x = 0.5$ and with the BDF method of order 2 in time with stepsize $\Delta t = 0.1$. The results for the case of two repelling pulses are shown in Fig. 40. One clearly observes the evolution of two pulses traveling in opposite directions with velocities $\bar{\mu}_1 = -\bar{\mu}_2 = -0.7966$, both being nicely captured by the method.

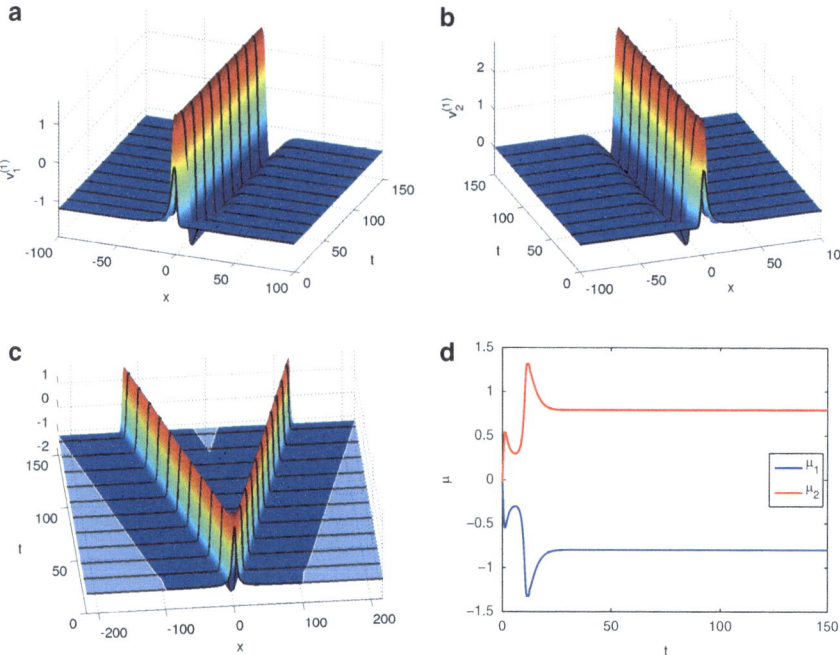

Fig. 40 Two repelling pulses in the FitzHugh-Nagumo system. (**a**)–(**b**): First components of the profiles v_1 and v_2, (**c**): first component of the superposition of v_1 and v_2, (**d**): time evolution of the velocities μ_1 and μ_2

(b) A situation with two colliding pulses occurs for the initial conditions

$$v_1^0(x) = u_2^- + \begin{pmatrix} w(x) \\ 0 \end{pmatrix}, \quad v_2^0(x) = \begin{pmatrix} w(-x) \\ 0 \end{pmatrix},$$

where w is a ramp function given by

$$w(x) = \begin{cases} 1 & , x \in [-100, -10], \\ \frac{1}{20}(10-x) & , x \in [-10, 10], \\ 0 & , x \in [10, 100]. \end{cases}$$

The initial positions are $g_1^0 = -100$ and $g_2^0 = 100$. Thus, for the decompose and freeze method with spatial domain $[-100, 100]$, these are completely separated at $t = 0$ and only influence each other through interpolated data. In contrast to all previous experiments, we choose the orthogonal phase condition in both frames. The bump function is $\varphi(x) = \mathrm{sech}\,(bx)$ with $b = 0.01$. We discretize in space with continuous piecewise linear finite elements with stepsize $\triangle x = 0.5$ and with the

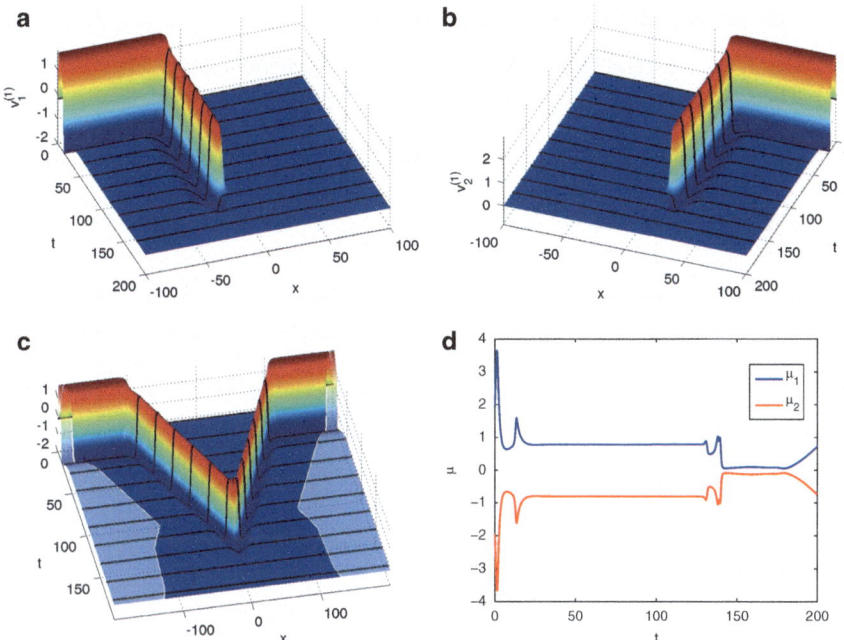

Fig. 41 Simulation of two colliding pulses in the FitzHugh-Nagumo system with the decompose and freeze method. (**a**)–(**b**): First components of the profiles v_1 and v_2, (**c**): first component of the superposition of v_1 and v_2, (**d**): time evolution of the velocities μ_1 and μ_2

BDF method of order 5 with stepsize $\Delta t = 0.1$. First the two pulses are generated from the opposite ramps, then they travel towards each other until their domains begin to overlap. Then the two pulses collide and finally cancel each other. A result of the simulation is shown in Fig. 41. When both pulses are extinguished to their stationary values, the PDAE system (141) becomes ill-posed, since the derivatives $v_{j,\xi}$ vanish and the velocities can no longer be determined from the phase condition. In experiments we observe such a blowup at $t \approx 180$ (see Fig. 41d) for the orthogonal phase condition and immediately after collision at $t \approx 140$ for the fixed phase condition (not shown).

5.3 Stability of the Decomposition System

In this subsection we present the main stability result for the decomposition system (143). We sketch only the main ideas in the rather technical and involved proof which can be found in the PhD thesis of S. Selle [63]. The result is related to the work of J.D. Wright [62, 69], who constructs manifolds for the (PDE) that are invariant for certain time intervals during which they attract the multi-structures.

We impose the following conditions.

(A1) Let $f \in C^2(\mathbb{R}^m, \mathbb{R}^m)$ and let $A \in \mathbb{R}^{m,m}$ be positive definite.
(A2) The system $u_t = Au_{xx} + f(u)$ has N traveling wave solutions

$$u_j(x,t) = w_j(x - c_j t), \quad j = 1, \ldots, N,$$

such that $c_1 < c_2 < \cdots < c_N$ and the limits $w_j^\pm = \lim_{x \to \pm\infty} w_j(x)$ satisfy $w_j^+ = w_{j+1}^-, j = 1, \ldots, N-1$.

(A3) For some $\beta > 0$, the constant coefficient operators $\Lambda_{j,\pm} := A\partial_{xx} + c_j \partial_x + C_{j,\pm}$ with $C_{j,\pm} = Df(w_j^\pm)$ for $j = 1, \ldots, N$ satisfy the **Spectral Condition SC** from Sect. 1.5 for the given β, cf. (33).

(A4) The variable coefficient operators $\Lambda_j = A\partial_{xx} + c_j \partial_x + Df(w_j)$ $j = 1, \ldots, N$, have the simple eigenvalue 0 and no further eigenvalues with $\operatorname{Re} s \geq -\beta$ and eigenfunctions in $L^2(\mathbb{R}, \mathbb{R}^m)$.

(A5) There exist $C_1, C_2, C_3 > 0$ such that $\varphi \in C^\infty(\mathbb{R}, \mathbb{R})$ satisfies for all $x \in \mathbb{R}$

$$C_1 e^{-\beta|x|} \leq \varphi(x) \leq C_2 e^{-\beta|x|},$$
$$|\varphi'(x)| \leq C_3 e^{-\beta|x|}.$$

(A6) The template functions $\hat{v}_j$, $j = 1, \ldots, N$, satisfy $\hat{v}_j + u_j^- - w_j \in H^2(\mathbb{R}, \mathbb{R}^m)$,

$$\left(\hat{v}_j + u_j^- - w_j, \hat{v}_{j,x}\right)_{L^2} = 0 \quad \text{and} \quad \left(w_{j,x}, \hat{v}_{j,x}\right)_{L^2} \neq 0.$$

Note that the same constant β appears in all conditions (A3)-(A5). In essence, $\beta > 0$ should be chosen such that $-\beta$ is an upper bound for the spectrum of all Λ_j's except zero. Then one can take a bump function with the asymptotic behavior of $e^{-\beta|x|}$, for example $\varphi(x) = \operatorname{sech}(\beta x)$, $x \in \mathbb{R}$.

Under these assumptions one can prove what is called *joint asymptotic stability* in [63], i.e. stability of the system (143) for initial data that lead to weak interaction, see Thm. 11 below. The result holds in weighted L^2 and H^1 spaces given by

$$L_b^2(\mathbb{R}, \mathbb{R}^m) = \{u : \cosh(b\cdot)u \in L^2(\mathbb{R}, \mathbb{R}^m)\} \text{ with norm } \|u\|_{L_b^2} = \|\cosh(b\cdot)u\|_{L^2}$$

and

$$H_b^1(\mathbb{R}, \mathbb{R}^m) = \{u : \cosh(b\cdot)u \in H^1(\mathbb{R}, \mathbb{R}^m)\} \text{ with norm } \|u\|_{H_b^1} = \|\cosh(b\cdot)u\|_{H^1},$$

where $b > 0$ must be chosen positive and sufficiently small.

Theorem 11 (Stability Theorem, [63]). *Assume (A1)–(A6). Then there exists $b_0 > 0$ so that for every $b_0 \geq b > 0$ there exist $\delta > 0$, $g_{min} > 0$ such that for all initial data v_j^0, g_j^0, satisfying*

Stability and Computation of Dynamic Patterns in PDEs

$$\left\| v_j^0 + u_j^- - w_j \right\|_{H^{1,b}} \leq \delta, \quad \left(\hat{v}_{j,x}, v_j^0 - \hat{v}_j \right)_{L^2} = 0,$$

$$g_1^0 < g_2^0 < \cdots < g_N^0, \quad g_{min} \leq \left| g_{j+1}^0 - g_j^0 \right|, \quad \text{for all } j = 1, \ldots, N,$$

the PDAE (143) has a unique global solution $v(t) = (v_1(t), \ldots, v_N(t))$, $\mu(t) = (\mu_1(t), \ldots, \mu_N(t))$, $g(t) = (g_1(t), \ldots, g_N(t))$ *for all* $t \geq 0$.

Moreover, there exist asymptotic phases $\tau_j^\infty \in \mathbb{R}$, $j = 1, \ldots, N$ *such that the solution converges exponentially fast with some rate* $0 < \varepsilon < \beta$,

$$\left\| v_j + u_j^- - w_j \right\|_{H^{1,b}} + \left| g_j(t) - c_j t - g_j^0 - \tau_j^\infty \right| + \left| \mu_j(t) - c_j \right| \leq C e^{-\varepsilon t}, \quad \forall j = 1, \ldots, N.$$

Consider the original PDE (136) with initial condition $u_0(x) = \sum_{j=1}^N v_j^0(x - g_j^0)$, where v_j^0 and g_j^0 satisfy the assumptions from Thm. 11. Then the theorem implies exponential convergence of the solution u to a (linear) superposition of the individual traveling waves with individual asymptotic phases:

$$\left\| u(\cdot, t) - \sum_j w_j(\cdot - c_j t - g_j^0 - \tau_j^\infty) \right\|_{H^{1,b}} \leq C e^{-\varepsilon t}.$$

In [69] results of this type are proved directly for the original system (136) by using analytic information about the single waves. On the contrary, Thm. 11 states a result about the 'blown-up' system (143) which is accessible to numerical computation.

Proof (A sketch of ideas, for details see [63, pp. 37–98]).

Step 1: First linearize the system (141) at the shifted exact waves $w_j - u_j^-$ (see (142)) and their speeds c_j for each $j = 1, \ldots, N$. This yields a system of the form

$$u_{j,t} = \Lambda_j u_j + \lambda_j(t) w_{j,x} + E_j(t) u + T_j(t) + N_j(t, u, r, \lambda),$$

$$r_{j,t} = \lambda_j(t),$$

$$0 = \left(\hat{v}_{j,x}, u_j \right)_{L^2},$$

for the unknowns $u_j = v_j - (w_j - u_j^-)$, $r_j(t) = g_j(t) - c_j t - g_j^0$, $\lambda_j = \mu_j - c_j$. When omitting the coupling terms E_j, T_j, and N_j, the system decouples with linear differential operators known from the analysis of single traveling waves in Sects. 1 and 2.

Step 2: All coupling terms E_j, T_j and N_j turn out to be nonlocal. The term $T_j(t)$ collects the nonlocal terms $f\left(\sum_{k=1}^N v_k(*)\right) - \sum_{k=1}^N f(v_k(*))$ from (140) obtained by inserting the exact traveling waves w_j shifted to the initial and well separated positions g_j^0, $j = 1, \ldots, N$. Due to assumption (A2), the individual waves w_j, $j = 1, \ldots, N$ are exponentially converging towards their limits.

Therefore, the influence of one wave on the other decays exponentially in time. This statement still holds in the weighted space H_b^1, if the weight b is taken sufficiently small (which is the reason for the smallness assumption on b in the theorem). In a sense this property expresses the well-known phenomenon of *convective (in)stability*, see [60].

The operator $E_j(t)$ is the linearization of the nonlinear coupling terms when shifted to the positions $c_j t + g_j^0$ of the individual traveling waves. Here the use of the weighted spaces with $b > 0$ implies that the u_j are exponentially located. An interplay with the separation of the positions as time increases, then shows exponential decay of the operator E_j in the weighted space.

Combining these considerations yields an estimate in the weighted space L_b^2 of the form

$$\|T_j(t)\|_{L_b^2} + \|E_j(t)\|_{L_b^2 \to L_b^2} \le C_L \exp\left(-C_L' g_{min}\right) \exp\left(-C_L'' t\right) \quad \text{for all } t \ge 0,$$

with positive constants C_L, C_L', C_L''. Therefore, these terms decay exponentially in time and exponentially with respect to the initial separation of patterns.

Step 3: The terms N_j contain the nonlinear and nonlocal terms in all variables that are at least of second order. Using the weighted norm one can show an estimate

$$\|N_j(t, u, r, \lambda)\|_{L_b^2} \le C_N \|u\|_{H_b^1} \left(\|u\|_{H_b^1} + |\lambda_j| + \exp\left(C_N' \|r\|\right) \|r\|\right)$$
$$+ C_N \exp\left(C_N' \|r\|\right) \|r\| \left(1 + \|u\|_{H_b^1}\right) \exp\left(-\gamma_N t - \gamma_N g_{min}\right) \quad \text{for all } t \ge 0.$$

Step 4: Consider the linear PDAE with T_j and N_j replaced by an inhomogeneity, i.e. the following coupled system ($j = 1, \ldots, N$),

$$u_{j,t} = \Lambda_j u_j + \lambda_j(t) w_{j,x} + E_j(t) u + k_j(t),$$
$$r_{j,t} = \lambda_j(t), \tag{145}$$
$$0 = \left(\hat{v}_{j,x}, u_j\right)_{L^2}.$$

This can be reduced to a PDAE of index 1 with the algebraic variables λ eliminated. For this inhomogeneous system one shows a variation of constants formula by first proving that the systems yield sectorial operators. The variation of constants formula is then used to prove an estimate of the form

$$\sup_{0 \le s \le t} e^{\varepsilon s} \|u(s)\|_{H_b^1} \le C \left(\|u^0\|_{H_b^1} + \sup_{0 \le s \le t} \|k(s)\|_{L_b^2}\right)$$

with a suitable $\varepsilon > 0$ for the solution $u = (u_1, \ldots, u_N)$ of the coupled linear problem (145).

Step 5: Finally, the estimates from Steps 2–4 are combined and yield global existence as well as the asserted exponential decay. □

5.4 Generalization to an Abstract Framework

The idea to decompose and freeze multi-structures can be combined with the general idea of freezing solutions in equivariant evolution equations. For this we consider the setting from Sect. 1.7, i.e.

$$u_t = F(u), \quad u(0) = u_0, \tag{146}$$

where $F : Y \subset X \to X$ with X is a Banach space and Y a dense subspace. The evolution equation (146) is assumed to be equivariant under the action of a Lie group G so that

$$a : G \to GL(X), \quad F(a(\gamma)u) = a(\gamma)F(u)$$

holds. To generalize the idea of a time-dependent partition of unity from Sect. 5.2, we use the abstract concept of a module E (a vector space with abelian multiplication) that acts on the state space X

$$\bullet : E \times X \to X, \quad (\varphi, u) \to \varphi \bullet u.$$

As a standard example consider $E = C^1_{\text{unif}}(\mathbb{R}^d)$ and $X = H^1(\mathbb{R}^d, \mathbb{R}^m)$ with the action of E on X given by multiplication.

We assume that the Lie group G also acts on E, denoting the action by

$$\mathfrak{a} : G \to GL(E), \quad \gamma \mapsto \mathfrak{a}(\gamma).$$

We require that both actions a and $\mathfrak{a}$ satisfy the identities:

$$\begin{aligned}\mathfrak{a}(\gamma)(\varphi\psi) &= (\mathfrak{a}(\gamma)\varphi)(\mathfrak{a}(\gamma)\psi), \\ a(\gamma)(\varphi \bullet u) &= (\mathfrak{a}(\gamma)\varphi) \bullet (a(\gamma)u)\end{aligned} \tag{147}$$

for all $\gamma \in G, \varphi, \psi \in E$ and $u \in X$.

Example 21 (Ginzburg-Landau equation in 1D). Reconsider Example 5, i.e.

$$u_t = \alpha u_{xx} + \delta u + \beta |u|^2 u + \gamma |u|^4 u, \quad x \in \mathbb{R}, \, t \geq 0, \tag{148}$$

where $u(x,t) \in \mathbb{C}$. Recall that (148) is equivariant under the action $a : G \to GL(X)$ of the Lie group $G = \mathbb{R} \times S^1 \ni (\tau, \theta)$ on X given by

$$[a(\tau,\theta)u](x) = e^{-i\theta}u(x-\tau), \quad \text{for all } (\tau,\theta) \in G, u \in X.$$

As above choose the Banach space $X = L^2(\mathbb{R},\mathbb{C})$ and the module $E = C^0_{\text{unif}}(\mathbb{R},\mathbb{R})$, which acts on X by multiplication, i.e. $(\varphi \bullet u)(x) = \varphi(x)u(x)$ for all $x \in \mathbb{R}$ and all $\varphi \in E, u \in X$.

The group G acts on the module E via the action $\mathfrak{a} : G \to \text{GL}(E)$ given by

$$[\mathfrak{a}(\tau,\theta)\varphi](x) = \varphi(x-\tau), \quad \text{for all } (\tau,\theta) \in G, \varphi \in E.$$

In this case, (147) follows from

$$\left[a(\tau,\theta)(\varphi \bullet u)\right](x) = \varphi(x-\tau)e^{-i\theta}u(x-\tau), \quad \forall x \in \mathbb{R},$$

for all $\varphi \in C^0_{\text{unif}}(\mathbb{R}), u \in L^2(\mathbb{R},\mathbb{C}), \tau \in \mathbb{R}, \theta \in [0,2\pi)$.

In the abstract framework, the idea of the decomposition (138) is generalized as follows

$$u(t) = \sum_{j=1}^{N} a(g_j(t))v_j(t), \tag{149}$$

where $g_j : [0,\infty) \to G$ denotes the time dependent location in the group G of the j-th profile $v_j : [0,\infty) \to Y$.

Assume an element $\varphi \in E$ such that the inverse of $\sum_j \mathfrak{a}(g_j)\varphi \in E$ with respect to the multiplication in E exists for all $g_1,\ldots,g_N \in G$. We denote this inverse $\left(\sum_j \mathfrak{a}(g_j)\varphi\right)^{-1}$ by $\frac{1}{\sum_j \mathfrak{a}(g_j)\varphi}$. Then a calculation, similar to (140) and (72) yields

$$\sum_{j=1}^{N} a(g_j)\left[v_{j,t} + a(g_j)^{-1}d[a(g_j)v_j]g_{j,t}\right] = u_t = F(u)$$

$$= \sum_{j=1}^{N}\left[F(a(g_j)v_j) + \frac{\mathfrak{a}(g_j)\varphi}{\sum_{k=1}^{N}\mathfrak{a}(g_k)\varphi} \bullet \left(F\left(\sum_{k=1}^{N}a(g_k)v_k\right) - \sum_{k=1}^{N}F(a(g_k)v_k)\right)\right]$$

$$= \sum_{j=1}^{N}a(g_j)\left[F(v_j) + \frac{\varphi}{\sum_{k=1}^{N}\mathfrak{a}(g_j^{-1}g_k)\varphi} \bullet \left(F\left(\sum_{k=1}^{N}a(g_j^{-1}g_k)v_k\right)\right.\right.$$

$$\left.\left. - \sum_{k=1}^{N}F(a(g_j^{-1}g_k)v_k)\right)\right]. \tag{150}$$

As in the derivation of the decompose and freeze method for multi-pulses and multi-fronts in Sect. 5.2, we now require that for each $j = 1,\ldots,N$ the summand on the left hand side and on the right hand side of (150) coincides. This yields the following nonlinear coupled system for the unknowns $v_j \in Y$, $\mu_j \in \mathscr{A}$, and $g_j \in G$, $j = 1, \ldots, N$:

$$v_{j,t} = F(v_j) - d[a(\mathbb{1})v_j]\mu_j + \frac{\varphi}{\sum_k \alpha(g_j^{-1}g_k)\varphi} \bullet \left(F\left(\sum_k a(g_j^{-1}g_k)v_k\right) \right.$$

$$\left. - \sum_k F(a(g_j^{-1}g_k)v_k) \right),$$

$$v_j(0) = v_j^0,$$

$$g_{j,t} = dL_{g_j}(\mathbb{1})\mu_j, \quad g_j(0) = g_j^0,$$

$$0 = \left(v_j - \hat{v}_j, d[a(\mathbb{1})\hat{v}_j]\lambda\right)_H \quad \forall \lambda \in T_\mathbb{1} G. \tag{151}$$

For the second summand in the first line we used the identity $a(g_j)^{-1}d[a(g_j)v_j]$ $g_{j,t} = d[a(\mathbb{1})v_j]\mu_j$, where $\mu_j = dL_{g_j}(g_j)^{-1}g_{j,t} \in T_\mathbb{1} G$.

As in Sect. 5.2 we obtain that a solution of (151) with initial data v_j^0, g_j^0, satisfying $u_0 = \sum_{j=1}^N a(g_j^0)v_j^0$, yields a solution to the original Cauchy problem (146) by setting

$$u(t) = \sum_{j=1}^N a(g_j(t))v_j(t) \quad \text{for all } t \geq 0. \tag{152}$$

Example 22 (Freezing pulse and front simultaneously in Ginzburg-Landau equation). As an example in $1D$ we consider the quintic-cubic Ginzburg-Landau equation in $1D$ from Example 21 again

$$u_t = \alpha u_{xx} + \left(\delta + \beta |u|^2 + \gamma |u|^4\right) u, \quad x \in \mathbb{R}, \ t \geq 0$$

with $u(x, t) \in \mathbb{C}$. The parameter values are $\alpha = 1$, $\delta = -0.1$, $\beta = 3 + i$, and $\gamma = -2.75 + i$. In this case one finds a multi-structure, consisting of a standing rotating pulse v_1 and a rotating front v_2 that travels to the right. In Fig. 42 we show the result obtained by the decompose and freeze method for this problem. One observes that the individual structures are well captured in their respective frames and the single speed correctly reproduced for the single waves, compare Example 5. Note that this is a case of weak interaction. However, for strong interactions, such as the collision of a rotating and a traveling pulse, the decompose and freeze method did not work properly.

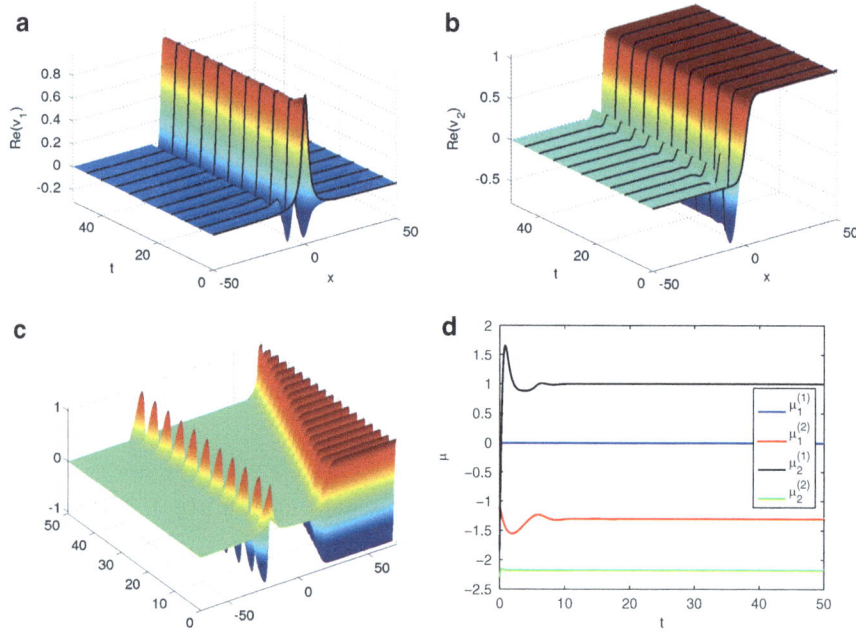

Fig. 42 Result of the decompose and freeze method for a multi-structure in the quintic-cubic Ginzburg-Landau equation, weak interaction of a standing and rotating pulse and a traveling front. Profile of the standing and rotating pulse v_1 (**a**), profile of the rotating and right traveling front v_2 (**b**), simulation of the nonfrozen equation (**c**), time-dependence of the derivatives of the group variables (**d**)

5.5 Multisolitons: Interaction of Spinning Solitons

We finish with numerical results of the method where we try to capture simultaneously two and more solitons in the $2D$ quintic-cubic complex Ginzburg-Landau equation from Example 13.

Example 23 (Quintic-cubic Ginzburg-Landau equation in 2D).

$$u_t = \alpha \Delta u + \delta u + \beta |u|^2 u + \gamma |u|^4 u, \quad (x,y) \in \mathbb{R}^2, \ u(x,y,t) \in \mathbb{C}. \qquad (153)$$

The parameter values are the same as in (115) for which single spinning solitons are known to exist. As initial data we take the sum of two such solitons, shifted a certain distance apart. If this distance is large enough, we have weak interaction and a multi-structure consisting of two (or more) spinning solitons stabilizes. The result of such a simulation is shown in Fig. 43. The first row shows the superposition of the profiles obtained from the decompose and freeze method at different time instances. The next row contains the single profiles v_1 (d) and v_2 (e) and the trace $\{\tau_j(t) : t \geq 0\}$ of the two group orbits $g_j(t) = (\theta_j(t), \tau_j(t)), j = 1,2$ from the reconstruction equation in (151). Figure 43g–i displays the time-dependence of all 6 velocities.

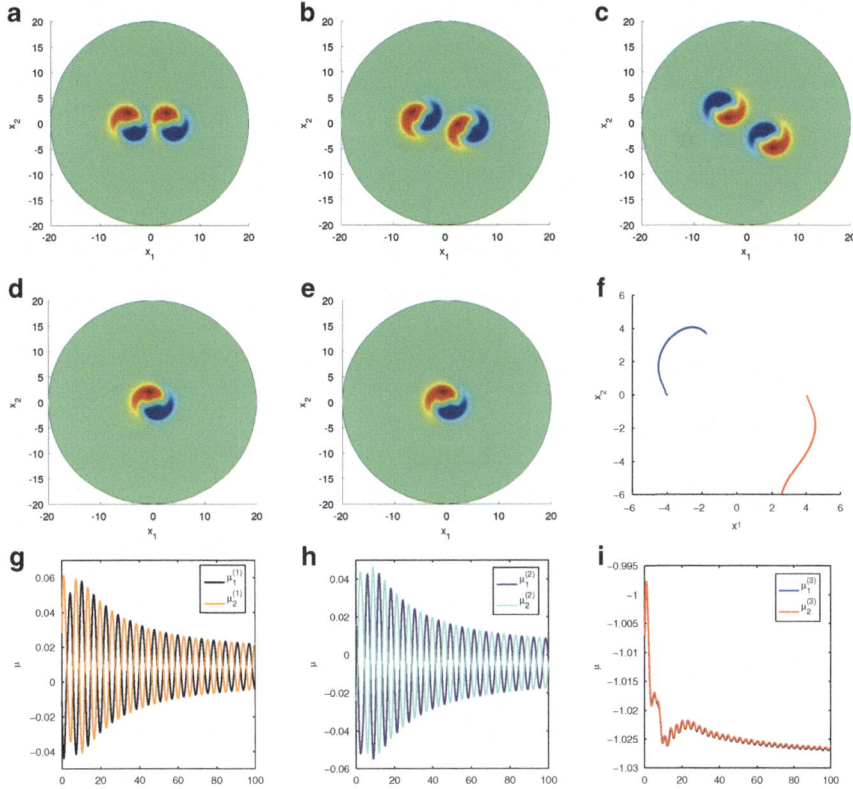

Fig. 43 Weak interaction of two spinning solitons. Initial location of the profiles at $\pm(4,0)$, real part of superposition at time $t = 0, 30, 150$ (**a**)–(**c**), real parts of profiles v_1 and v_2 at time $t = 150$ (**d**), (**e**), position of the centers of the profiles v_1, v_2 from $t = 0$ to $t = 500$ calculated by solving the reconstruction equation (**f**), time evolution of translational velocities $\mu_1^{1/2}(t)$ and $\mu_2^{1/2}(t)$ in x-direction (**g**) and in y-direction (**h**), evolution of angular velocities (**i**). The colorbar is scaled to $[-1.65, 1.65]$. Solution by Comsol Multiphysics with piecewise linear finite elements, Neumann boundary conditions, $\Delta x = 0.5$, $\Delta t = 0.1$, BDF of order 2

The translational velocities μ_j^1, μ_j^2 converge to zero and the angular velocities to their limiting values. However, the convergence is oscillatory and very slow (oscillations become invisible at $t \approx 1000$). Therefore, we show details in the interval $0 \le t \le 100$.

Figure 44 shows a case of strong interaction of two spinning solitons with pictures selected as in Fig. 43. The solution converges to single soliton (c), which is represented by the decompose and freeze method as the superposition of two single but deformed solitons (d), (e). The two group orbits apparently trace a circle, and velocities slowly decay as in the case of weak interaction. (b) (Strong interaction of 2 spinning solitons) Finally, we consider the strong interaction of 3 spinning solitons, see Fig. 45. Initially, the solitons are put on the vertices of

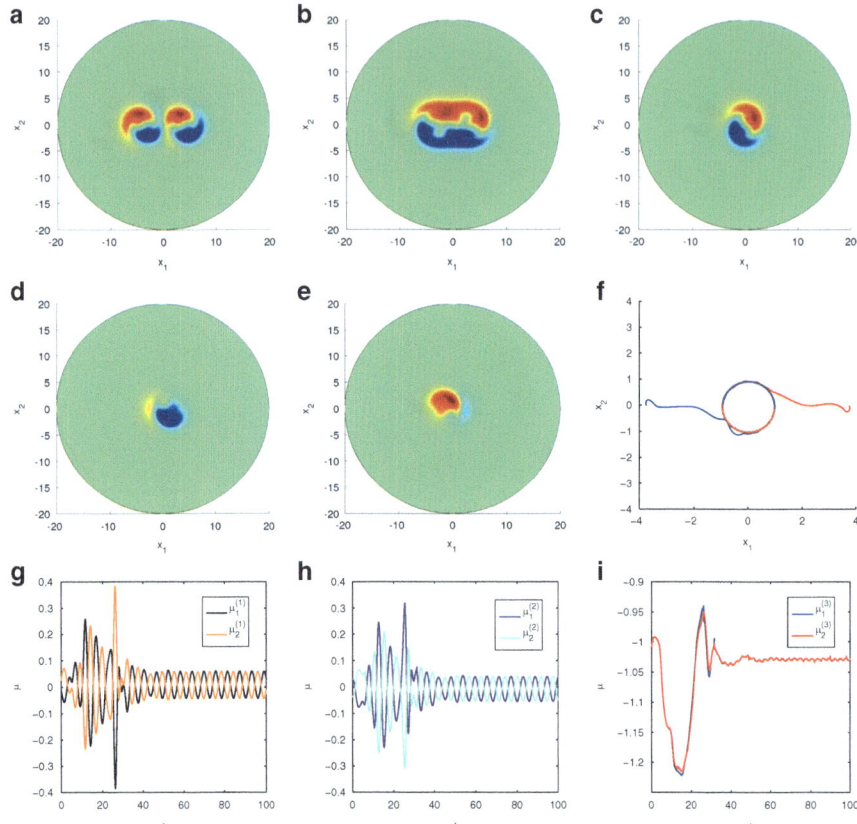

Fig. 44 Strong interaction of two spinning solitons. Initial location of profiles at $\pm(3.75, 0)$, real part of superposition at time $t = 0, 7.2, 36$ (**a**)–(**c**), real part of profiles v_1 and v_2 at time $t = 150$ (**d**), (**e**), position of centers for the profiles v_1, v_2 for $0 \leq t \leq 500$ (**f**), evolution of translational velocities $\mu_1^{1/2}(t)$ and $\mu_2^{1/2}(t)$ in x-direction (**g**) and y-direction (**h**), angular velocities (**i**), further data are as in Fig. 43

an equilateral triangle. The behavior is quite similar to the two-solitons case. Translational velocities oscillate rapidly for a long time before tending to zero, and the traces of the group orbits, after a sharp turn, seem to follow a common circle with different phases.

5.6 Summary

A summary of this section is the following:

- Excitable reaction diffusion systems in 1D show multi-structures composed of fronts and pulses.

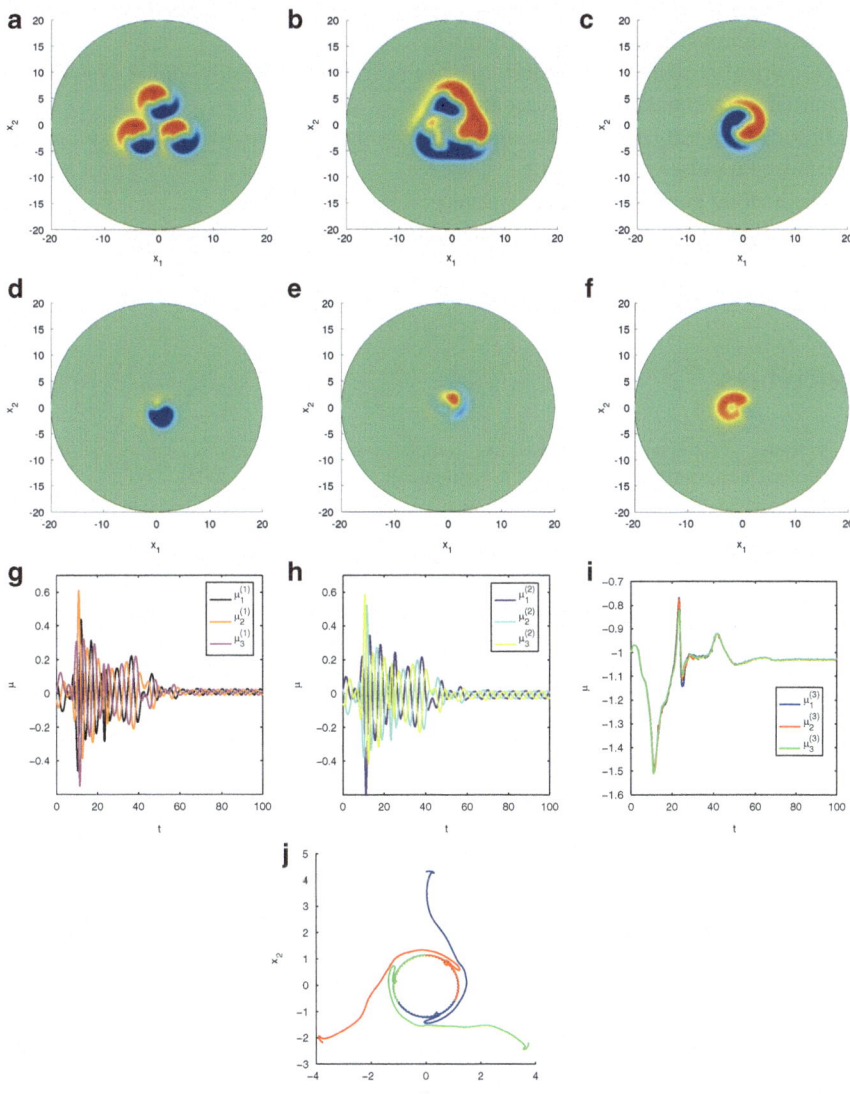

Fig. 45 Strong interaction of three spinning solitons. Centers are put initially on an equilateral triangle with radius of circumcircle 3.75, real part of superposition at times $t = 0$, 7.2, 36 (**a**)–(**c**), real parts of profiles v_1, v_2 and v_3 at time $t = 150$ (**d**)–(**f**), evolution of translational velocities $\mu_1^{1/2}(t)$, $\mu_2^{1/2}(t)$ and $\mu_3^{1/2}(t)$ in x-direction (**g**), y-direction (**h**) and evolution of angular velocities (**i**), reconstruction of the group orbits for the profiles v_1, v_2 and v_3 for $0 \leq t \leq 500$ (**j**). The colorbar is scaled to $[-1.8, 1.8]$ while further data are as in Fig. 43

- The freezing method is extended to a 'decompose and freeze' method to capture solutions consisting of multi-structures.
- Numerical solution of a system of nonlinear and nonlocal coupled systems of partial differential algebraic equations.
- Proof of stability for the decomposition method in case of weakly interacting fronts and pulses.
- The method generalizes to equivariant evolution equations.
- Numerical computations with freezing multi-structures in dimensions ≥ 2 are in initial state, no theory available.

References

1. S. Ahuja, I.G. Kevrekidis, C.W. Rowley, Template-based stabilization of relative equilibria in systems with continuous symmetry. J. Nonlinear Sci. **17**, 109–143 (2007)
2. J. Alexander, R. Gardner, C.K.R.T. Jones, A topological invariant arising in the stability analysis of travelling waves. J. Reine Angew. Math. **410**, 167–212 (1990)
3. H.W. Alt, *Lineare Funktionalanalysis*, 3rd edn. (Springer, Berlin, 1999)
4. W. Arendt, C.J.K. Batty, M. Hieber, F. Neubrander, *Vector-Valued Laplace Transforms and Cauchy Problems*. Monographs in Mathematics, vol. 96 (Birkhäuser, Basel, 2001)
5. J.M. Arrieta, M. López-Fernández, E. Zuazua, Approximating travelling waves by equilibria of nonlocal equations. Asymptot. Anal. **78**, 145–186 (2012)
6. D. Barkley, A model for fast computer simulation of waves in excitable media. Physica D **49**, 61–70 (1991)
7. P.W. Bates, C.K.R.T. Jones, Invariant manifolds for semilinear partial differential equations, in *Dynamics Reported*. Dynam. Report. Ser. Dynam. Systems Appl. vol. 2 (Wiley, Chichester, 1989), pp. 1–38
8. W.-J. Beyn, The numerical computation of connecting orbits in dynamical systems. IMA J. Numer. Anal. **10**(3), 379–405 (1990)
9. W.-J. Beyn, J. Lorenz, Stability of traveling waves: Dichotomies and eigenvalue conditions on finite intervals. Numer. Funct. Anal. Optim. **20**, 201–244 (1999)
10. W.-J. Beyn, J. Lorenz, Nonlinear stability of rotating patterns. Dyn. PDEs **5**, 349–400 (2008)
11. W.-J. Beyn, J. Rottmann-Matthes, Resolvent estimates for boundary value problems on large intervals via the theory of discrete approximations. Numer. Funct. Anal. Optim. **28**(5–6), 603–629 (2007)
12. W.-J. Beyn, V. Thümmler, Freezing solutions of equivariant evolution equations. SIAM J. Appl. Dyn. Syst. **3**(2), 85–116 (2004)
13. W.-J. Beyn, V. Thümmler, Phase conditions, symmetries, and PDE continuation, in *Numerical Continuation Methods for Dynamical Systems*, ed. by B. Krauskopf, H. Osinga, J. Galan-Vioque. Series in Complexity (Springer, Berlin, 2007), pp. 301–330
14. W.-J. Beyn, V. Thümmler, Dynamics of patterns in nonlinear equivariant PDEs. GAMM Mitteilungen **32**(1), 7–25 (2009)
15. W.-J. Beyn, S. Selle, V. Thümmler, Freezing multipulses and multifronts. SIAM J. Appl. Dyn. Syst. **7**, 577–608 (2008)
16. L.Q. Brin, Numerical testing of the stability of viscous shock waves. Math. Comp. **70**, 1071–1088 (2001)
17. L.Q. Brin, K. Zumbrun, Analytically varying eigenvectors and the stability of viscous shock waves. Math. Contemp. **22**, 19–32 (2002)
18. C. Cattaneo, Sulla conduzione del calore. Atti Semin. Mat. Fis. Univ. Modena **3**, 83–101 (1948)

19. P. Chossat, R. Lauterbach, *Methods in Equivariant Bifurcations and Dynamical Systems*. Advanced Series in Nonlinear Dynamics, vol. 15 (World Scientific, Singapore, 2000)
20. W.A. Coppel, *Dichotomies in Stability Theory*. Lecture Notes in Mathematics, vol. 629 (Springer, Berlin, 1978)
21. L.-C. Crasovan, B.A. Malomed, D. Mihalache, Spinning solitons in cubic-quintic nonlinear media. Pramana J. Phys. **57**, 1041–1059 (2001)
22. J.W. Demmel, L. Dieci, M.J. Friedman, Computing connecting orbits via an improved algorithm for continuing invariant subspaces. SIAM J. Sci. Comput. **22**(1), 81–94 (electronic) (2000)
23. K.-J. Engel, R. Nagel, *One-Parameter Semigroups for Linear Evolution Equations*. Graduate Texts in Mathematics (Springer, Berlin, 2000)
24. J.W. Evans, Nerve axon equations. I. Linear approximations. Indiana Univ. Math. J. **21**, 877–885 (1971/1972)
25. J.W. Evans, Nerve axon equations. II. Stability at rest. Indiana Univ. Math. J. **22**, 75–90 (1972/1973)
26. J.W. Evans, Nerve axon equations. III. Stability of the nerve impulse. Indiana Univ. Math. J. **22**, 577–593 (1972/1973)
27. J.W. Evans, Nerve axon equations. IV. The stable and the unstable impulse. Indiana Univ. Math. J. **24**(12), 1169–1190 (1974/1975)
28. M.J. Field, *Dynamics and Symmetry*. ICP Advanced Texts in Mathematics, vol. 3 (Imperial College Press, London, 2007)
29. P.C. Fife, J.B. McLeod, The approach of solutions of nonlinear diffusion equations to travelling front solutions. Arch. Rat. Mech. Anal. **65**, 335–361 (1977)
30. R. FitzHugh, Impulses and physiological states in theoretical models of nerve membrane. Biophys. J. **1**, 445–466 (1961)
31. A.J. Foulkes, V.N. Biktashev, Riding a spiral wave; numerical simulation of spiral waves in a comoving frame of reference. Phys. Rev. E (3) **81**, 046702 (2010)
32. M.J. Friedman, E.J. Doedel, Numerical computation and continuation of invariant manifolds connecting fixed points. SIAM J. Numer. Anal. **28**(3), 789–808 (1991)
33. S. Froehlich, P. Cvitanović, Reduction of continuous symmetries of chaotic flows by the method of slices. Comm. Nonlinear Sci. Numer. Simulat. **17**, 2074–2084 (2012)
34. A. Ghazaryan, Y. Latushkin, S. Schecter, Stability of traveling waves for degenerate systems of reaction diffusion equations. Indiana Univ. Math. J. **60**(2), 443–472 (2011)
35. I.C. Gohberg, M.G. Kreĭn, *Introduction to the Theory of Linear Nonselfadjoint Operators*. Translations of Mathematical Monographs, vol. 18 (American Mathematical Society, Providence, 1969)
36. M. Golubitsky, I. Stewart, *The Symmetry Perspective*. Progress in Mathematics, vol. 20 (Birkhäuser, Basel, 2002)
37. S. Hastings, On travelling wave solutions of the Hodgkin-Huxley equations. Arch. Ration. Mech. Anal. **60**, 229–257 (1976)
38. D. Henry, *Geometric Theory of Semilinear Parabolic Equations*. Lecture Notes in Mathematics, vol. 840 (Springer, Berlin, 1981)
39. S. Hermann, G.A. Gottwald, The large core limit of spiral waves in excitable media: A numerical approach. SIAM J. Appl. Dyn. Syst. **9**, 536–567 (2010)
40. A.L. Hodgkin, A.F. Huxley, A quantitative description of membrane current and its application to conduction and excitation in nerve. J. Physiol. **117**, 500–544 (1952)
41. J. Humpherys, K. Zumbrun, An efficient shooting algorithm for Evans function calculations in large systems. Physica D **220**, 116–126 (2006)
42. J. Humpherys, B. Sandstede, K. Zumbrun, Efficient computation of analytic bases in Evans function analysis of large systems. Numer. Math. **103**, 631–642 (2006)
43. J. Keener, J. Sneyd, *Mathematical Physiology. I: Cellular Physiology*, 2nd edn. (Springer, New York, 2009)
44. Y.J. Kim, A.E. Tzavaras, Diffusive N-waves and metastability in the Burgers equation. SIAM J. Math. Anal. **33**(3), 607–633 (electronic) (2001)

45. G. Kreiss, H.O. Kreiss, N.A. Petersson, On the convergence of solutions of nonlinear hyperbolic-parabolic systems. SIAM J. Numer. Anal. **31**(6), 1577–1604 (1994)
46. V. Ledoux, S. Malham, V. Thümmler, Grassmannian spectral shooting. Math. Comp. **79**, 1585–1619 (2010)
47. S. Malham, J. Niesen, Evaluating the Evans function: Order reduction in numerical methods. Math. Comp. **261**, 159–179 (2008)
48. R.M. Miura, Accurate computation of the stable solitary waves for the FitzHugh-Nagumo equations. J. Math. Biol. **13**, 247–269 (1982)
49. K.J. Palmer, Exponential dichotomies and transversal homoclinic points. J. Differ. Equat. **55**(2), 225–256 (1984)
50. A. Pazy, *Semigroups of Linear Operators and Applications to Partial Differential Equations (2. corr. print)*. Applied Mathematical Sciences, vol. 44 (Springer, Berlin, 1983)
51. R.L. Pego, M.I. Weinstein, Eigenvalues, and instabilities of solitary waves. Philos. Trans. Roy. Soc. Lond. Ser. A **340**, 47–94 (1992)
52. J. Rottmann-Matthes, Computation and Stability of Patterns in Hyperbolic-Parabolic Systems. PhD thesis, Shaker Verlag, Aachen (2010)
53. J. Rottmann-Matthes, Linear stability of traveling waves in first-order hyperbolic PDEs. J. Dyn. Differ. Equat. **23**(2), 365–393 (2011)
54. J. Rottmann-Matthes, Stability and freezing of nonlinear waves in first order hyperbolic PDEs. J. Dyn. Differ. Equat. **24**(2), 341–367 (2012)
55. J. Rottmann-Matthes, Stability and freezing of waves in non-linear hyperbolic-parabolic systems. IMA J. Appl. Math. **77**(3), 420–429 (2012)
56. J. Rottmann-Matthes, Stability of parabolic-hyperbolic traveling waves. Dyn. Part. Differ. Equat. **9**(1), 29–62 (2012)
57. C.W. Rowley, I.G. Kevrekidis, J.E. Marsden, K. Lust, Reduction and reconstruction for self-similar dynamical systems. Nonlinearity **16**(4), 1257–1275 (2003)
58. K.M. Saad, A.M. El-shrae, Numerical methods for computing the Evans function. ANZIAM J. Electron. Suppl. **52** (E), E76–E99 (2010)
59. B. Sandstede, Stability of traveling waves, in *Handbook of Dynamical Systems*, ed. by B. Fiedler, vol. 2 (North Holland, Amsterdam, 2002), pp. 983–1055
60. B. Sandstede, A. Scheel, Absolute and convective instabilities of waves on unbounded and large bounded domains. Physica D **145**(3–4), 233–277 (2000)
61. B. Sandstede, A. Scheel, C. Wulff, Dynamics of spiral waves on unbounded domains using center manifold reductions. J. Differ. Equat. **141**, 122–149 (1997)
62. A. Scheel, J.D. Wright, Colliding dissipative pulses – the shooting manifold. J. Differ. Equat. **245**(1), 59–79 (2008)
63. S. Selle, Decomposition and Stability of Multifronts and Multipulses. PhD thesis, University of Bielefeld, Bielefeld (2009)
64. G.W. Stewart, J.G. Sun, *Matrix Perturbation Theory*. Computer Science and Scientific Computing (Academic, Boston, 1990)
65. V. Thümmler, Numerical Analysis of the Method of Freezing Traveling Waves. PhD thesis, Bielefeld University (2005)
66. V. Thümmler, Numerical approximation of relative equilibria for equivariant PDEs. SIAM J. Numer. Anal. **46**, 2978–3005 (2008)
67. V. Thümmler, The effect of freezing and discretization to the asymptotic stability of relative equilibria. J. Dyn. Differ. Equat. **20**, 425–477 (2008)
68. A. Volpert, V.A. Volpert, V.A. Volpert, *Traveling Wave Solutions of Parabolic Systems*. Translations of Mathematical Monographs, vol. 140 (AMS, Providence, 1994)
69. J.D. Wright, Separating dissipative pulses: The exit manifold. J. Dyn. Differ. Equat. **21**(2), 315–328 (2009)

Continuous Decompositions and Coalescing Eigenvalues for Matrices Depending on Parameters

Luca Dieci, Alessandra Papini, Alessandro Pugliese, and Alessandro Spadoni

Abstract This contribution is a summary of four lectures delivered by the first author at the CIME Summer school in June 2011 at Cetraro (Italy). Preparation of those lectures was greatly aided by the other authors of these lecture notes.

Our goal is to present some classical, as well as some new, results related to decompositions of matrices depending on one or more parameters, with particular emphasis being paid to the case of coalescing eigenvalues (or singular values) for matrices depending on two or three parameters. There is an extensive literature on this subject, but a systematic collection of relevant results is lacking, and this provided the impetus for writing the lecture notes.

During the last 15 years, Dieci has had several collaborators on the topics under scrutiny. Besides the coauthors of these lectures, the collaboration with the following people is gratefully acknowledged: Timo Eirola (Helsinki University of Technology, Finland), Jann-Long Chern (National Central University, Taiwan), Mark Friedman (University of Alabama, Huntsville) and Maria Grazia Gasparo (University of Florence, Italy).

L. Dieci (✉)
School of Mathematics, Georgia Institute of Technology, Atlanta, GA 30332, USA
e-mail: dieci@math.gatech.edu

A. Papini
Dip. di Ingegneria Industriale, Università di Firenze, Florence, Italy
e-mail: alessandra.papini@unifi.it

A. Pugliese
Dip. di Matematica, Università degli studi di Bari "Aldo Moro", Bari, Italy
e-mail: alessandro.pugliese@uniba.it

A. Spadoni
Laboratory of Wave Mechanics and Multi-field Interactions, EPFL, Lausanne, Switzerland
e-mail: alex.spadoni@epfl.ch

Notation. The conjugate transpose (Hermitian) of a matrix is written as $A^* = \bar{A}^T$. A matrix $U \in \mathbb{C}^{n \times n}$ is called unitary if $U^*U = I$, whereas a matrix $Q \in \mathbb{R}^{n \times n}$ is called orthogonal if $Q^T Q = I$; we will use the term orthonormal to identify a matrix (real or complex valued) whose columns are orthonormal. With $\sigma(A)$ we will indicate the set of eigenvalues of a matrix A (repeated by their multiplicities).

A matrix valued function smoothly depending on a real parameter will be written as $A \in \mathscr{C}^k(\mathbb{R}, \mathbb{R}^{n \times n})$ (or $A \in \mathscr{C}^k(\mathbb{R}, \mathbb{C}^{n \times n})$, and similarly $A \in \mathscr{C}^k(\mathbb{R}, \mathbb{R}^{m \times n})$, etc.); here, $k \geq 0$ and typically $k > 1$. Further, if A is periodic of minimal period τ, we will write $A \in \mathscr{C}_\tau^k(\mathbb{R}, \mathbb{C}^{m \times n})$, etc. Ω will indicate a region of $\mathbb{R}^2$ or of $\mathbb{R}^3$, diffeomorphic to the unit ball, and we will write $A \in \mathscr{C}^k(\Omega, \mathbb{C}^{n \times n})$ for a $\mathscr{C}^k$ matrix valued function on Ω, $k \geq 0$, or also $A \in \mathscr{C}^\omega$. Coordinates in Ω will be given as $x = (x_1, x_2)$, or $x = (x_2, x_2, x_3)$ as appropriate. $\mathbb{S}^2$ will always be the unit 2-sphere, the boundary of the unit ball in $\mathbb{R}^3$.

1 Background and Motivation

In this chapter, we lay down the questions examined in these notes and present some applications which have motivated our study.

1.1 Familiar Linear Algebra Results

The spectral theorem (also called Schur decomposition) for symmetric (in $\mathbb{R}$) or Hermitian (in $\mathbb{C}$) matrices, and the singular value decomposition (SVD) of a general matrix, are certainly two of the best known and most important results in linear algebra. Let us recall them (e.g., see [1]).

Theorem 1.1 (Spectral Theorem).

($\mathbb{R}$) *Given $A \in \mathbb{R}^{n \times n}$, $A = A^T$, then there exists orthogonal $Q \in \mathbb{R}^{n \times n}$: $Q^T A Q = \Lambda$, $Q^T Q = I$, $\Lambda = \mathrm{diag}(\lambda_1, \ldots, \lambda_n)$. The eigenvalues can be assumed to be ordered, say: $\lambda_1 \geq \cdots \geq \lambda_n$.*

($\mathbb{C}$) *Given $A \in \mathbb{C}^{n \times n}$, $A = A^*$, then there exists unitary $Q \in \mathbb{C}^{n \times n}$: $Q^* A Q = \Lambda$, $Q^* Q = I$, $\Lambda = \mathrm{diag}(\lambda_1, \ldots, \lambda_n)$, $\lambda_i \in \mathbb{R}$. The eigenvalues can be taken ordered, say: $\lambda_1 \geq \cdots \geq \lambda_n$.*

Theorem 1.2 (SVD, Real Case).

(SVD-1) *(Full SVD) Given $A \in \mathbb{R}^{m \times n}$, $m \geq n$, then there exist orthogonal $U \in \mathbb{R}^{m \times m}$, $V \in \mathbb{R}^{n \times n}$, and $\Sigma \in \mathbb{R}^{m \times n}$: $U^T A V = \Sigma$, $\Sigma = \begin{bmatrix} S \\ 0 \end{bmatrix}$, $S = \mathrm{diag}(\sigma_1, \ldots, \sigma_n)$. The usual convention is that the singular values are nonnegative and ordered: $\sigma_1 \geq \cdots \geq \sigma_n \geq 0$.*

(SVD-2) *(Reduced SVD) Given $A \in \mathbb{R}^{m \times n}$, $m \geq n$, then there exist orthonormal $U \in \mathbb{R}^{m \times n}$, $U^T U = I_n$, orthogonal $V \in \mathbb{R}^{n \times n}$, and $\Sigma \in \mathbb{R}^{n \times n}$: $U^T A V = \Sigma$, $\Sigma = \mathrm{diag}(\sigma_1, \ldots, \sigma_n)$. As before, $\sigma_1 \geq \cdots \geq \sigma_n \geq 0$.*

Similarly, for the SVD of $A \in \mathbb{C}^{m \times n}$, where now U and V are complex valued, V is unitary, and U is unitary (SVD-1) or simply orthonormal (SVD-2).

Remark 1.3. In Theorem 1.2, U gives the Schur form for AA^T and V gives the Schur form for $A^T A$; AA^T and $A^T A$ share n eigenvalues: $\sigma_1^2, \ldots, \sigma_n^2$.

An important question, relative to the decompositions of Theorems 1.1 and 1.2, is: "To what extent are the decompositions unique?" For example, what about uniqueness of Q in the Spectral Theorem when all eigenvalues are distinct? Below, we summarize the relevant results about uniqueness in a form which will be useful later. The proof of Theorem 1.4 is left as an exercise.

Theorem 1.4 (Uniqueness). *For the spectral decomposition of a symmetric/Hermitian matrix, we have:*

($\mathbb{R}$) *If $\lambda_1 > \lambda_2 > \cdots > \lambda_n$, then Q is unique up to changes of signs of its columns:*

$$Q \equiv Q \begin{bmatrix} \pm 1 & & \\ & \ddots & \\ & & \pm 1 \end{bmatrix}.$$

($\mathbb{C}$) *If $\lambda_1 > \cdots > \lambda_n$, then Q is unique up to a **phase matrix**: $Q \equiv Q\Phi$, $\Phi = \mathrm{diag}(e^{i\phi_j}, \phi_j \in \mathbb{R}, j = 1, \ldots, n)$.*

For the SVD, we have similar results.

(SVD-2) *For the reduced SVD, with distinct singular values, $\sigma_1 > \cdots > \sigma_n \geq 0$, in the real case U and V are unique up to joint changes of signs of their columns: $U \equiv UD$, $V \equiv VD$, $D - \mathrm{diag}(\pm 1)$. In the complex case, U and V are unique up to a **phase matrix**: $U \equiv U\Phi$, $V \equiv V\Phi$, $\Phi = \mathrm{diag}(e^{i\phi_j}, \phi_j \in \mathbb{R}, j = 1, \ldots, n)$.*

(SVD-1) *For the full SVD, even with distinct and positive singular values, $\sigma_1 > \cdots > \sigma_n > 0$, writing $U = [U_1 \ U_2]$, U_1 and V have the same degree of uniqueness identified in (SVD-2). However, we must now account for extra freedom in the complementary part of U_1; we have $U_2 \rightarrow U_2 Q$, where Q is $\mathbb{R}^{m-n, m-n}$ (or $\mathbb{C}^{m-n, m-n}$) and is orthogonal (unitary).*

In all cases, if the eigenvalues (singular values) are not distinct, then there is a more severe lack of uniqueness. For example, in the case of the spectral theorem for symmetric matrices, if we have $\lambda_1 = \lambda_2 > \lambda_3 > \cdots > \lambda_n$, then the first two columns of Q can be modified with any 2×2 orthogonal matrix $\hat{Q}$, namely

$$Q \rightarrow Q \begin{bmatrix} \hat{Q} & 0 \\ 0 & I_{n-2} \end{bmatrix}.$$

1.2 Our Concern

The degree of uniqueness (or lack thereof) is indicative of whether or not we may be able to enforce some smooth variation of the decompositions in case we work with matrix valued functions, rather than matrices. Indeed, one of our concerns is precisely to understand the degree of smoothness of the eigenspaces (or singular spaces) when we have matrix valued functions, rather than just matrices. Moreover, we will want to distinguish between generic properties (that is, what we should expect to have) and non-generic ones. Typical questions follow, relatively to the spectral decomposition, with similar questions relatively to the SVD.

1. Suppose we have a function A, symmetric, depending on a real parameter. Suppose further that A has some degree of smoothness, say $A \in \mathscr{C}^k(\mathbb{R}, \mathbb{R}^{n \times n}))$. Can we say that we can choose also the eigenvalues and eigenvectors to be $\mathscr{C}^k$? When can we (not)?
2. Further, suppose that we have a symmetric function $A \in \mathscr{C}^k(\mathbb{R}, \mathbb{R}^{n \times n})$ which is also periodic in the parameter. When can we choose the eigenspaces to be periodic? With same period or different period? And, if there is a smooth (perhaps periodic) path, how can we compute it?
3. What if the symmetric function A now depends on two parameters? Are there new phenomena occurring?
4. What if $A \in \mathscr{C}^k(\mathbb{R}, \mathbb{C}^{n \times n})$ and is Hermitian? Does something change? What about if A depends on two or three parameters?

In these notes we will address the above questions and qualify what can be said and when. We will use concepts (and motivation) from differential geometry, dynamical systems, continuation techniques, as well as matrix theory and perturbation theory. As reference for general geometrical concepts (genericity, transversality), see [2, 3]. As general references for concepts and motivation from dynamical systems and theory of continuation, see [4–9]. As general references on matrix theory, see [1, 10–12]. As references of works specifically concerned with theory and perspectives on coalescing eigenvalues, see [13–16]. For further reading and computational approaches to 1-d continuation, see also [17–21].

For a representative (if biased) collection of works concerned with smoothness of decompositions and their application (1-parameter case), we refer to [22–29] for theoretical developments. For applications and motivations for functions of one parameter, see [30–37], as well as the following works more specifically concerned with the SVD [38–44].

For the case of two parameters, we refer to [45–47] and references there. For three parameters, we have used [48, 49]. For other general multiparameter studies, please refer to [50, 51] for important theoretical works, and to [52–54] for a sample of multi-parameter continuation techniques.

Also, there is an extensive literature in the Physics and Chemical Physics literature, ultimately related with matrices depending on two and three parameters, given the intimate connection of coalescing eigenvalues of symmetric (Hermitian) matrices and potential energy surfaces; see [55–60], where [55, 56] are truly pioneering works on the subject. Further, there are also several relevant works connected to random matrix models; e.g., see [61–63].

Finally, the problem is of interest in Mechanical Engineering given its relation to wave propagation; e.g., see [64–67].

1.2.1 Outline of Lecture Notes

In the next part of this first chapter, we look at some applications and motivating examples. In Chap. 2, we consider the 1-parameter case, give some general smoothness results on spectral decompositions and SVDs, some periodicity results, and present an algorithm for continuation of an SVD path. In Chap. 3, we consider the 2-parameter case for the symmetric eigenvalue problem and the SVD. In this case, we discuss genericity and coalescing points and present algorithmic techniques for localizing generic coalescing points. Finally, in Chap. 4 we look at the Hermitian eigenvalue problem in the 3-parameter case. Again we consider coalescing points and highlight the importance of the so-called geometric (Berry) phase. Algorithmic developments and examples conclude the chapter and these notes.

1.3 Some Applications

There are many situations where it is convenient (indeed necessary) to make use of the underlying smooth dependence on parameters in order to devise appropriate numerical methods and/or to draw inferences on what one observes. A rich class of problems where matrices depending on parameters arise is in dynamical system studies, and we will look at the problem of fold bifurcation in Chap. 2.

Here below, we look at two prototypical situations: (a) How to set up smooth boundary conditions for connecting orbits computation, and (b) A classic engineering structure made up of coupled oscillators or pistons engine, as toy models of a vibrating beam.

1.3.1 Some Problems from Dynamical Systems

In studies of dynamical systems, there are many instances where parameter dependent matrices arise. The most common instance is in stability studies for dynamical systems (differential equations) that depend on parameters, say when continuing paths of equilibria, or periodic orbits, where one is often also interested in building smooth projections onto specific subspaces, say onto the dominant eigenspace. But there are many other instances where the problem arise. For example, in studies of differential algebraic equations (DAEs), where the problem is connected to having manifold constraints (or simply index theory), but also in continuation techniques and when one needs to build smooth bases.

Remark 1.5. In these notes, we will always assume that we have the matrix to be decomposed. For example, this could be the Jacobian along a (computed) trajectory, though we notice that there is also interest in the case where the matrix to be decomposed is not explicitly known (such is the case for the fundamental matrix solution of a linearized system), but we do not consider this situation.

1.3.2 Building Smooth Bases

Suppose we have a smooth full rank matrix valued function $A \in \mathscr{C}^k(\mathbb{R}, \mathbb{R}^{m \times n})$, $m > n$, and need a smooth orthonormal basis for $\mathscr{R}(A)^\perp$. Below are two instances where this problem arises.

Example 1.6 (Moving coordinate system). Let

$$\Gamma_0 = \{u \in \mathbb{R}^n : u = g(\theta, \lambda_0) \ \theta \in [a, b]\}$$

be an invariant curve of a dynamical system $\dot{x} = f(x, \lambda)$ for a given value λ_0. If the curve is hyperbolic, then for λ close to λ_0, the system has an invariant curve Γ. [For some details on this problem, see [5]; a similar situation occurs also for invariant tori; e.g., see [68, 69].]

To exploit closeness of Γ to Γ_0, we parametrize all points $x \in \mathbb{R}^n$ close to Γ_0 as functions of θ, in terms of corrections from the old Γ_0:

$$x = g(\theta, \lambda_0) + \big(c_1(\theta)b_1(\theta) + \cdots + c_{n-1}(\theta)b_{n-1}(\theta)\big).$$

Here, $\{b_1, \ldots, b_{n-1}\}$ span the normal space to Γ_0. In other words, we would form $A(\theta) = g_\theta$ which spans the tangent space, and we want an orthonormal representation for the orthogonal complement of $A(\theta)$.

Example 1.7 (Null space problem). Frequently, one needs to solve a constrained minimization problem

$$\min_{x \in \mathbb{R}^n} F(x), \ \text{s.t.} \ Bx = b, \ B \in \mathbb{R}^{n \times m}, n < m,$$

where B is full rank (all constraints are active). Any feasible x must be of the form $x = x_p + y$, where x_p is a particular solution of the constraints, and $y \in \mathscr{N}(B)$.

When the constraints are parameter dependent, it is desirable to be able to obtain a smooth representation for $\mathscr{N}(B)$, so that solutions in the null space can be continued exploiting closeness to each other for small variation of the parameter(s).

We observe that Examples 1.6 and 1.7 are the same problem.

Now, a textbook solution for Example 1.7 proceeds as follows. Consider the QR factorizations of B^T, $B^T = QR$, obtained by using Householder matrices; then,

$$B^T = QR = [Q_1 \ Q_2] \begin{bmatrix} R_1 \\ 0 \end{bmatrix}, \quad \text{and thus} \quad Q_2 = <\mathcal{N}(B)> .$$

The caveat is that this technique does not usually give a smooth basis for $\mathcal{N}(B)$; e.g., see [70].

An instance similar to the previous two examples arises in the context of computing connecting orbits.

Example 1.8 (Heteroclinic connections and continuation of invariant subspaces). We have the system

$$\dot{x} = f(x, \lambda), \quad (x, \lambda) \in (\mathbb{R}^n, \mathbb{R}^p)$$

and let x_- and x_+ be hyperbolic fixed points; that is, the eigenvalues of the Jacobian f_x evaluated at $x_\pm$, call these $f_x^\pm$, are not on the imaginary axis. Let E_-^u, E_+^s, be unstable, stable, subspaces of f_x^-, f_x^+, of dimensions n_-, n_+.

It is well understood (e.g., see [71]) that if $p = n - n_- - n_+ + 2$, and the unstable and stable manifolds of x_- and x_+ intersect transversally, then there is a branch of connecting orbits. We want to approximate these (heteroclinic) orbits numerically.

The most successful approach goes through a reformulation as a two-point boundary value problem (TPBVP):

(a) Truncate time: $T_- \leq t \leq T_+$;
(b) Prescribe appropriate boundary conditions (BCs) at T_- and T_+ and solve the resulting TPBVP.

The key issue is how to prescribe the BCs. Ever since the original work of Beyn (again, see [71]), the most successful mean to provide BCs goes through the following *geometrical idea*:

(i) We know that a heteroclinic connection must leave x_- along the tangent plane to its unstable manifold

$$x(T_-) - x_- \in E_-^u .$$

(ii) We know that the connection must enter x_+ along the tangent plane to its stable manifold

$$x(T_+) - x_+ \in E_+^s .$$

(iii) Therefore, to enforce (i)–(ii), the idea is to use so-called *projected* BCs

$$P_-^s(x(T_-) - x_-) = 0, \quad P_+^u(x(T_+) - x_+) = 0,$$

where P_-^s is the orthogonal projection onto the stable subspace of x_- (similarly for P_+^u).

The present concern is how to build the above smooth projectors. Note that smoothness is needed surely for the theory, but also because otherwise numerical procedures may experience convergence difficulties.

To clarify, let the branch of connecting orbits be parametrized by τ (this could be arc length): $x_\pm(\tau)$, $\lambda(\tau)$, and let

$$A(\tau) = f_x^-(\tau)$$

(similarly for $f_x^+(\tau)$). We want a smooth projector onto $E_-^s(\tau)$.

If we had a smooth function Q such that (for all τ of interest)

$$A(\tau) = Q(\tau) T(\tau) Q^T(\tau), \quad Q(\tau) = [Q_1(\tau), Q_2(\tau)]$$

$$T(\tau) = \begin{bmatrix} T_{11}(\tau) & T_{12}(\tau) \\ 0 & T_{22}(\tau) \end{bmatrix}, \quad \sigma(T_{11}) \in \mathbb{C}^-, \quad \sigma(T_{22}) \in \mathbb{C}^+,$$

then we could use the projection

$$P_-^s(\tau) = Q_1(\tau) Q_1^T(\tau).$$

This problem fits into the framework of a smooth block Schur decompositions, which will be reviewed in Chap. 2. Presently, we remark that several algorithmic studies have clearly shown that techniques which exploit the underlying smoothness lead to better results than traditional linear algebra techniques; e.g., see [33, 34] and see also [35, 37, 72] for issues arising for large problems. Finally, see [36] for extensions to computation and continuation of equilibrium-to-periodic and periodic-to-periodic connections.

1.3.3 Coupled Oscillators

An important class of problems giving symmetric (Hermitian) parameter dependent matrices is from structural engineering.

In many situations of interest (see [73]) a typical structure (say, a beam) is modeled by a finite element discretization and this will lead to a system of second order differential equations of the type $M\ddot{x} + C\dot{x} + Kx = F$, where K and M are often positive semi-definite (M is typically invertible and often it is diagonal). Of particular interest is to understand how the structure behaves in response to a certain excitation, and the free modes of vibrations of the system need to be analyzed. It is particularly revealing to consider the case of no damping, $C = 0$, which serves also as a good approximation to the small damping case.

Fig. 1 System of coupled oscillators

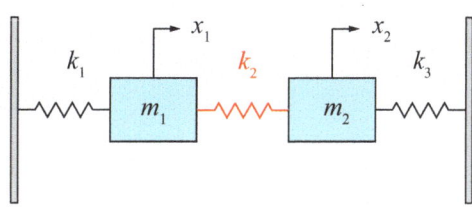

Fig. 2 Forces acting on the first mass

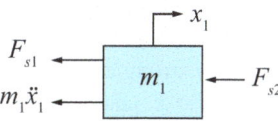

As a prototype for this situation, we consider a simple model consisting of coupled oscillators, see Fig. 1. The structure of the problem is exemplified by Fig. 2, whereby the forces acting on the first mass are shown.

As a consequence of this type of structure, force balance and Hooke's law ($F = -kx$) render the following system of homogeneous coupled equations for the system of Fig. 1:

$$m_1 \ddot{x}_1 + (k_1 + k_2)x_1 - k_2 x_2 = 0,$$
$$m_2 \ddot{x}_2 + (k_2 + k_3)x_2 - k_2 x_1 = 0, \tag{1}$$

where k_2 is the coupling variable, or more compactly as the system

$$\begin{bmatrix} m_1 & 0 \\ 0 & m_2 \end{bmatrix} \begin{bmatrix} \ddot{x}_1 \\ \ddot{x}_2 \end{bmatrix} + \begin{bmatrix} k_1 + k_2 & -k_2 \\ -k_2 & k_2 + k_3 \end{bmatrix} \begin{bmatrix} x_1 \\ x_2 \end{bmatrix} = 0,$$

which by the ansatz $\begin{bmatrix} x_1 \\ x_2 \end{bmatrix} = e^{i\omega t} \begin{bmatrix} v_1 \\ v_2 \end{bmatrix}$ gives the eigenvalue problem

$$\left(-\omega^2 \begin{bmatrix} m_1 & 0 \\ 0 & m_2 \end{bmatrix} + \begin{bmatrix} k_1 + k_2 & -k_2 \\ -k_2 & k_2 + k_3 \end{bmatrix} \right) \begin{bmatrix} v_1 \\ v_2 \end{bmatrix} = 0.$$

It is of interest to study the eigenvalues of this problem as the various parameters change, in particular to explore if and when the eigenvalues coincide. In Fig. 3 we show the simplified structure afforded in the limiting case of no coupling ($k_2 = 0$). In this case, it is rather trivial to solve the system explicitly and obtain:

$$x_1 = A_1 \cos(\omega_1 t) + B_1 \sin(\omega_1 t), \quad \omega_1^2 = k_1/m_1,$$
$$x_2 = A_2 \cos(\omega_2 t) + B_2 \sin(\omega_2 t), \quad \omega_2^2 = k_3/m_2.$$

Fig. 3 Uncoupled system: $k_2 = 0$

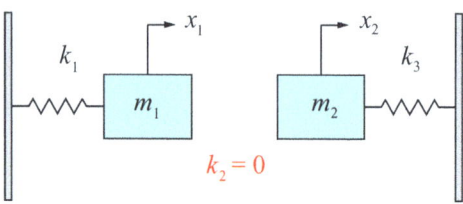

In a similar way, in Fig. 4 we show the structure in the limiting case of infinitely strong coupling ($k_2 = +\infty$). In this case, we really have only one mass, $x_1 = x_2 \equiv x$, and it is again trivial to solve the system explicitly:

$$x = A\cos(\omega t) + B\sin(\omega t), \quad \omega^2 = \frac{k_1 + k_3}{2(m_1 + m_2)}.$$

In the uncoupled case, since the eigenfrequencies (the values ω_1 and ω_2) do not influence each other, it is entirely possible that they will become equal by varying just one of the various parameters; of course, in the rigid case there is only one frequency. We illustrate in Fig. 5, obtained varying just k_3.

Much more interesting is the case when there is a non extreme value of the coupling term k_2. When the coupling is strong, then the eigenfrequencies stay away from each other; see left figure below. However, when the coupling is weak, then the eigenfrequencies approach each other and seem to coincide for some parameter value, though effectively they do not: they get close, but then steer away from each other; see right figure below. This phenomenon is known as *veering*.

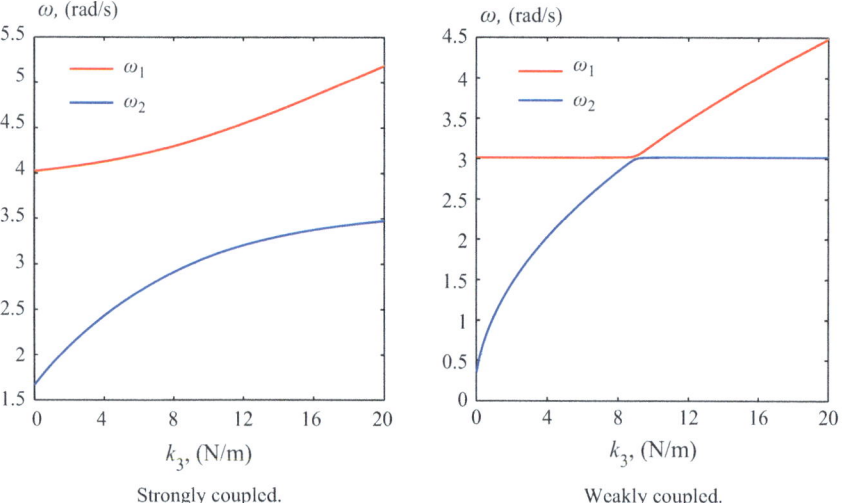

Strongly coupled.   Weakly coupled.

The most intriguing aspect of the behavior of a weakly coupled system is associated to the eigenvectors. In Fig. 6, we show the two components (labeled

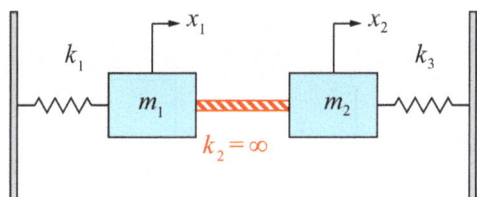

Fig. 4 Rigid system: $k_2 \to \infty$

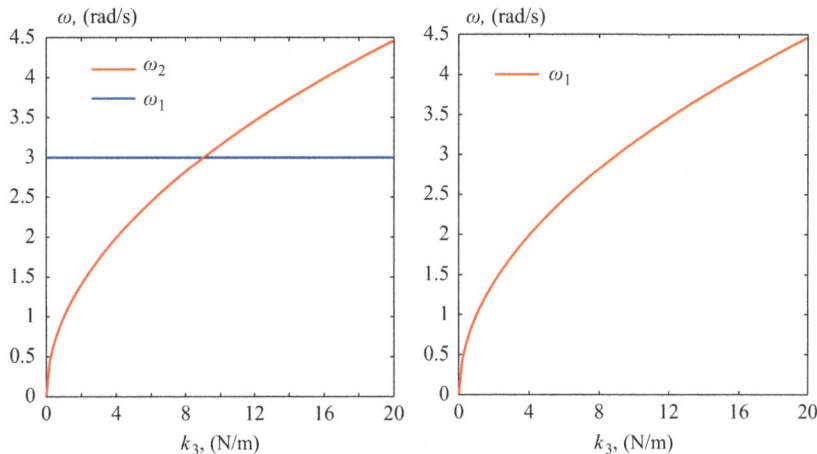

Fig. 5 Uncoupled (*left*) and rigid (*right*) cases

v_1 and v_2) of the two eigenvectors associated to the two (distinct) eigenvalues in this weakly coupled case. As it can be observed, the components of the eigenvectors seem to undergo an exchange with each other through a veering point; this is a so-called *mode exchange*.

There is actually nothing particularly special about having a system made up by just two classical mass-spring oscillators. The same type of system is obtained considering a classical type of a piston-engine as in Fig. 7.

The system of differential equations again rewrites much like in (1):

$$m_1 \ddot{y}_1 + (k_1 + k_{c_1})y_1 - k_{c_1} y_2 = 0,$$
$$m_2 \ddot{y}_2 + (k_2 + k_{c_1})y_2 - k_{c_1} y_1 = 0,$$

where now k_{c_1} is the coupling variable. But, more importantly, there is nothing special about having just two pistons. A very similar situation is encountered when we couple several pistons (oscillators) together. For example, consider the piston-engine system with N pistons of Fig. 8.

Fig. 6 Eigenvectors components associated to veering phenomenon

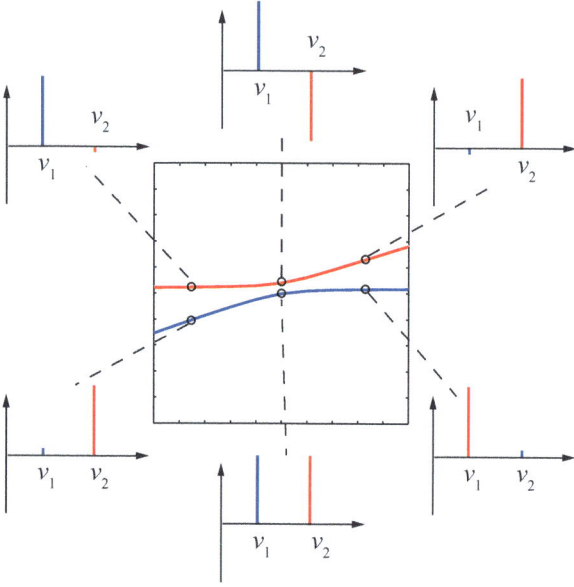

Fig. 7 Two pistons engine

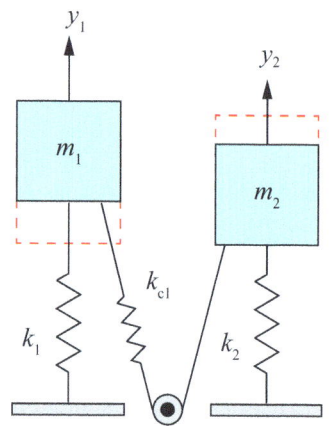

Taking (for simplicity) identical masses all equal to 1, this configuration gives the following symmetric, tridiagonal, positive definite system:

$$\ddot{y} + \begin{bmatrix} k_1 + k_{c_1} & -k_{c_1} & 0 & \cdots & \\ -k_{c_1} & k_2 + k_{c_1} + k_{c_2} & -k_{c_2} & & \ddots \\ 0 & -k_{c_2} & k_3 + k_{c_2} + k_{c_3} & \ddots & \\ & & \ddots & & -k_{c_{N-1}} \\ & & & -k_{c_{N-1}} & k_N + k_{c_{N-1}} \end{bmatrix} y = 0, \ y = \begin{bmatrix} y_1 \\ \vdots \\ y_N \end{bmatrix},$$

which—as before—leads to the eigenvalue problem $(A - \omega^2 I)v = 0$.

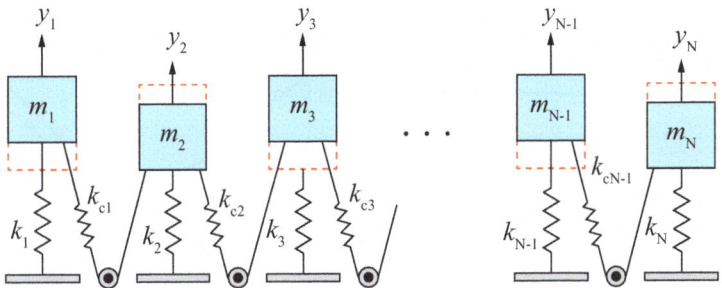

Fig. 8 N pistons engine

Of course, this system now depends on several parameters, and thus so do the eigenvalues ω^2, but for the sake of illustration let us consider the (realistic) situation in which the coupling terms $R = \frac{k_{c_j}}{k_j}$, are all equal (for $j = 1, \ldots, N$), and the stiffness constant k_j of each piston depart from a nominal stiffness $\bar{k}$ by the value Δk_j, $j = 1, \ldots, N$. Further, assuming that each stiffness perturbation term Δk_j is a random value of mean 0, indicating with ε' the root mean square standard deviation of the perturbations, at first approximation we get

$$\Delta k = k\varepsilon', \quad \Delta k = \begin{bmatrix} \Delta k_1 \\ \vdots \\ \Delta k_N \end{bmatrix}, \quad k = \begin{bmatrix} k_1 \\ \vdots \\ k_N \end{bmatrix}.$$

So doing, we have reduced our problem to two parameters, R and ε', and thus the eigenvalues ω are just $\omega(R, \varepsilon')$. The first parameter controls the strength of the coupling, and is the key one to monitor.

For strong coupling (large R), the eigenvalues do not cross. For weak coupling, the eigenvalues get close to each other, though they do not cross either: they all veer away from one another at $\varepsilon' = 0$. We illustrate in Fig. 9, obtained for $N = 10$ pistons, varying ε'; the first figure is for $R = 0.5$ (strong coupling), the second for $R = 0.01$.

We are ready to summarize what we learned from the above example:

(i) If the oscillators (pistons) are coupled, then we have an unreduced[1] symmetric tridiagonal (and positive definite) structure, and thus no eigenvalue can be equal. [The verification of this statement is an interesting exercise, and it can be proved from properties of Sturm polynomials.] Therefore, if the eigenvalues of two sub-systems of coupled oscillators coalesce, then the two sub-systems must be uncoupled, in other words the system matrix must have a reduced tridiagonal structure. Indeed, in this case we observed that coalescing of eigenvalues can occur by varying one parameter (the internal frequency of

---

[1] No sub-diagonal element is 0.

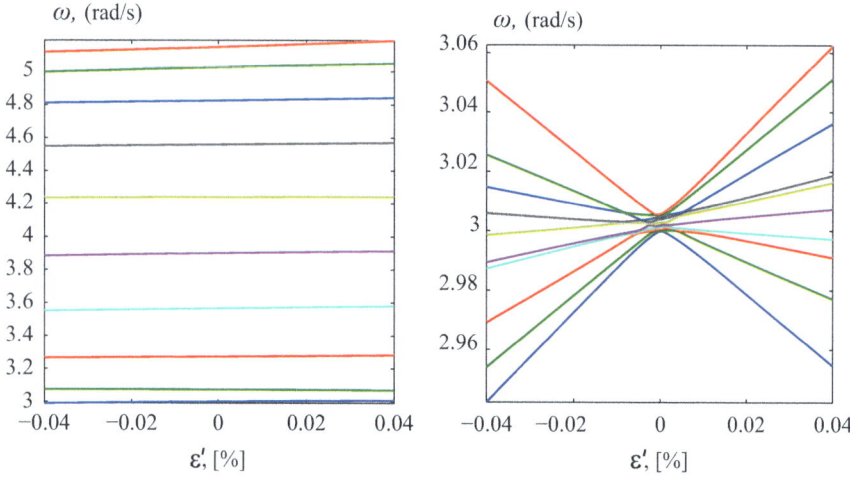

Fig. 9 $R = 0.5$ (*left*), and $R = 0.01$ (*right*)

a subsystem). As we will see, this is somewhat unusual. For these types of uncoupled systems, there does not seem to be anything peculiar happening at a point where eigenvalues coalesce, in the sense that the eigenvalues and eigenvectors vary smoothly through the coalescing point.
(ii) Having a small coupling term produces the most intriguing effect of replacing the coalescing phenomenon by the veering effect. More dramatic is the exchange between the eigenvectors.
(iii) We did not consider the case when there is friction in these systems of coupled oscillators. Of course, the problem is no longer necessarily reducible to a symmetric eigenproblem, and all modes are damped. Still, interesting models of friction (practically used in engineering studies) continue to lead to a model where the internal frequencies can be obtained by solving a symmetric eigenproblem and so the analysis of the eigenvalues of a symmetric matrix depending on parameters remains the mathematical object of study. See below.

1.3.4 Coupled Oscillators with Friction

Consider the damped coupled oscillators model

$$M\ddot{x} + 2C\dot{x} + Kx = 0, \qquad (2)$$

where M, and K are symmetric positive definite, and C is also symmetric (it is also positive definite if all modes are damped). Using the change of variable

$x \to M^{1/2}x$, and further letting $C \to M^{-1/2}CM^{-1/2}$ and $K \to M^{-1/2}KM^{-1/2}$, (2) rewrites as the system

$$\ddot{x} + 2C\dot{x} + Kx = 0. \tag{3}$$

Now, the idea is to write the fundamental matrix solution X of (3) in the form $X = ZY$, where Z will be accounting for the damping coming from C, and Y will account for the oscillatory part. We have the following result.

Lemma 1.9. *Consider the fundamental matrix solution X of* (3)*. Let $X = ZY$, where Z satisfies $\dot{Z} + CZ = 0$. Then, Y satisfies*

$$\ddot{Y} + (K - C^2)Y = 0,$$

if and only if $KC = CK$. In particular, the given factorization is feasible whenever $C = \alpha I + \beta K$.

Proof. Differentiating twice the relation $X = ZY$ and substituting in (3), we obtain

$$\ddot{X} + 2C\dot{X} + KX = 0 \iff \ddot{Y} + (Z^{-1}KZ - Z^{-1}C^2Z)Y = 0,$$

and since $\dot{Z} = -CZ$, and clearly $e^{Ct}C = Ce^{Ct}$, we get that $Z^{-1}C^2Z = C^2$, and thus we have

$$\ddot{X} + 2C\dot{X} + KX = 0 \iff \ddot{Y} + (Z^{-1}KZ - C^2)Y = 0.$$

It remains to verify that $Z^{-1}KZ = K$ if and only if $KC = CK$. But this is clear: If $KC = CK$, then also $e^{Ct}K = Ke^{Ct}$ and thus $ZK = KZ$; conversely, if $e^{Ct}K = Ke^{Ct}$, for all t, then it must be that $KC = CK$.

The validity of the final statement on the specific case of $C = \alpha I + \beta K$ is obvious. □

As a consequence of Lemma 1.9, we have the following situation relatively to the system (2), which justifies studying a symmetric eigenproblem to find the oscillatory component of the eigenmodes.

Example 1.10. Consider (2), with friction coefficients given by (e.g., see [73])

$$C = \alpha M + \beta K.$$

This case fits precisely in the situation of Lemma 1.9. Indeed, (2) can be rewritten as the new problem $\ddot{x} + 2\tilde{C} + \tilde{K}x = 0$, with $\tilde{C} = M^{-1/2}(\alpha M + \beta K)M^{-1/2} = \alpha I + \beta \tilde{K}$, and $\tilde{K} = M^{-1/2}KM^{-1/2}$.

2 Smoothness and Periodicity: One Parameter

In this chapter, we consider smooth matrix valued functions depending on one parameter, $A \in \mathscr{C}^k(\mathbb{R}, \mathbb{C}^{m \times n})$, $k \geq 1$ (but also $k = \omega$, the analytic case, is of interest). Our purpose in this chapter is threefold.

(a) In Sect. 2.1, we will review some results on the degree of smoothness of the factors in several decompositions of A, which, under some non-degeneracy assumptions, will be carried out by obtaining differential equations for the sought factors. Then, we'll proceed to weaken the assumptions in three ways: (1) we will work with block decompositions, (2) give weaker sufficient conditions to get smooth decompositions, possibly with some loss of smoothness, and (3) see what we should expect generically.
(b) In Sect. 2.2, we will consider the issue of periodicity of the factors when the functions depend periodically on the parameter.
(c) Finally, in Sect. 2.3, we will give an algorithm which computes an SVD-path. This 1-d continuation technique will be the backbone of other continuation techniques to be introduced in Chaps. 3 and 4.

2.1 Smoothness Results

This section is based on the work [26], where complete proofs for the results below may be found.

2.1.1 QR Factorization

Let us begin with the case of the QR factorization, which we will review precisely according to the previous plan: First we'll give a typical result under some nondegeneracy assumptions (full rank), then we'll weaken the assumptions and pay for some lack of smoothness, and finally we will see what we can expect in the generic case.

The next theorem can be proved by careful analysis of the Gram-Schmidt process.

Theorem 2.1 (QR Factorization).

(i) *Let $A \in \mathscr{C}^k(\mathbb{R}, \mathbb{R}^{m \times n})$, $m \geq n$, $k \geq 0$, be of full rank n. Then, A has a $\mathscr{C}^k$ QR decomposition: $A = QR$, where $Q \in \mathscr{C}^k(\mathbb{R}, \mathbb{R}^{m \times n})$ is orthonormal, $Q^T Q = I$, and $R \in \mathscr{C}^k(\mathbb{R}, \mathbb{R}^{n \times n})$ is upper triangular. Further, given a reference decomposition—say at $t = 0$—$A(0) = Q_0 R_0$, then Q and R going through this reference decomposition are unique.*
(ii) *If $A \in \mathscr{C}^\omega(\mathbb{R}, \mathbb{R}^{m \times n})$, $m \geq n$, then A has a $\mathscr{C}^\omega$ QR decomposition.*

Similarly for $A \in \mathscr{C}^{k,\omega}(\mathbb{R}, \mathbb{C}^{m \times n})$.

What happens if A is rank deficient (and not analytic)? What type of singular behavior can we handle?

Example 2.2. Consider $A = \begin{bmatrix} t^k|t| \\ t^d \end{bmatrix}$, $d, k \in \mathbb{N}, d \leq k+1 \Rightarrow A \in \mathscr{C}^k$. If $d \leq k$, we have the QR-factorization

$$A(t) = \begin{bmatrix} t^{k-d}|t|/\sqrt{1+t^{2(k-d+1)}} \\ 1/\sqrt{1+t^{2(k-d+1)}} \end{bmatrix} t^d \sqrt{1+t^{2(k-d+1)}},$$

and so $Q \in \mathscr{C}^{k-d}$. If $d = k+1$ we have

$$A(t) = \begin{bmatrix} \operatorname{sgn}(t)/\sqrt{2} \\ 1/\sqrt{2} \end{bmatrix} t^d \sqrt{2},$$

and Q is not even continuous.

Motivated by Example 2.2, we realize that there is a subtle interplay between the rate at which A loses rank and the degree of smoothness of A. Guided by this consideration, the following result gives sufficient conditions for A to admit a smooth QR-factorization, albeit with some loss of smoothness.

Theorem 2.3. *Let $A \in \mathscr{C}^k(\mathbb{R}, \mathbb{C}^{m \times n})$, $m \geq n$, $k \geq 1$. Assume that there exists $d \leq k$:*

$$\limsup_{\tau \to 0} \frac{1}{\tau^{2d}} \det(A^*A)(t+\tau) > 0, \quad \forall t. \tag{4}$$

Then, given any QR-factorization $A(t_0) = Q(t_0)R(t_0)$ at a point t_0 where A has full rank, there exists a $\mathscr{C}^{k-d}$ QR-factorization of A satisfying this initial condition. If $R(t_0)$ has real diagonal, the QR-factorization becomes unique, if we require the diagonal of R to be real.

Remarks 2.4. (i) As Example 2.2 shows, Theorem 2.3 is sharp for Q. Taking $A = \begin{bmatrix} t^k|t| & 1 \\ t^d & 0 \end{bmatrix}$, $d \leq k+1$, shows that the result is sharp also for the R factor.

(ii) If $m = n$, then it suffices that the matrix $A_{1:n-1}$ satisfies the assumptions of Theorem 2.3 (since the n-th column of Q is essentially determined by the previous columns).

Although it is quite easy to give a function A which loses rank at several points (and even more degenerate situations are possible), rather than in the pathological behavior of a given function A it is often much more appropriate to consider the entire function space (for us, $\mathscr{C}^k(\mathbb{R}, \mathbb{R}^{m \times n})$), endow it with the proper topology (for us, the "fine topology"), and consider what are the *generic* properties one should expect. Recall that a property is generic if it holds for a set which is of second Baire category (see [3,4]). Whenever a property is generic, this means that we can perturb a given function not satisfying this property into one which does satisfy it, hence a generic statement gives a convenient mean to characterize what has to be expected within a certain class. For example, in the class of rectangular matrices, generically a matrix $A \in \mathbb{R}^{m \times n}$, $m > n$, has full rank.

With this in mind, we are ready to see what one should generically expect for the QR-factorization of a function A.

Theorem 2.5 (Generic QR). *Generically, a function $A \in \mathscr{C}^k(\mathbb{R}, \mathbb{R}^{m \times n})$, $m \geq n$, $k \geq 1$, has a $\mathscr{C}^k$ QR-factorization. Similarly for generic $A \in \mathscr{C}^k(\mathbb{R}, \mathbb{C}^{m \times n})$.*

Sketch of Proof

(i) Consider the stratified variety of matrices of rank $r < n$. In the real case, each stratum is a smooth submanifold of codimension $(m-r)(n-r)$. In particular, if $m > n$, then $r \leq n-1$ and the codimension of each stratum is at least 2.

(ii) As a consequence, generically, a one-parameter function does not meet these strata, so it has full rank for all t and the result follows from Theorem 2.1.

(iii) If $m = n$ a generic $A \in \mathscr{C}^k(\mathbb{R}, \mathbb{R}^{n \times n})$ is such that the function made up by its first $(n-1)$ columns, $[a_1 \ldots a_{n-1}]$, is of full rank for all t and, thus A has a $\mathscr{C}^k$ QR decomposition (see Remark 2.4(ii)).

2.1.2 Schur Decomposition and SVD

Here we consider the eigendecomposition (or Schur decomposition) of a symmetric/Hermitian matrix depending on one parameter, as well as the SVD of a general matrix valued function. As anticipated, we will first derive differential equations models for the sought factors, under the assumption of distinct eigenvalues (singular values). Then, we will work with block decompositions, and also consider weaker sufficient conditions to get smooth decomposition with some loss of smoothness, and finally we will review what we should expect generically.

Let us begin with the spectral decomposition of a Hermitian function (the case of real symmetric function is left as an exercise).

Theorem 2.6 (Hermitian case, distinct eigenvalues). *Any $A \in \mathscr{C}^k(\mathbb{R}, \mathbb{C}^{n \times n})$, $k \geq 1$, $A^* = A$, with simple eigenvalues, is diagonalizable with a unitary $\mathscr{C}^k$-matrix Q: for all t we have $Q^*AQ = \Lambda$, where $Q^*Q = I$, and $\Lambda = \mathrm{diag}(\lambda_1, \ldots, \lambda_n)$. The factors Q and Λ passing through a reference decomposition at t_0: $A(t_0) = Q_0 \Lambda_0 Q_0^*$ are unique, except for a smooth phase factor function: $Q \to Qe^{\Phi}$, $\Phi = \mathrm{diag}(\phi_k, k = 1, \ldots, n)$, where each ϕ_k is a $\mathscr{C}^k$ real valued function.*

Proof. We give a proof which relies on devising differential equations for the factors, given initial conditions $A(t_0) = Q_0 \Lambda_0 Q_0^*$.

Let us formally differentiate $Q^*(t)A(t)Q(t) = \Lambda(t)$, $\Lambda = \mathrm{diag}(\lambda_1, \ldots, \lambda_n)$, to obtain

$$\dot{\Lambda} = Q^*\dot{A}Q + \Lambda Q^*\dot{Q} + \dot{Q}^*Q\Lambda.$$

Let $H = Q^*\dot{Q}$, and notice that $H^* = -H$, so that we get the system of differential algebraic equation (DAEs)

$$\begin{aligned}\dot{\Lambda} &= Q^*\dot{A}Q + \Lambda H - H\Lambda, \\ \dot{Q} &= QH.\end{aligned} \quad (5)$$

From this, we immediately get $\dot{\lambda}_i = (Q^*\dot{A}Q)_{ii}$. Further, we use the algebraic equations (i.e., the part relative to the off diagonal entries in $\dot{\Lambda}$) to determine H: H_{ii}'s must be purely imaginary (and smooth), but otherwise arbitrary, e.g. we can set $H_{ii} = 0, i = 1, \ldots, n$. Since the eigenvalues are distinct, then we also get

$$H_{ij} = \frac{(Q^*\dot{A}Q)_{ij}}{\lambda_j - \lambda_i}, \ i \neq j.$$

With this, the description of the system of differential equations Q and Λ is complete, and any Q and Λ satisfying the given initial conditions and the given differential equations, provide a $\mathscr{C}^k$ eigendecomposition of A, as stated. The uniqueness statement is a consequence of the above construction. □

The next result is about the SVD. Under the assumption of full rank, and distinct singular values, again differential equations for the factors can be derived; a similar result holds in the real case (and it is left as an exercise).

Theorem 2.7 (SVD: full rank and distinct singular values). *Let $A \in \mathscr{C}^k(\mathbb{R}, \mathbb{C}^{m\times n})$, $m \geq n$, $k \geq 1$. If A has full rank, and distinct singular values, then it has a $\mathscr{C}^k$ singular value decomposition: $A = U\Sigma V^*$, where $U \in \mathscr{C}^k(\mathbb{R}, \mathbb{C}^{m\times m})$ and unitary ($U^*U = I_m$), $V \in \mathscr{C}^k(\mathbb{R}, \mathbb{C}^{n\times n})$ and unitary ($V^*V = I_n$), and $\Sigma \in \mathscr{C}^k(\mathbb{R}, \mathbb{R}^{m\times n})$ with structure $\Sigma = \begin{bmatrix} S \\ 0 \end{bmatrix}$, $S = \operatorname{diag}(\sigma_1, \ldots, \sigma_n)$, and ordered singular values, $\sigma_1 > \cdots > \sigma_n$.*

Given a reference SVD at a value t_0: $A(t_0) = U_0 \Sigma_0 V_0^*$, the $\mathscr{C}^k$ functions U, V, Σ, of an SVD passing through this reference decompositions enjoy the following uniqueness properties: Σ is unique; writing $U = [U_1, U_2]$, where $U_1 \in \mathbb{C}^{m\times n}$, then U_1 and V are unique except for a joint transformation of the type $U \to Ue^{i\Phi}$, and $V \to Ve^{i\Phi}$, with $\Phi = \operatorname{diag}(\phi_k, k = 1, \ldots, n)$ and each ϕ_k is a $\mathscr{C}^k$ real valued function; finally, U_2 is unique within the equivalence class $U_2 \to U_2 Z$, where Z is any $\mathscr{C}^k(\mathbb{R}, \mathbb{C}^{m-n,m-n})$ unitary function: $Z^*Z = I_{m-n}$.

Proof. Again we resort to a differential equation model. Given an initial SVD at t_0: $A(t_0) = U_0\Sigma_0 V_0^*$, we seek $\mathscr{C}^k$ unitary U and V, and real Σ such that for all t: $A(t) = U(t)\Sigma(t)V^*(t)$. Formally differentiating this relation, we now get globally well defined differential equations whose solutions will render the sought factors.

We have:
$$U^*\dot{A}V = (U^*\dot{U})\Sigma + \dot{\Sigma} + \Sigma(\dot{V}^*V).$$

Let us set
$$H := U^*\dot{U}, \qquad K := V^*\dot{V},$$

and note that both H and K are skew-Hermitian. Thus, we obtain the system of DAEs
$$\dot{\Sigma} = U^*\dot{A}V - H\Sigma + \Sigma K,$$
$$\dot{U} = UH, \qquad (6)$$
$$\dot{V} = VK.$$

From this, using the structure of Σ, we immediately get
$$\dot{\sigma}_i = (U^*\dot{A}V)_{ii} - H_{ii}\sigma_i + \sigma_i K_{ii}, \quad i = 1, \ldots, n.$$

Next we use the algebraic part in (6) to obtain expressions for H and K. From the 0-structure in the equation for $\dot{\Sigma}$, and for $i \neq j$, $i, j = 1, \ldots, n$, we get:
$$0 = (U^*\dot{A}V)_{ij} - H_{ij}\sigma_j + \sigma_i K_{ij}$$
$$0 = (U^*\dot{A}V)_{ji} - H_{ji}\sigma_i + \sigma_j K_{ji},$$

and thus (since H and K are skew)
$$H_{ij} = \frac{\sigma_j(U^*\dot{A}V)_{ij} + \sigma_i(U^*\dot{A}V)_{ji}}{\sigma_j^2 - \sigma_i^2}$$
$$K_{ij} = \frac{\sigma_j(U^*\dot{A}V)_{ji} + \sigma_i(U^*\dot{A}V)_{ij}}{\sigma_j^2 - \sigma_i^2}.$$

Also,
$$H_{ij} = -\bar{H}_{ji} = \frac{(U^*\dot{A}V)_{ij}}{\sigma_j}, \quad i = n+1, \ldots, m, \; j = 1, \ldots, n.$$

Now, to have $\dot\sigma_i$ real, we need

$$(H_{ii} - K_{ii}) = \mathrm{Im}((U^*\dot A V)_{ii})/\sigma_i$$

which leaves n degrees of freedom; e.g., we can choose $K_{ii} = 0$, and determine H_{ii}. The bottom right $(m-n)\times(m-n)$ block of H is not determined and we may set it to 0 (or another smooth skew matrix valued function). The statement on uniqueness is a consequence of the above proof. □

Remark 2.8. An interesting observation is that—when $m = n$—then one singular value can become 0 (i.e., A is allowed to lose rank), the other singular values remaining distinct. This is because the last column of U (and V) is effectively determined by the previous $(n-1)$ ones. However, we note that in order to retain smoothness, σ_n must be allowed to become negative if it goes to zero, hence we obtain a so-called **signed SVD**.

2.1.3 Block Decompositions

There are some clear shortcomings in the previously derived differential equations models. For one thing, the previous models break down if the stated non-degeneracy assumptions are not satisfied; namely, if the eigenvalues or singular values of A coalesce at some value of t. Moreover, even for simple eigenvalues, numerical difficulties can be expected in case two or more eigenvalues (singular values) become close. In such cases, it may be better to compute the subspaces relative to clusters of eigenvalues (singular values).

We present two such results, relatively to a block Schur decomposition and block SVD.

Theorem 2.9 (Block-Schur). *Let $A \in \mathscr{C}^k(\mathbb{R}, \mathbb{C}^{n\times n})$ and suppose that the eigenvalues of A can be clustered in p groups $\Lambda_1, \ldots, \Lambda_p$, each containing n_j eigenvalues ($j = 1, \ldots, p$, and $n_1 + \cdots + n_p = n$), which stay disjoint for all t. Then, A has a $\mathscr{C}^k$ block Schur decomposition, the blocks corresponding to these groups. That is, there exists $Q \in \mathscr{C}^k(\mathbb{R}, \mathbb{C}^{n\times n})$, unitary ($Q^*Q = I$), such that*

$$Q^*AQ = \begin{bmatrix} R_{11} & R_{12} & \cdots & R_{1p} \\ 0 & R_{22} & \ddots & \vdots \\ \vdots & \ddots & \ddots & \vdots \\ 0 & \cdots & 0 & R_{pp} \end{bmatrix}, \quad \sigma(R_{jj}) = \Lambda_j, \ j = 1, \ldots, p.$$

The factor Q is unique within the class of block-diagonal and unitary transformations: $Q \to QZ$, $Z = \mathrm{diag}(Z_{jj},\ j = 1, \ldots, p)$, $Z_{jj} \in \mathscr{C}^k(\mathbb{R}, \mathbb{C}^{n_j\times n_j})$, $Z_{jj}^*Z_{jj} = I_{n_j}$, $j = 1, \ldots, p$. Finally, if A is Hermitian, then we obtain a block diagonal form.

Proof for Two Blocks. We consider the case of two blocks, so that we have $\Lambda_1(t) = \{\lambda_1(t), \ldots, \lambda_m(t)\}$ and $\Lambda_2(t) = \{\lambda_{m+1}(t), \ldots, \lambda_n(t)\}$ disjoint for all t, and want to find unitary Q and block upper triangular R, as smooth as A, such that $Q^*(t)A(t)Q(t) = R(t)$ for all t. Here $R = \begin{bmatrix} R_{11} & R_{12} \\ 0 & R_{22} \end{bmatrix}$. As previously, we derive differential equations models for the factors.

Let $A(t_0) = Q_0 R_0 Q_0^*$ be a reference decomposition at t_0 so that $\Lambda(R_{jj}(t_0)) = \Lambda_j(t_0)$, $j = 1, 2$. We differentiate $R = Q^*AQ$, let $H := Q^*\dot{Q}$, and obtain the system of DAE's

$$\dot{R} = Q^*\dot{A}Q + RH - HR,$$
$$\dot{Q} = QH,$$
$$R_{21} = 0, \quad H^* = -H.$$

To arrive at the sought system of differential equations, rewrite the first equation above in block form

$$\begin{bmatrix} \dot{R}_{11} & \dot{R}_{12} \\ 0 & \dot{R}_{22} \end{bmatrix} = \begin{bmatrix} (Q^*\dot{A}Q)_{11} & (Q^*\dot{A}Q)_{12} \\ (Q^*\dot{A}Q)_{21} & (Q^*\dot{A}Q)_{22} \end{bmatrix} +$$

$$\begin{bmatrix} R_{11} & R_{12} \\ 0 & R_{22} \end{bmatrix} \begin{bmatrix} H_{11} & H_{12} \\ -H_{12}^* & H_{22} \end{bmatrix} - \begin{bmatrix} H_{11} & H_{12} \\ -H_{12}^* & H_{22} \end{bmatrix} \begin{bmatrix} R_{11} & R_{12} \\ 0 & R_{22} \end{bmatrix}$$

and thus we must have

$$R_{22}H_{12}^* - H_{12}^*R_{11} = (Q^*\dot{A}Q)_{21},$$

and since the eigenvalues of R_{11} and R_{22} are distinct, we get $H_{12} \in \mathbb{C}^{m \times (n-m)}$. Note that the blocks H_{11} and H_{22} are not uniquely determined; we may set them both to 0, or any other skew-Hermitian function of appropriate size. Setting them to 0, we will take $H = \begin{bmatrix} 0 & H_{12} \\ -H_{12}^* & 0 \end{bmatrix}$, and have found the sought differential equations for Q, R_{11}, R_{12}, R_{22}. The statement on uniqueness is again a consequence of the proof, and the statement when A is Hermitian is a simple verification left as an exercise. □

Remark 2.10. Theorem 2.9 gives theoretical validation to the construction of *smooth BCs* we saw in Example 1.8. Also, note that the diagonal blocks R_{jj}, $j = 1, \ldots, p$, in Theorem 2.9 are generally not triangular.

In a very similar way, we can obtain smoothness results for block SVDs. We present one such result for real valued function A, and leave its proof as an exercise.

Theorem 2.11 (Block SVD). *Let $A \in \mathscr{C}^k(\mathbb{R}, \mathbb{R}^{m \times n})$, $m \geq n$, $k \geq 1$, be full rank and with p disjoint groups of singular values for all t: $\Sigma_1, \ldots, \Sigma_p$, each containing*

n_j singular values ($j = 1, \ldots, p$, and $n_1 + \cdots + n_p = n$). Then, A has a $\mathscr{C}^k$ block SVD. That is, there exist $U \in \mathscr{C}^k(\mathbb{R}, \mathbb{R}^{m \times m})$, orthogonal, and $V \in \mathscr{C}^k(\mathbb{R}, \mathbb{R}^{n \times n})$, orthogonal, such that

$$U^T A V = \begin{bmatrix} S_{11} & 0 & \cdots & 0 \\ 0 & S_{22} & \ddots & \vdots \\ \vdots & \ddots & \ddots & \vdots \\ 0 & \cdots & 0 & S_{pp} \\ 0 & \cdots & 0 & 0_{m-n} \end{bmatrix},$$

where further $S_{jj} = S_{jj}^T$ and positive definite, $\sigma(S_{jj}) = \Sigma_j$, $j = 1, \ldots, p$. Partitioning $U = [U_1, U_2]$ where $U_1 \in \mathbb{R}^{m \times n}$, then U_1 and V are determined within the class of block-diagonal and orthogonal transformations: $U \to UZ$, $V \to VZ$, $Z = \mathrm{diag}(Z_{jj}, \, j = 1, \ldots, p)$, $Z_{jj} \in \mathscr{C}^k(\mathbb{R}, \mathbb{R}^{n_j \times n_j})$, $Z_{jj}^T Z_{jj} = I_{n_j}$, $j = 1, \ldots, p$. The factor U_2 is determined up to a right orthogonal transformation $U_2 \to U_2 W$, $W \in \mathscr{C}^k(\mathbb{R}, \mathbb{R}^{(m-n) \times (m-n)})$, $W^T W = I_{m-n}$.

Remarks 2.12. (i) Note that we are requiring positive definite structure for the blocks S_{jj} in Theorem 2.11; this could be relaxed.

(ii) Taking just one block in Theorem 2.11 gives $A = U \begin{bmatrix} S \\ 0 \end{bmatrix} V^T$, with S positive definite. This writing is sufficient to validate the goal of obtaining smooth bases for $(\mathrm{range}(A))^\perp$ as well as for $\mathrm{null}(A^T)$ which we saw in Examples 1.6 and 1.7.

2.1.4 Degenerate Cases: Multiple Eigenvalues and Singular Values

Results such as Theorem 2.6 or Theorem 2.7 required the assumption of distinct eigenvalues, respectively singular values. It is natural to inquire what can happen when this assumption is violated. In particular, insofar as smoothness is concerned, under what types of conditions can we expect to obtain some smoothness? What is known?

As a general principle, it is simple to appreciate that, when eigenvalues (singular values) coalesce, the eigenspaces (singular vectors) undergo a more obviously pathological behavior than the eigenvalues. In general, not even for a function $A \in \mathscr{C}^\omega$, can we continue an eigenvector past a point where its associated eigenvalue coincides with some other eigenvalue, simply because we may fail to have a basis of eigenvectors, as the following example highlights.

Example 2.13.

$$A(t) = \begin{bmatrix} t & 1 \\ 0 & 0 \end{bmatrix}, \quad A \in \mathscr{C}^\omega$$

$$\lambda_1 = t, \; \lambda_2 = 1, \; \lambda_{1,2} \in \mathscr{C}^\omega$$

but

$$v_1 = \begin{bmatrix} 1 \\ 0 \end{bmatrix}, \quad v_2 = \begin{bmatrix} -1/t \\ 1 \end{bmatrix}, \, t \neq 0,$$

and obviously v_2 cannot be continued at $t = 0$, where we have a Jordan block of size 2.

Naturally, we may suspect that difficulties arise because A is not diagonalizable. But in fact, even when we restrict to Hermitian matrices, things are not trivial.

Example 2.14 ([13]). Consider the following function:

$$A(t) = e^{-1/t^2} \begin{bmatrix} \cos(2/t) & \sin(2/t) \\ \sin(2/t) & -\cos(2/t) \end{bmatrix}, \, t \neq 0, \, A(0) = 0.$$

Clearly $A \in \mathscr{C}^\infty$, and so are the eigenvalues: $\lambda_{1,2}(t) = \pm e^{-1/t^2}$ for $t \neq 0$ and $\lambda_{1,2}(0) = 0$. For $t \neq 0$, the eigenvectors are the columns of Q:

$$Q(t) = \begin{bmatrix} \cos(1/t) & \sin(1/t) \\ \sin(1/t) & -\cos(1/t) \end{bmatrix},$$

and obviously the eigenvectors cannot be continuously continued at $t = 0$. Geometrically, as t approaches 0, Q rotates faster and faster and it selects all possible directions infinitely many times. An important observation here is that not only $\lambda_1(0) = \lambda_2(0)$, but in fact the eigenvalues coalesce at all orders:

$$0 = \lambda_1^{(k)}(0) = \lambda_2^{(k)}(0), \, \forall k.$$

To recap, in this Example, the function is symmetric, the eigenvalues—albeit coalescing—retain all smoothness of A, but the eigenvectors are not even continuous.

Before stating a theorem giving sharp sufficient conditions in order to maintain (some) smoothness of the eigendecomposition in the Hermitian case, let us recall what is known about this case.

($\mathscr{C}^1$) A major step was achieved by Rellich (see [14]), who proved that
 "$\mathscr{C}^1$ Hermitian matrices have $\mathscr{C}^1$ eigenvalues".
 This result is optimal, in two ways:

 (a) If A is not Hermitian things can go awry. E.g., even $A \in \mathscr{C}^1$ and diagonalizable does not give $\mathscr{C}^1$ eigenvalues; see [13, Example 5.9].
 (b) In general, one cannot expect anything more that $\mathscr{C}^1$ eigenvalues. Indeed, the following example (see [74]) shows that "*not even $\mathscr{C}^\infty$ Hermitian functions have $\mathscr{C}^2$ eigenvalues*".:

$$A(t) = \begin{bmatrix} e^{-(\alpha+\beta)/|t|} & e^{-\beta/|t|}\sin(1/t) \\ e^{-\beta/|t|}\sin(1/t) & -e^{-(\alpha+\beta)/|t|} \end{bmatrix}, \; t \neq 0, \text{ and } A(0) = 0,$$

with $\alpha, \beta > 0$, so that $A \in \mathscr{C}^\infty$. The eigenvalues are

$$\lambda_\pm(t) = \pm e^{-\beta/|t|}\left(e^{-2\alpha/|t|} + \sin^2(1/t)\right)^{\frac{1}{2}},$$

and these are in $\mathscr{C}^1$ but not in $\mathscr{C}^2$, if $\alpha \geq \beta$.

($\mathscr{C}^\omega$) In the real analytic case, things are actually much friendlier, as the following fundamental result shows.

Theorem 2.15 (Hermitian and analytic). *If $A \in \mathscr{C}^\omega(\mathbb{R}, \mathbb{C}^{n \times n})$ is Hermitian, then it has analytic eigendecomposition. In other words: $A = Q\Lambda Q^*$, where Q is unitary and analytic and Λ is diagonal and analytic.*

We refer to [13] for a proof of this result, but it is instructive to have an informal understanding of why the result is true.

(a) In general, multiple eigenvalues lead to a local expansion in fractional powers of the function describing the eigenvalue; this is called a Newton–Puiseux series, e.g. see [16]. For example, an eigenvalue of multiplicity m at $t = 0$ will have a local expansion of the form $\lambda(t) = \sum_{k=0}^\infty t^{k/m} b_k$, with $b_0 = \lambda(0)$. Similarly, if A is diagonalizable (surely the case for a Hermitian function), the associated eigenvectors will also have a fractional powers expansion.

(b) An analytic function $A \in \mathscr{C}^\omega(\mathbb{R}, \mathbb{C}^{n \times n})$, Hermitian, is writeable as $A(t) = \sum_{k=0}^\infty t^k A_k$, where $A_k^* = A_k$ for all k. Since A is analytic, also the characteristic polynomial $p(\lambda, t)$ will have coefficients which are analytic functions of t. If—say at $t = 0$—we have a multiple eigenvalue of algebraic multiplicity m, then that eigenvalue will have an algebraic singularity and it can be expressed as a Newton–Puiseux series around $t = 0$: $\lambda(t) = \sum_{k=0}^\infty t^{k/m} b_k$. However, for all $t \in \mathbb{R}$, the eigenvalues of $A(t)$ must be real since A is Hermitian, and fractional powers of t become complex valued for t near 0. This means that in the Newton–Puiseux series we can only have integer powers. That is, $\lambda(t)$ is an analytic function of t. Similar considerations (or see [13]) justify that the eigenvectors also admit a regular expansion in powers of t.

Our next task is to give a smoothness result, relatively to coalescing eigenvalues, under realistic sufficient conditions in the $\mathscr{C}^k$ case. Motivated by Example 2.14 and the results above, the idea is that "the order of contact of the eigenvalues should not exceed smoothness of A". To properly define order of contact, we use the following.

Definition 2.16. Let $A \in \mathscr{C}^k(\mathbb{R}, \mathbb{C}^{n \times n})$, Hermitian. We say that $A \in W_p$ if there exist a labeling of the eigenvalues, $\lambda_1, \ldots, \lambda_n$, such that for any t and $i \neq j$ we have

$$\liminf_{\tau \to 0} \frac{|\lambda_i(t+\tau) - \lambda_j(t+\tau)|}{|\tau^p|} > 0. \tag{7}$$

The value p is called order of contact of the eigenvalues, and it is the smallest value for which (7) is satisfied (for all t).

To appreciate the meaning of (7), we make two observations:

(i) if the eigenvalues remain distinct for all t, then $p = 0$;
(ii) if we had two eigenvalues equal at some given t, and differentiable there as often as we would like to, then—relatively to this pair of eigenvalues—the value of p in (7) would be nothing else but the index of the first differing term in the Taylor expansions of these two eigenvalues.

We are ready to recall the following result from [26], to which we refer for its proof.

Theorem 2.17. *Let $A \in \mathscr{C}^k(\mathbb{R}, \mathbb{C}^{n \times n}) \cap W_p$, $p \leq k$, A Hermitian. Then, there exists $Q \in \mathscr{C}^{k-p}$, and real diagonal $\Lambda \in \mathscr{C}^k$ such that $A = Q \Lambda Q^*$. Similarly in the real case.*

In other words, in the $\mathscr{C}^k$ case, as long as the order of contact of eigenvalues does not exceed that of the underlying degree of smoothness, the eigenvalues remain as smooth as the original function regardless of their coalescing, whereas the eigenvectors lose smoothness in the amount determined by the order of contact at points where eigenvalues coalesce. It should be appreciated that the assumptions of Theorem 2.17—and the reason why such result holds—imply that coalescing of eigenvalues occur at isolated values of t. As consequence, the loss of differentiability is also a localized phenomenon.

For completeness, we now recall a different type of result (see [12, 28]), which, albeit less refined than Theorem 2.17, is often rather useful in practice to ascertain if a multiple eigenvalue splits into several branches of distinct eigenvalues. We present it in the form of a remark, relatively to a first order split (splitting at higher orders is handled in a similar way, and we leave the case of a 2nd or higher order split as an exercise).

Remark 2.18 (Multiple Eigenvalues Split). Suppose that the Hermitian function $A(\cdot)$ has an eigenvalue of multiplicity k at (say) $t = 0$, call it λ_0. Then, we have a k-dimensional invariant subspace for $A(0)$ which can be represented in terms of k orthonormal columns: $A(0)U_0 = U_0 D_0$, where $D_0 = \lambda_0 I_k$, and $U_0^* U_0 = I_k$. Now, since A is smooth, by writing (for t near 0) $A(t) = A(0) + t\dot{A}(0) + \mathscr{O}(t^2)$, we consider $A(t)U_0 = A(0)U_0 + t\dot{A}(0)U_0 + \mathscr{O}(t^2)$; so, we are lead to look at the eigenvalues of $\lambda_0 I_k + t(U_0^* \dot{A}(0) U_0) + \mathscr{O}(t^2)$. If $(U_0^* \dot{A}(0) U_0)$ has k distinct eigenvalues, then—locally—the multiple eigenvalue λ_0 splits into k distinct eigenvalues, which (since they are t-dependent roots of the characteristic polynomial) determine k differentiable branches through the coalescing point.

The situation for the SVD is similar to the case of a Hermitian function, by virtue of the relation of the SVD of A to the eigendecompositions of A^*A and AA^*. In particular, if A is analytic, as a consequence of Theorem 2.15, it is immediate that it will have an analytic SVD, where the singular values necessarily must be allowed to possibly change sign and ordering (along the diagonal of Σ) if they go through zero or cross each other.

In the smooth case, however, sufficient conditions to retain (some) smoothness for the SVD factors are slightly different than those for the Hermitian case, because we need to account for the possibility that the null space of A changes dimension, and for the rate at which A may lose rank. As before, we will typically also have only a *signed* SVD (the singular values may become negative in order to retain smoothness) and with singular values not necessarily ordered. The result below is again from [26], to which we refer for its proof (essentially, it consists of putting together the results on smooth QR factorization and Hermitian eigenproblem, and making sure to take smooth square roots).

Theorem 2.19 (SVD: rank deficient case, multiple singular values). *Let $A \in \mathscr{C}^k(\mathbb{R}, \mathbb{C}^{m \times n})$, $m \geq n$, $k \geq 1$. Assume that the singular values can be labeled so that at any point the order of coalescing of singular values is at most $p \leq k$, that is (see Definition 2.16)*

$$\liminf_{\tau \to 0} \frac{|\sigma_i(t+\tau) - \sigma_j(t+\tau)|}{|\tau^p|} > 0,$$

and that for every t (see (4))

$$\limsup_{\tau \to 0} \frac{1}{\tau^{2d}} \det(A^*A)(t+\tau) > 0.$$

Also assume that $\mathrm{rank}(A) \geq n - 1$ for all t (i.e., only one of the singular values can become zero). Then, there exists a $\mathscr{C}^{k-\max(d,p)}$ singular value decomposition of A: $A = U \Sigma V^*$, with U and V unitary and $\mathscr{C}^{k-\max(d,p)}$ whereas the singular values in Σ can be taken to be $\mathscr{C}^k$ functions.

Remark 2.20. For completeness, we report on an interesting variation of the above result, which is concerned with the SVD in case the $\mathscr{C}^k$ function A has **constant rank**. The following result was proven in [27].

Let $A \in \mathscr{C}^k(\mathbb{R}, \mathbb{C}^{m \times n})$, $m \geq n$ and $k \geq 0$, $\mathrm{rank}(A(t)) = n - r$ for all t, r fixed: $0 \leq r \leq n - 1$. Then, there exist unitary $U \in \mathscr{C}^k(\mathbb{R}, \mathbb{C}^{m \times m})$ and $V \in \mathscr{C}^k(\mathbb{R}, \mathbb{C}^{n \times n})$ such that

$$U^*(t)A(t)V(t) = \begin{bmatrix} S_+ & 0 \\ 0 & 0 \end{bmatrix},$$

where $S_+ \in \mathscr{C}^k(\mathbb{R}, \mathbb{C}^{(n-r) \times (n-r)})$ is Hermitian positive definite.

Naturally, the above S^+ can be further decomposed as in Theorem 2.17.

Recap. Whenever some of the eigenvalues of a Hermitian function (or singular values of a general function) coalesce, we should anticipate a loss of smoothness of the unitary factors in the eigendecomposition (respectively, singular value decomposition), although the eigenvalues (respectively, the singular values) are expected to retain the same degree of smoothness of the underlying function.

2.1.5 Generic Properties

Just as we argued about the QR factorization, it is often more meaningful to focus on what one should expect *generically*, that is when considering the topological properties of the entire function space. This way, we will be able to properly distinguish between what properties should a typical function have and what instead are degenerate situations from which we can perturb away.

For example, a simple—yet fundamental—result in linear algebra states that generically a Hermitian matrix (not a matrix function) $A \in \mathbb{C}^{n \times n}$ has distinct eigenvalues. [This is simply because from an eigendecomposition of A: $A = U\Lambda U^*$, $U^*U = I$, $\Lambda = \text{diag}(\lambda_i, i = 1, \ldots, n)$, we can trivially perturb the diagonal of Λ to make all of its entries distinct.] Indeed, in the general class of matrices in $\mathbb{C}^{n \times n}$, the property of having distinct eigenvalues is generic (similar argument as above using the Schur form of a matrix A: $A = URU^*$, $U^*U = I$, R upper triangular; now we can perturb (in $\mathbb{C}$) the diagonal of R). However, when considering matrix valued functions, we cannot expect that having smooth eigenvalues be a generic property, not even in the analytic case.

Example 2.21. We leave the verification of the following claim as an exercise.

Any real perturbation ϵB of $A(t) = \begin{bmatrix} 0 & 1 \\ t & 0 \end{bmatrix}$ will be non-diagonalizable in a neighborhood of $t = 0$.

In other words, whereas each single matrix $A(t)$ (for given t) may be effectively perturbed into a diagonalizable one, the entire family cannot.

In the Hermitian case, things are friendlier. Of course, we already know that analytic Hermitian functions admit analytic eigendecompositions (see Theorem 2.15), but actually we can claim more, and also in the $\mathscr{C}^k$ case.

Example 2.22. Let $A(t) = \begin{bmatrix} t & 0 \\ 0 & 0 \end{bmatrix}$. The perturbed problem $\begin{bmatrix} t & \epsilon \\ \epsilon & 0 \end{bmatrix}$ has simple eigenvalues $\frac{1}{2}(t \pm \sqrt{t^2 + 4\epsilon^2})$ for every t.

Example 2.23. Consider the symmetric function $A(t) = \begin{bmatrix} a(t) & b(t) \\ b(t) & c(t) \end{bmatrix}$, $\forall t \in \mathbb{R}$. A trivial computation shows that the eigenvalues of A are identical if and only if A is diagonal with equal diagonal terms. Thus, to have identical eigenvalues at some value of t, we must satisfy:

$$a - c = 0, \quad b = 0.$$

Now, this is a system of two scalar equations in the single variable t, and within the class of smooth functions a, b, c, generically the 0-sets of $(a - c)$ and of b are given by isolated, and different, values of t, hence generically A does not have equal eigenvalues. As a consequence, it is a generic property that smooth symmetric functions have a smooth orthogonal eigendecomposition (see Theorem 2.6).

Example 2.23 is actually indicative of the general situation, as the following result summarizes.

Theorem 2.24. *Generically, a symmetric $A \in \mathscr{C}^k(\mathbb{R}, \mathbb{R}^{n \times n})$, $k \geq 1$, has a $\mathscr{C}^k$ orthogonal eigendecomposition.*

Sketch of Proof. For a complete proof see [26], but the following is sufficient to appreciate why the result holds true.

(i) We can write any symmetric matrix $A \in \mathbb{R}^{n \times n}$ in terms of its eigenvectors and eigenvalues (outer product expansion): $A = \sum_{i=1}^{n} \lambda_i u_i u_i^T$, so that in general we have $n(n-1)/2 + n = n(n+1)/2$ degrees of freedom (the first number, $n(n-1)/2$ specifies an orthogonal matrix, the remaining n degrees of freedom are the eigenvalues).

(ii) Whenever two eigenvalues are equal, say $\lambda_1 = \lambda_2 = \lambda$, then the number of degrees of freedom drops to $n(n-1)/2 + n - 1$, which seemingly gives a codimension of 1 for the dimension of the set of symmetric matrices with two equal eigenvalues. However, observe that

$$Q \begin{bmatrix} \lambda & 0 \\ 0 & \lambda \end{bmatrix} Q^T = \begin{bmatrix} \lambda & 0 \\ 0 & \lambda \end{bmatrix}$$

where Q is $(2,2)$ and orthogonal, hence it is fully specified by one angular coordinate, and so we actually have one fewer degree of freedom. Therefore, the dimension of the manifold of symmetric matrices with two equal eigenvalues is $n(n-1)/2 - 1 + n - 1$, and thus its codimension in the set of symmetric matrices is 2. As a consequence, a generic one-parameter function does not meet this manifold and therefore has simple eigenvalues.

Remark 2.25. As a consequence of the argument in the proof of Theorem 2.24, the property of having two coalescing eigenvalues of a symmetric function will necessitate the function to depend on two real parameters in order to be generic.

Remark 2.26. There is an analogous result to Theorem 2.24 for the case of a $\mathscr{C}^k$ Hermitian function. Now the dimension of the set of matrices with equal eigenvalues is even higher, it is 3. This is easy to appreciate by realizing that a $(2,2)$ Hermitian matrix is of the form $A = \begin{bmatrix} a & c+ib \\ c-ib & d \end{bmatrix}$ (with a, b, c, real valued) and A has equal eigenvalues if and only if it is a multiple of the identity, and thus we must have three relations satisfied: $a - d = 0$, $b = 0$, $c = 0$. As a consequence, generically, again a function of one or two real parameters does not meet the manifold of matrices with two (or more) equal eigenvalues. Indeed, we will need three real parameters in order for two eigenvalues of a Hermitian function coalescing to be a generic property.

A very similar situation to the symmetric (Hermitian) eigenproblem occurs for the SVD. In the present context, we need to emphasize once more that the singular values cannot be required to remain positive.

Theorem 2.27. *Generically, a function $A \in \mathscr{C}^k(\mathbb{R}, \mathbb{R}^{m \times n})$, $k \geq 1$, $m \geq n$, (or taking values in $\mathbb{C}^{m \times n}$) has a $\mathscr{C}^k$ singular value decomposition.*

Sketch of Proof

(i) Generically, the singular values are distinct (since their squares are the eigenvalues of $A^T A$, which is symmetric).
(ii) In the real case, if $m > n$ generically A has full rank (see Theorem 2.5). If $m = n$, since generically the singular values are distinct, hence at most one vanishes, Remark 2.8 validates the claim.
(iii) In the complex case, A is generically of full rank for $m \geq n$, and has distinct singular values, so the result follows from Theorem 2.7.

Remark 2.28. An important **consequence** of the above genericity results is that it is conceptually appropriate to devise computational procedures for smooth eigendecompositions of $\mathscr{C}^k$ one-parameter symmetric (Hermitian) functions (or SVDs), by relying on the differential equations models put forward in Theorems 2.6 and 2.7. See Sect. 2.3 and Chaps. 3 and 4.

2.2 Periodicity Results

The results in this section are based on [27], where complete proofs and further periodicity and smoothness results can be found. We consider the following problem.

- Given $A \in \mathscr{C}_1^k(\mathbb{R}, \mathbb{R}^{n \times n})$, symmetric. Assume that A has a smooth eigendecomposition (perhaps with some loss of smoothness). What can be said about the periodicity of the eigenvalues? Of the eigenvectors? What can be said in the case of $A \in \mathbb{C}^{n \times n}$, Hermitian? Similarly, what can be said about periodicity of the SVD factors for a function A taking values in $\mathbb{R}^{m \times n}$ (or $\mathbb{C}^{m \times n}$)? Naturally, similar questions can be phrased for block-decompositions.

Beside being mathematically natural and interesting questions, the above problems turn out to be very relevant when we will consider functions of 2 and 3 parameters, as we will do in Chaps. 3 and 4.

Next, before considering the general case (including the more challenging cases of coalescing eigenvalues or singular values), let us recall two fundamental results which provide insight into our general concern.

(i) *Floquet theory*. This well known result is concerned with the fundamental matrix solution of a time-periodic linear system (e.g., see [5]).

> Let X be the fundamental matrix solution of a 1-periodic linear system: $\dot{X} = B(t)X$, $X(0) = I$, $B(t+1) = B(t)$ for all t, and B is real valued. Then, we can write:
>
> $$\begin{aligned} \mathbb{C}: \quad & X(t) = P(t)e^{Ct}, \ P(\cdot) \in \mathbb{R}^{n \times n}, \ C \in \mathbb{C}^{n \times n}, \ P(t+1) = P(t), \\ \mathbb{R}: \quad & X(t) = P(t)e^{Ct}, \ P(\cdot), C \in \mathbb{R}^{n \times n}, \ P(t+2) = P(t), \end{aligned} \quad (8)$$
>
> for all t.

In other words, the fundamental matrix solution factors as the product of a time periodic function P and an exponential factor. If we insist that both factors be real valued, then the function P may have twice the period of the matrix B of the coefficients. [It is insightful to look at the proof of (8) in [5] and appreciate why one may need to double the period in the real case. This is due to the need to take the logarithm of the matrix $X(1)$, which albeit surely invertible (hence it has a logarithm), in general has only a complex valued logarithm.]

(ii) *Sibuya's theory*. In [22], Sibuya gave a key result on periodicity of eigendecompositions. To be precise, he was concerned with periodicity of a block diagonalization of a function with two disjoint groups of eigenvalues and effectively proved the following result.

> Let $A \in \mathscr{C}_1^k(\mathbb{R}, \mathbb{C}^{n \times n})$ and assume that $\sigma(A(t)) = \Lambda_1(t) \cup \Lambda_2(t)$, where $\Lambda_1 \cap \Lambda_2 = \emptyset$ for all t. Then, there exist invertible $S \in \mathscr{C}_1^k(\mathbb{R}, \mathbb{C}^{n \times n})$, such that for all t:
>
> $$S^{-1}(t)A(t)S(t) = \begin{bmatrix} E_{11} & 0 \\ 0 & E_{22} \end{bmatrix},$$
>
> and $\sigma(E_{ii}) = \Lambda_i$, $i = 1, 2$. Moreover, if the function A is real valued, and each complex conjugate pair of eigenvalues belong to one of the groups Λ_1 or Λ_2, then S and E_{ii}, $i = 1, 2$, can be taken real valued and in general will be of period 2: $S, E_{11}, E_{22} \in \mathscr{C}_2^k(\mathbb{R}, \mathbb{R}^{n \times n})$.

It is rather straightforward to refine Sibuya's result to several distinct groups of eigenvalues, and to further specialize these types of results to the case of all eigenvalues being distinct, to the case of symmetric (Hermitian) functions, and to the SVD as well. For completeness, we present one such extension (which we will need later), in the form of a Theorem and leave its verification as an exercise.

Theorem 2.29. *Let $A \in \mathscr{C}_1^k(\mathbb{R}, \mathbb{C}^{n \times n})$, with $k \geq 0$. Assume that $\sigma(A(t)) = \Lambda_1(t) \cup \cdots \cup \Lambda_p(t)$, where $\Lambda_i(t) \cap \Lambda_j(t) = \emptyset$ for all t and $i \neq j$, $i, j = 1, \ldots, p$. Then there exist invertible $S \in \mathscr{C}_1^k(\mathbb{R}, \mathbb{C}^{n \times n})$, and unitary $Q \in \mathscr{C}_1^k(\mathbb{R}, \mathbb{C}^{n \times n})$ such that*

$$S^{-1}(t)A(t)S(t) = \begin{bmatrix} E_{11} & 0 & \ldots & 0 \\ 0 & E_{22} & \ldots & 0 \\ \vdots & \ddots & \ddots & \vdots \\ 0 & \ldots & 0 & E_{pp} \end{bmatrix} \equiv E(t),$$

$$Q^*(t)A(t)Q(t) = \begin{bmatrix} T_{11} & T_{12} & \ldots & T_{1p} \\ 0 & T_{22} & \ldots & T_{2l} \\ \vdots & \ddots & \ddots & \vdots \\ 0 & \ldots & 0 & T_{pp} \end{bmatrix} \equiv T(t),$$

where each $T_{ij} \in \mathscr{C}_1^k(\mathbb{R}, \mathbb{C}^{n_i \times n_j})$ and $E_{ii} \in \mathscr{C}_1^k(\mathbb{R}, \mathbb{C}^{n_i \times n_i})$ with $\sigma(E_{ii}) = \sigma(T_{ii}) = \Lambda_i$, $i = 1, \ldots, p$, and $n_1 + \cdots + n_p = n$.

In the real case, if we assume that $\det(\Lambda_i) \in \mathbb{R}$ for all i, then the previous statements are true for real invertible $S \in \mathscr{C}_2^k$, and orthogonal $Q \in \mathscr{C}_2^k$. Now, in general, we have that each $T_{ij} \in \mathscr{C}_2^k$ and $E_{ii} \in \mathscr{C}_2^k$.

In either the real or complex case, if A is symmetric, or Hermitian, then $T_{ij} = 0$, $i \neq j$, and T_{ii}, $i = 1, \ldots, p$, are also symmetric or Hermitian.

Finally, in the real case, if A has only simple eigenvalues (real or complex conjugate), then can choose $S \in \mathscr{C}_2^k$ such that we have real $E_{ii} \in \mathscr{C}_1^k$. Likewise, in the symmetric case, we have real $T_{ii} \in \mathscr{C}_1^k$, though in general the $\mathscr{C}^k$ orthogonal function Q is 2-periodic.

Results for the SVD are similar, and we quote one such result without proof.

Let $A \in \mathscr{C}_1^k(\mathbb{R}, \mathbb{C}^{m \times n})$, respectively $\mathbb{R}^{m \times n}$, $m \geq n$, $\text{rank}(A) = n - r$, $\forall t$, $0 \leq r \leq n - 1$. Then, A has the SVD:

$$A = U \begin{bmatrix} S_+ & 0 \\ 0 & 0 \end{bmatrix} V^*,$$

with U and V unitary (orthogonal), S_+ symmetric (Hermitian) positive definite and further

$$\mathbb{C}: \quad U \in \mathscr{C}_1^k, \ V \in \mathscr{C}_1^k, \ S_+ \in \mathscr{C}_1^k$$
$$\mathbb{R}: \quad U \in \mathscr{C}_2^k, \ V \in \mathscr{C}_2^k, \ S_+ \in \mathscr{C}_2^k.$$

The function S_+ can be further decomposed as appropriate using Theorem 2.29.

To recap, we have that for distinct groups of eigenvalues/singular values, Sibuya's theory guarantees that eigendecompositions and SVD results retain a similar flavor to those from Floquet theory. The next remark summarizes one of the key consequences of this.

Remark 2.30. Of particular relevance to us is the fact that a smooth 1-periodic Hermitian (symmetric) function A with distinct eigenvalues admits an eigendecomposition where the eigenvalues retain full smoothness and are 1-periodic; further, in the complex case, the unitary function of eigenvectors will also be smooth and

Fig. 10 Example 2.31. Eigenvalues

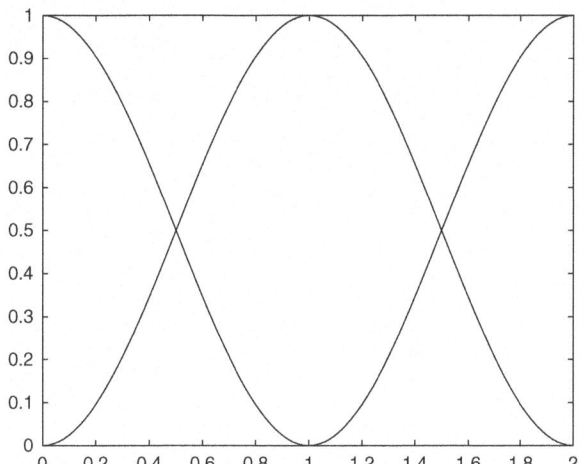

1-periodic, whereas in the real case (i.e., the case of symmetric A) the orthogonal factor will also be smooth, but possibly 2-periodic.

The question now becomes: what happens if eigenvalues/singular values coalesce? Let us focus just on the case of a symmetric (Hermitian) function, since this is the problem in which we are interested. Extensions for the SVD will be immediate. So, here below we will restrict to the case:

$$A \in \mathscr{C}_1^{k,\omega}(\mathbb{R}, \mathbb{C}^{n \times n}), \; A^* = A \quad \text{or} \quad A \in \mathscr{C}_1^{k,\omega}(\mathbb{R}, \mathbb{R}^{n \times n}), \; A^T = A.$$

To arrive at an appropriate periodicity result when eigenvalues coalesce, the following simple observations are useful.

(i) The eigenvalues are roots of the characteristic polynomial and the coefficients of the characteristic polynomial are 1-periodic (since A is).
(ii) As a consequence of (i), one can surely label the eigenvalues so that they are 1-periodic functions of t, and continuous. However, in general, with such labeling they are not smooth functions, they are just continuous whenever they coalesce during one period, as the next example clarifies.

Example 2.31. Consider the symmetric analytic function $A \in \mathscr{C}_1^\omega$ (see Fig. 10 for a plot of the eigenvalues):

$$A(t) = \begin{bmatrix} 1 - \frac{1}{2}\sin^2 \pi t & -\frac{1}{4}\sin 2\pi t \\ -\frac{1}{4}\sin 2\pi t & \frac{1}{2}\sin^2 \pi t \end{bmatrix}.$$

We make the following observations:

1. We can take (see Theorem 2.15) analytic eigenvalues and eigenvectors: $\lambda_1(t) = \frac{1+\cos \pi t}{2}$ and $\lambda_2(t) = \frac{1-\cos \pi t}{2}$,

$$Q = \begin{bmatrix} \cos\frac{\pi}{2}t & \sin\frac{\pi}{2}t \\ -\sin\frac{\pi}{2}t & \cos\frac{\pi}{2}t \end{bmatrix}, \quad Q^T A Q = \begin{bmatrix} \lambda_1 & 0 \\ 0 & \lambda_2 \end{bmatrix}.$$

With this, we note that $\lambda_{1,2} \in \mathscr{C}_2^\omega$, and $Q \in \mathscr{C}_4^\omega$.

2. Or, we could also take a different labeling:

$$\tilde{\lambda}_1 = 1/2 + \begin{cases} 1/2 \cos \pi t, & 0 \le t \le 1/2, \text{ or } 3/2 \le t \le 2; \\ -1/2 \cos \pi t, & 1/2 \le t \le 3/2; \end{cases}$$

$$\tilde{\lambda}_2 = 1/2 + \begin{cases} -1/2 \cos \pi t, & 0 \le t \le 1/2, \text{ or } 3/2 \le t \le 2; \\ 1/2 \cos \pi t, & 1/2 \le t \le 3/2. \end{cases}$$

Now $\tilde{\lambda}_{1,2} \in \mathscr{C}_1^0$, and $\tilde{Q}$ such that $\tilde{Q}^T A \tilde{Q} = \begin{bmatrix} \tilde{\lambda}_1 & 0 \\ 0 & \tilde{\lambda}_2 \end{bmatrix}$ is discontinuous!

So, in Example 2.31 we have that when we retain smoothness—and the eigenvalues coalesced over one period—the eigenvalues have increased (doubled) their period, and the eigenvectors have further doubled the eigenvalues period; we could have kept the eigenvalues merely continuous and of period 1, but at the expense of discontinuous eigenvectors. We also make a seemingly innocent observation: as soon as the eigenvalues are chosen (labeled) so as to retain the desired smoothness, there is no further control (freedom) on the periodicity of the eigenvalues and eigenvectors. In other words, the appropriate mindframe is to proceed in two steps: first to decompose A smoothly in order to retain smooth eigenvalues, then to study the period of the eigenvalues and of the associated eigenvectors.

Now, it is natural to suspect that it is the coalescing of the eigenvalues which prompted the increase in period in Example 2.31; however, this is not entirely correct as the next example shows.

Example 2.32. Take the following function (see Fig. 11 for a plot of the eigenvalues):

$$A(t) = Q(t) D(t) Q^T(t), \quad D(t) = \text{diag}\left(\cos^2 \pi t, \frac{1}{2} \sin^2 \pi t\right),$$

$$Q(t) = \begin{bmatrix} \cos \pi t & \sin \pi t \\ -\sin \pi t & \cos \pi t \end{bmatrix}, \quad \forall t.$$

Now we have that $A, D \in \mathscr{C}_1^\omega$, while $Q \in \mathscr{C}_2^\omega$. So, the eigenvalues coalesced and retained full smoothness and period 1, the eigenvectors retained full smoothness but doubled their period.

As anticipated, there is a subtle conflict between minimizing the period and retaining smoothness of eigenvalues. In order to keep smooth eigenvalues, we may need to increase their period. It is also clear that periodicity for the smooth eigenvalues is related to whether or not the eigenvalues coalesce. **But**, what really matters is the relative position of the smooth eigenvalues after 1 unit of time, rather

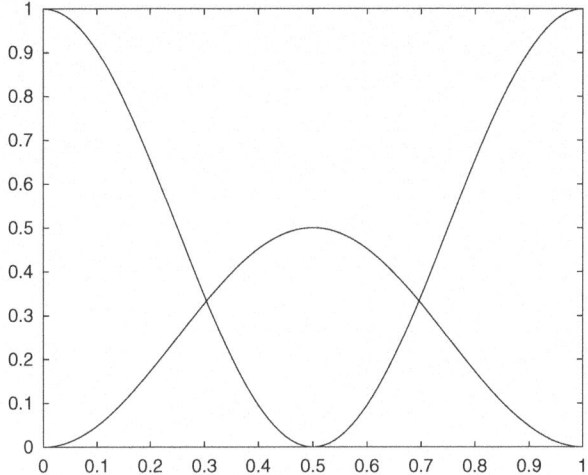

Fig. 11 Example 2.32. Eigenvalues

than whether or not they coalesced (see Example 2.32). To avoid trivialities, let us henceforth assume that no two eigenvalues are identical for all t.

To understand what is going on, let us begin by assuming that we have a smooth decomposition of A (recall, that A is here assumed to be Hermitian, respectively symmetric in the real case). That is, we have unitary U (respectively, orthogonal Q) such that $U^*AU = D$, respectively $Q^TAQ = D$, where D is the diagonal function of the smooth eigenvalues. [For example, such smooth decomposition is guaranteed to exist under the assumptions of Theorem 2.17 (smooth case) or Theorem 2.15 (analytic case).]

Now, suppose we label the (smooth) eigenvalues $\lambda_1, \ldots, \lambda_n$, and that they are distinct and ordered at time $t = 0$: $\lambda_1(0) > \cdots > \lambda_n(0)$. Since A is periodic, the set of eigenvalues at time $t = 1$ must coincide with the set of eigenvalues at time $t = 0$. However, in general, we can only say that for any $j = 1, \ldots, n$, we have $\lambda_j(1) = \lambda_k(0)$, for some value of $k = 1, \ldots, n$. This means that we have a permutation matrix P that performs the transition from $t = 0$ to $t = 1$:

$$\begin{bmatrix} \lambda_1(1) \\ \vdots \\ \lambda_n(1) \end{bmatrix} = P \begin{bmatrix} \lambda_1(0) \\ \vdots \\ \lambda_n(0) \end{bmatrix}. \tag{9}$$

Since P is a permutation, there is a (least) integer p such that $P^p = I$. Therefore, after p units of time, the eigenvalues will all have returned to their initial ordering. Of course, it is entirely possible that some eigenvalues have period smaller than p, but the entire diagonal matrix of eigenvalues will have period p. Therefore, p is the (integer) period of the smooth eigenvalues matrix.

There are two natural questions: (i) What is p? How large can it be? (ii) What is the period of the matrix of the (smooth) eigenvectors? The answer to (i) is the content of the following example.

Example 2.33. Suppose that P is an irreducible permutation matrix.[2] Then (and this verification is left as an exercise for the reader), P is cyclic: $P = \begin{bmatrix} 0 & 1 & 0 & \cdots & 0 \\ 0 & 0 & 1 & & \vdots \\ \vdots & & \ddots & \ddots & \\ 0 & & \cdots & 0 & 1 \\ 1 & & \cdots & 0 & 0 \end{bmatrix}$. If such an irreducible P is of size n, then the period is easily given by $p = n$.

Now, when the permutation matrix P of (9) is not in irreducible form, we perform its *irreducible decomposition*: $E^T P E = \begin{bmatrix} P_1 & & & \\ & P_2 & & \\ & & \ddots & \\ & & & P_k \end{bmatrix}$, where E is itself a permutation matrix and each P_j is an irreducible permutation of size n_j, $j = 1, \ldots, k$, and $n_1 + \cdots + n_k = n$. Since each P_j has period n_j, then P has period given by the least common multiple of the n_j's. In other words:

$$p = \mathrm{lcm}(n_1, \ldots, n_k). \tag{10}$$

At this point, we can understand how large can p in (10) be. Letting $n_1, \ldots, n_l$ be a partition of n: $n_1 + \cdots + n_l = n$, and $\mu(n) = \mathrm{lcm}(n_1, \ldots, n_l)$, we let $\mu^*(n)$ be the maximum value of $\mu(n)$ over all partitions of n. Then, p can be as large as $\mu^*(n)$, and we notice that $\log(\mu^*(n)) \approx \sqrt{n \log n}$, quite a large value!

For completeness, we remark that the above result on periodicity of the eigenvalues is sharp. [In [27], there are examples of As of stated smoothness with D taking all possible periods of the type p/q, $(p, q) = 1$, and $1 \leq p \leq \mu^*(n)$.]

Next, consider the function of eigenvectors of A, which we can assume is such that $U^* A U = D$ with D of the form

$$D = \mathrm{diag}(D_1, D_2, \ldots, D_k), \quad D_j = \mathrm{diag}(\lambda_j^{(l)}, l = 1, \ldots, n_j), \quad j = 1, \ldots, k,$$

so that pairs of coalescing eigenvalues belong to the same diagonal block. In other words, we can partition $U = \begin{bmatrix} U_1 & U_2 & \ldots & U_k \end{bmatrix}$ (similarly, Q in the real symmetric case) conformally to D's partitioning:

$$A U_j = U_j D_j, \quad D_j = \mathrm{diag}(\lambda_j^{(l)}, l = 1, \ldots, n_j), \quad j = 1, \ldots, k.$$

---

[2]Recall that a matrix M is irreducible if there does not exist a permutation matrix E for which $E^T M E$ is in block upper triangular form: $E^T M E = \begin{bmatrix} M_{11} & M_{12} \\ 0 & M_{22} \end{bmatrix}$. In particular, a permutation matrix P is irreducible if there does not exist another permutation matrix E for which $E^T P E = \begin{bmatrix} P_{11} & 0 \\ 0 & P_{22} \end{bmatrix}$ with P_{11} and P_{22} permutation matrices of smaller size.

With these preparations, the following result on periodicity of the function of the eigenvectors was proved in [27].

Theorem 2.34 (Periodicity of Eigenvectors). *Let $A = A^* \in \mathscr{C}_1^k(\mathbb{R}, \mathbb{C}^{n \times n})$, and let there exist $U \in \mathscr{C}^l(\mathbb{R}, \mathbb{C}^{n \times n})$ ($l \leq k$) such that $U^* A U = \mathrm{diag}(D_1, D_2, \ldots, D_k)$, where $D_1, \ldots, D_k$ are diagonal, and for all t $\sigma(D_i) \cap \sigma(D_j) = \emptyset$, $i \neq j$, with each $D_j \in \mathscr{C}_{p_i}^k(\mathbb{R}, \mathbb{R}^{n_j \times n_j})$. $1 \leq p_j \leq n_j$, $j = 1, \ldots, k$. Then, we can take each U_j of period p_j, and hence U of period $p = \mathrm{lcm}(p_1, p_2, \cdots, p_l)$.*

If the function A is symmetric real valued, and $Q \in \mathscr{C}^l(\mathbb{R}, \mathbb{R}^{n \times n})$ ($l \leq k$) is such that $Q^T A Q = \mathrm{diag}(D_1, D_2, \ldots, D_k)$, then each Q_j can be taken of period $2 p_j$, $j = 1, \ldots, k$, and Q of period $2 p$.

Remarks 2.35. (i) The proof of Theorem 2.34 is technical, the key step requiring to show that a given set of smooth orthonormal columns, say U_j above: $A U_j = U_j D_j$, can be smoothly modified into a smooth periodic orthonormal set of period p_j. For details, see [27].
(ii) The case of $A \in \mathscr{C}^\omega$ is actually easier, and one can take the unitary (respectively, orthogonal) eigenvector function analytic with period given by $p = \mathrm{lcm}(p_1, \ldots, p_l)$ (respectively, orthogonal with period $2p$).
(iii) In general, the result on periodicity of Q is optimal.
(iv) It must be stressed that, in the real case, the function of smooth eigenvectors **may** have twice the period of the function of the eigenvalues. However, the result does **not** say that it **must** have twice the period. In particular, in case the eigenvalue are distinct, the function D will have period 1, and the function of eigenvectors, Q, will have period 1 or 2. But, it is not clear whether or not the period will double; see Problem 2.36 below and Chap. 3.

Problem 2.36. You have to show that for a given continuous and symmetric function A, continuous 1-periodic and 2-periodic eigen-decompositions cannot coexist. More precisely, show the following result.

Let $A \in \mathscr{C}_1^0(\mathbb{R}, \mathbb{R}^{n \times n})$, $A^T = A$. Suppose there exists orthogonal Q such that

$$A(t) = Q(t) D(t) Q^T(t), \quad \forall t,$$

with $D \in \mathscr{C}_1^0(\mathbb{R}, \mathbb{R}^{n \times n})$ diagonal, with distinct eigenvalues, and $Q \in \mathscr{C}_2^0(\mathbb{R}, \mathbb{R}^{n \times n})$, with

$$Q(t+1) = Q(t) S, \quad \forall t \in \mathbb{R},$$

where $S = \mathrm{diag}(\pm 1)$, but $S \neq I_n$. Then, there is no matrix function V such that: $V \in \mathscr{C}_1^0(\mathbb{R}, \mathbb{R}^{n \times n})$, orthogonal, and $V^T(t) A(t) V(t) = D(t)$ for all $t \in \mathbb{R}$.

[Hint: Argue by contradiction.]

For the SVD, one has very much the same type of periodicity results. Again, as we saw before, it is worth stressing that—in case the singular values go through zero and/or they coalesce—we must allow for an SVD with unordered and possibly negative singular values.

2.3 Computing an SVD Path

We complete this chapter by presenting an algorithm of predictor-corrector type which computes a smooth path for the SVD of a one-parameter function A taking values in $\mathbb{R}^{m \times n}$, $m \geq n$. We seek U orthonormal (in $\mathbb{R}^{m \times n}$) and V orthogonal, such that $A = U \Sigma V^T$. Naturally, a similar algorithm can be used to compute a smooth path for the eigendecomposition of a symmetric function.

The algorithm presented here was originally introduced in [42] and it underwent several levels of later fine tuning, though the basic description below is adequate for our present purposes and it suffices to understand the key aspects of the technique used for the numerical experiments of Chaps. 3 and 4.

After the explanation of the algorithm, we will illustrate the technique using it to continue curves of equilibria of a dynamical system. Recall that when finding a smooth SVD the singular values may become negative.

The basic idea of our approach is to compute an SVD at a given step, using standard linear algebra software, and then recover smoothness by techniques of least variation with respect to some suitably **predicted** factors. Important aspects that we want to guarantee are:

(i) the stepsize h (i.e., where to factor) must be chosen **adaptively**, based on the variation of the factors themselves;
(ii) the end result must be an "exact" SVD.

A high level description of our algorithm follows for a typical step from the point t_j to the point $t_{j+1} = t_j + h_j$ (the very first time, h_0 is an input value, then it is adjusted adaptively). Given an SVD at the value t_j: $A(t_j) = U_j \Sigma_j V_j^T$, we want an SVD at t_{j+1}, $A(t_{j+1}) = U_{j+1} \Sigma_{j+1} V_{j+1}^T$, along the smooth SVD path.

Skleton of the "SVD Path" Algorithm

(a) Predict the singular values from the differential equations (6). For example, we use

$$\Sigma^{\mathrm{pred}} = \Sigma_j + \mathrm{diag}(U_j^T (A(t_{j+1}) - A(t_j)) V_j) = \Sigma_j + h_j \dot{\Sigma}_j = O(h_j^2),$$

and for the singular vectors:

$$U^{\mathrm{pred}} = U_j + h_j \frac{U_j - U_{j-1}}{t_j - t_{j-1}}, \text{ and } V^{\mathrm{pred}} = V_j + h_j \frac{V_j - V_{j-1}}{t_j - t_{j-1}};$$

for $j = 0$ the initial factors at t_0 are trivially used as predictors.
Detect if we are near (or at) a possible crossing of singular values:
– Two singular values, σ_i and σ_{i+1}, are declared "close to coalesce" at t_j, if

$$\frac{|\sigma_{i+1}(t_j) - \sigma_i(t_j)|}{|\sigma_i(t_j)| + 1} < \text{TOL};$$

– Declare possible crossing if either of these holds:

$$\frac{\sigma_{i+1}^{\text{pred}} - \sigma_i^{\text{pred}}}{\sigma_{i+1}(t_j) - \sigma_i(t_j)} < 0, \quad \frac{|\sigma_{i+1}^{\text{pred}} - \sigma_i^{\text{pred}}|}{|\sigma_i^{\text{pred}}| + 1} < \text{TOL},$$

where TOL is an input tolerance (we used 10^{-4}).

If a possible crossing is detected, then we group the "coalescing" singular values into a (2×2) block.

(b) Compute an algebraic SVD $A(t_{j+1}) = U^c \Sigma^c (V^c)^T$, and then **correct** it by solving an orthogonal Procrustes problem, to recover smoothness.

– The solution of the orthogonal Procrustes problem is trivial relatively to columns of U^c and V^c corresponding to distinct singular values (it is at most a change of sign). But relatively to a (2×2) block we need to bring the corresponding computed singular vectors $U^c(:, i : i+1)$ and $V^c(:, i : i+1)$ as close as possible to the exact $U_{j+1}(:, i : i+1)$ and $V_{j+1}(:, i : i+1)$. We proceed as follows.

– When σ_i^c and σ_{i+1}^c are close but numerically well separated, apart from the signs, exchange columns of U^c and V^c, and reorder singular values if (a criterion of [39])

$$|u_{11}| + |u_{22}| < |u_{12}| + |u_{21}|,$$

where $u = U^{\text{pred}}(:, i : i+1)^T U^c(:, i : i+1)$.

– If σ_i and σ_{i+1} do coalesce at t_{j+1} (say within 10^{-13}, when working in double precision), solve the following orthogonal Procrustes problem:

$$\min_{Z^T Z = I} \|U^c(:, i : i+1)Z - U^{\text{pred}}(:, i : i+1)\|_F^2$$
$$+ \|V^c(:, i : i+1)Z - V^{\text{pred}}(:, i : i+1)\|_F^2.$$

It is well known that the solution is given by $Z = vu^t$, where u and v are the orthogonal factors of the SVD decomposition of the 2×2 matrix:

$$U^{\text{pred}}(:, i : i+1)^T U^c(:, i : i+1) + V^{\text{pred}}(:, i : i+1)^T V^c(:, i : i+1).$$

(c) As we proceed, we will **unroll** the block if

$$\frac{|\sigma_{i+1}(t_{j+1}) - \sigma_i(t_{j+1})|}{|\sigma_i(t_{j+1})| + 1} > \text{TOL}.$$

(d) We trigger accurate localization of crossings and zeros (using the secant method) when in the interval $[t_j, t_{j+1}]$, for some i or k, we have either

$$(\sigma_{i+1}(t_j) - \sigma_i(t_j))(\sigma_{i+1}(t_{j+1}) - \sigma_i(t_{j+1})) \leq 0, \quad \text{(crossing)}$$

$$\text{or} \quad \sigma_k(t_j)\sigma_k(t_{j+1}) \leq 0, \quad \text{(zero)}.$$

(e) The stepsize selection mechanism is based on the "predictor error". We have predicted values U^{pred}, Σ^{pred}, V^{pred}, and *corrected* values U_{j+1}, Σ_{j+1}, V_{j+1}. We compute the weighted error factors $\rho_U = \|U^{\text{pred}} - U_{j+1}\|_1/\epsilon_a$, $\rho_V = \|V^{\text{pred}} - V_{j+1}\|_1/\epsilon_a$, and $\rho_\sigma = \max_{1 \leq i \leq n} \frac{|\sigma_i^{\text{pred}} - \sigma_i(t_{j+1})|}{\epsilon_r |\sigma_i(t_{j+1})| + \epsilon_a}$, where ϵ_r and ϵ_a are relative and absolute error tolerances. [In practice, we use $\epsilon_r = \epsilon_a = 100\text{EPS}$, where EPS is the machine precision.] Then, we set $\rho = \max(\rho_\sigma, \rho_U, \rho_V)$ and $h_{\text{new}} = h_j/\sqrt{\rho}$. If $\rho \leq 1.5$, the step is accepted, otherwise it is rejected. In all cases, h_{new} is used as the new steplength.

2.3.1 Curves of Equilibria

Consider computing curves of equilibria of smooth vector fields

$$f : \mathbb{R}^n \times \mathbb{R} \to \mathbb{R}^n : f(x, \alpha) = 0.$$

This is a standard problem, very well studied, with reliable software for its solution (e.g., see [18, 21]). In their essential aspects, existing methods effectively follow the path following approach of [6], outlined hereafter.

- Suppose the sought curve is parametrized by arc-length:

$$f(x(s), \alpha(s)) = 0, \quad \dot{x}^T(s)\dot{x}(s) + (\dot{\alpha}(s))^2 = 1,$$

where $\begin{bmatrix} \dot{x}(s) \\ \dot{\alpha}(s) \end{bmatrix}$ is the tangent vector: $f_x \dot{x} + f_\alpha \dot{\alpha} = 0$.

In reality, it is cumbersome to use the arc-length equations to find a next point on the curve. The adopted practice is to use the so-called *pseudo* (or approximate) arc-length. That is, given (x_0, α_0) and Δs, we seek (x_1, α_1) such that

$$f(x_1, \alpha_1) = 0, \quad \dot{x}_0^T(x_1 - x_0) + \dot{\alpha}_0(\alpha_1 - \alpha_0) = \Delta s.$$

- To solve the pseudo arc-length system above, we use a quasi-Newton procedure.

  - For $k = 0, 1, \ldots$, solve

  $$\begin{pmatrix} f_x & f_\alpha \\ \dot{x}_0^T & \dot{\alpha}_0 \end{pmatrix}_{x_1^{(0)}, \alpha_1^{(0)}} \begin{pmatrix} \Delta x \\ \Delta \alpha \end{pmatrix} = -\begin{pmatrix} f(x_1^{(k)}, \alpha_1^{(k)}) \\ \dot{x}_0^T (x_1^{(k)} - x_0) + \dot{\alpha}_0 (\alpha_1^{(k)} - \alpha_0) - \Delta s \end{pmatrix}$$

  and update the iterates

  $$\begin{pmatrix} x_1^{(k+1)} \\ \alpha_1^{(k+1)} \end{pmatrix} = \begin{pmatrix} x_1^{(k)} \\ \alpha_1^{(k)} \end{pmatrix} + \begin{pmatrix} \Delta x \\ \Delta \alpha \end{pmatrix}.$$

  The initial guess is provided by the tangent approximation (or a secant approximation to it):

  $$\begin{pmatrix} x_1^{(0)} \\ \alpha_1^{(0)} \end{pmatrix} = \begin{pmatrix} x_0 \\ \alpha_0 \end{pmatrix} + \Delta s \begin{pmatrix} \dot{x}_0 \\ \dot{\alpha}_0 \end{pmatrix}.$$

The present goal is to see how a smooth SVD path can be used to accomplish the previous tasks. Using the SVD for computing branches of equilibria of vector fields, and bifurcation points, has not been particularly exploited in works on the subject. It appeared in the work [41], but then one cannot find it anymore. Why? Most likely, because existing techniques seem to work just fine. But possibly also because, by using the standard linear algebra SVD, it is nearly **impossible** to land exactly on a fold or on a branch point. Indeed, the authors of [41] worked with a modified Newton method on the rectangular problem $[f_x, f_\alpha]$, and monitored when singular values of a standard algebraic SVD became small. As we understand, this is not a reliable practice: We must monitor a signed SVD!

Our approach is to work with the enlarged matrix, $M := \begin{pmatrix} f_x & f_\alpha \\ \dot{x}_0^T & \dot{\alpha}_0 \end{pmatrix}$, by finding a smooth SVD of f_x only. Notice that if we have $f_x = U \Sigma V^T$, then

$$\begin{pmatrix} f_x & f_\alpha \\ \dot{x}_0^T & \dot{\alpha}_0 \end{pmatrix} \begin{pmatrix} \Delta x \\ \Delta \alpha \end{pmatrix} = \begin{pmatrix} c \\ d \end{pmatrix}$$

becomes simply

$$\begin{pmatrix} \Sigma & U^T f_\alpha \\ \dot{x}_0^T V & \dot{\alpha}_0 \end{pmatrix} \begin{pmatrix} V^T \Delta x \\ \Delta \alpha \end{pmatrix} = \begin{pmatrix} U^T c \\ d \end{pmatrix}$$

which has a very simple structure and inexpensive solution process.

The tasks are therefore the following.

1) Compute a smooth SVD of f_x at points on the curve of equilibria. This we do according to the smooth SVD continuation method previously described.
2) Get the tangents and continue the path.

   - To get the tangent(s), we need to solve

   $$\left(\Sigma \, , \, U^T f_\alpha \right) \begin{pmatrix} y \\ \dot\alpha \end{pmatrix} = 0$$

   and normalize. If Σ is invertible, then this is easy to do (paying some attention to the signs). If Σ is singular, but $f_\alpha \notin \text{range}(f_x)$ then this is also easy. [This is a *Fold* point.] If Σ is singular, and $\text{rank}(M) = n$, then there are two tangents. [This is a *branch* point.]

3) Locate folds. To locate folds (or possible branch points) we monitor the singular values of f_x going through 0, a process we know how to do.
4) Locate simple branch points ($\text{rank}(M) = n$). Switch branches and continue new branches. Note that branch points can happen with $f_\alpha \in \text{range}(f_x)$, and a single singular value of f_x at 0. But can also happen with two singular values of f_x at 0.

   - To switch branches, we find an approximate vector in the null space of $[f_x, f_\alpha]$ orthogonal to the tangent we already have and take a small step along that direction.

Example 2.37. We illustrate the above algorithm on a test problem from the AUTO manual (see [21]):

$$f(x,\alpha) = \begin{bmatrix} x_1(1-x_1)-3x_1x_2 \\ -\frac{1}{4}x_2+3x_1x_2-3x_2x_3-\alpha(1-e^{-5x_2}) \\ -\frac{1}{2}x_3+3x_2x_3 \end{bmatrix}.$$

The continuation is initialized at $(x_0, \alpha_0) = ((1,0,0),0)$, and $\alpha \in [0, 0.6]$. The result of our computation is shown in Fig. 12. There is a branch point on the initial curve, and further folds and branch points on the bifurcating branches.

3 Two Parameters. Coalescing Eigenvalues and Singular Values

In this chapter, we consider matrix valued functions depending on two parameters, $A \in \mathscr{C}^k(\Omega, \mathbb{R}^{m \times n})$, $k \geq 1$, $m \geq n$, and more specifically restrict to either the case of the SVD for general A, or the eigendecomposition for symmetric A, $A = A^T$.

Fig. 12 Curves of equilibria

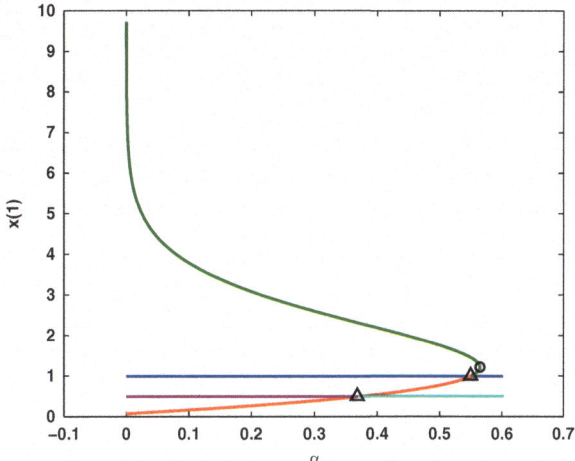

Naively, one may treat this case of two parameters by considering it as a family of one parameter problems, namely by freezing one parameter and considering the relevant decompositions with respect to the other, thereby (in principle) obtaining a parametrized family of decompositions. However, now in general we cannot expect to have smooth decompositions in case the eigenvalues coalesce, and so this approach is bound to be flawed (see below). In fact, in the present context, there are fundamental obstacles caused by coalescing eigenvalues (singular values) and—in general—there is no smooth decomposition through a coalescing point. For this reason, and further motivated by practical applications, our emphasis here is on **locating coalescing points** for the eigenvalues of a symmetric function, respectively singular values of a general function, that is values of $x \in \Omega$ where the eigenvalues (singular values) coincide. Insofar as the SVD is concerned, given that we will not try to continue an SVD through a coalescing point, in the present chapter the SVD will always refer to be the one with ordered (though not necessarily non-negative) singular values.

The presentation in this chapter is based on our works [45–47]. From the application point of view, the issue of coalescing of eigenvalues for two-parameter matrices appear in many contexts, at times in disguise; for example, in the Chemical Physics literature, where goes by the name of *conical intersections* (e.g., see [15]), and in Quantum Physics, where they have also been called *diabolical points* (see [75]), but also in the structural dynamics literature in mechanical engineering where it is relevant to robustness properties of beam and plates and to wave propagation (e.g., see Chap. 1 and [65, 67]).

In Sect. 3.1 we will first clarify what kind of coalescing point we can hope to detect, and then in Sect. 3.2 we review theoretical results on the symmetric eigenproblem and the SVD for one and several coalescing points. In Sect. 3.3 we will discuss a general localization and refinement algorithm to locate coalescing points. Finally, in Sect. 3.4 we will give some results relatively to the case of coalescing points which persist in 1 parameter (along a curve).

The following result (see [50,51]) will be quite useful in order to restrict attention to regions where coalescing of eigenvalues (singular values) takes place. It is an extension of Theorem 2.9 to the case of a function depending on two parameters. We state it in the case of a rectangular region R, but it holds whenever R is a closed simply connected region with piecewise smooth boundary.

Theorem 3.1 (Block-Diagonalization). *Let R be a closed rectangular region in $\mathbb{R}^2$. Suppose that the eigenvalues of $A \in \mathscr{C}^k(R, \mathbb{R}^{n \times n})$, $k \geq 0$, can be labeled so that they belong to two disjoint sets for all $x \in R$: $\lambda_1(x), \ldots, \lambda_{n_1}(x)$ in $\Lambda_1(x)$ and $\lambda_{n_1+1}(x), \ldots, \lambda_n(x)$ in $\Lambda_2(x)$, $\Lambda_1(x) \cap \Lambda_2(x) = \emptyset$, $\forall x \in R$, with complex conjugate eigenvalues belonging to the same group. Then, there exists $M \in \mathscr{C}^k(R, \mathbb{R}^{n \times n})$, invertible, such that*

$$M^{-1}(x)A(x)M(x) =: S = \begin{bmatrix} S_1(x) & 0 \\ 0 & S_2(x) \end{bmatrix}, \forall x \in R,$$

where $S_1 \in \mathscr{C}^k(R, \mathbb{R}^{n_1 \times n_1})$, $S_2 \in \mathscr{C}^k(R, \mathbb{R}^{(n-n_1) \times (n-n_1)})$, and the eigenvalues of $S_i(x)$ are those in $\Lambda_i(x)$, for all $x \in R$ and $i = 1, 2$.

Further, let Γ be a simple closed curve in R, parametrized as a $\mathscr{C}^p$ ($p \geq 0$) function γ in the variable t, so that $\gamma : t \in \mathbb{R} \to R$ is $\mathscr{C}^p$ and 1-periodic. Let $m = \min(k, p)$, and let M_γ be $M(\gamma(t))$, $t \in \mathbb{R}$. Then, $M_\gamma \in \mathscr{C}_1^m(\mathbb{R}, \mathbb{R}^{n \times n})$.

Remarks 3.2. (1) Theorem 3.1 can be refined to any number of disjoint groups of eigenvalues. In particular, in the case of distinct and real eigenvalues, M diagonalizes A (in case A has distinct, but possibly complex conjugate eigenvalues, then there will be small 2×2 bumps along the diagonal of $M^{-1}AM$).
(2) If A is also symmetric, then M can be taken orthogonal, and we obtain a block-Schur decomposition. In this case, M is unique up to smooth orthogonal transformations of each block. In the case of distinct eigenvalues, an orthogonal M is unique (for given eigenvalues ordering) up to sign changes of its columns.
(3) Theorem 3.1 has an immediate generalization to the case of a block-SVD, and we leave the details of such statement as an exercise. (For the uniqueness part, it will be useful to recall Theorems 1.4 and 2.11.)

3.1 Two Parameters: Preliminaries

To begin with, let us recall the gist of Remark 2.25: "The property of having two eigenvalues of a symmetric function to be equal necessitates two parameters in order to be generic" (and similarly for the singular values, see below). But, of course, this does not yet say that—or when—we should expect to observe two equal eigenvalues, nor whether a coalescing of eigenvalues should be an isolated phenomenon. To address these concerns, let us be guided by the considerations in the following example.

Example 3.3 (Regular Values, Transversality, Genericity). Let us begin with a quick refresher of some important concepts in differential geometry (e.g., see [3]).

(a) Suppose we look at the 0-set of a $\mathscr{C}^k$ function, $k \geq 2$, $f : \Omega \to \mathbb{R}$, that is we look at the values in S_0: $S_0 := \{x \in \Omega : f(x) = 0\}$. The Morse-Sard theorem tells us that 0 is expected to be a **regular value** of f, that is $\nabla f(x) \neq 0$ for $x \in S_0$. Further, the preimage $f^{-1}(0)$ is a $\mathscr{C}^k$ sub-manifold of Ω, that is a $\mathscr{C}^k$ curve.
(b) Now suppose we have two $\mathscr{C}^k$ functions from Ω to $\mathbb{R}$, f_1 and f_2, which vanish at a point $\xi_0 \in \Omega$: $f_1(\xi_0) = f_2(\xi_0) = 0$, $\nabla f_1(\xi_0) \neq 0$, $\nabla f_2(\xi_0) \neq 0$, and let 0 be a regular value for f_1 and f_2. Thus, we have two $\mathscr{C}^k$ curves intersecting at ξ_0. Within the class of $\mathscr{C}^k$ curves, generically the intersection is **transversal**, that is the tangent vectors to the two curves at ξ_0 are not multiple of each other. This means that the Jacobian $\begin{bmatrix} \nabla f_1(x) \\ \nabla f_2(x) \end{bmatrix}_{\xi_0}$ is invertible.

Now—in a similar way to what we did in Remark 2.25—let us highlight that having a pair of equal (and nonzero) singular values is a **codimension 2** phenomenon. Take $A = \begin{bmatrix} a_{11} & a_{12} \\ a_{21} & a_{22} \end{bmatrix}$. Having equal (and nonzero) singular values means

$$A \in \left\{ \sigma \begin{bmatrix} \cos(\theta) & \pm\sin(\theta) \\ \sin(\theta) & \mp\cos(\theta) \end{bmatrix} : \sigma > 0, \theta \in \mathbb{R} \right\}.$$

Therefore, to have equal singular values, A needs to be a multiple of an orthogonal matrix, that is it is a codimension 2 phenomenon. Note that having the two singular values equal to 0 is actually a codimension 4 phenomenon! Indeed, in this case A must be the 0-matrix! Naturally, the codimension tells us how many parameters we will need in order to observe the phenomenon, under genericity conditions. Namely, to have a pair of coalescing (and nonzero) singular values—for coefficients which are $\mathscr{C}^k$ functions—is a generic property when A depends on two parameters.

With the above in mind, let us now clarify what kind of coalescing we should expect to see.

Example 3.4 (Conical Intersections). Consider first the following simple example (see [13]), highlighting that there is a complete loss of smoothness even for analytic A (cfr. with Theorem 2.15):

$$A(x) = I + \begin{bmatrix} x_1 & x_2 \\ x_2 & -x_1 \end{bmatrix}; \quad \lambda_\pm(x_1, x_2) = 1 \pm \sqrt{x_1^2 + x_2^2}.$$

Obviously, the eigenvalues $\lambda_\pm$ are continuous, but not even differentiable at the origin (and explicit computation shows that the eigenvectors are not even continuous there!).

The above matrix A is in a simplified form, but the type of singularity of the eigenvalues is not specific to this simple form of A. To clarify, take the general symmetric function

$$A(x) = \begin{bmatrix} a(x) & b(x) \\ b(x) & d(x) \end{bmatrix}, \quad x = (x_1, x_2) \in \Omega,$$

and let ξ_0 be a point where the eigenvalues of A coalesce. Then, we will have the two curves given by the following zero sets $(a - d)(x) = 0$ and $b(x) = 0$, which (generically) intersect transversally at ξ_0. Write the (continuous) eigenvalues as

$$\lambda_\pm(x) = \frac{a(x) + d(x)}{2} \pm \frac{1}{2}\sqrt{(a(x) - d(x))^2 + 4b(x)^2} =: \frac{(a+d)(x)}{2} \pm \frac{1}{2}\sqrt{h(x)}.$$

Let us expand h about ξ_0: $h(x) = h(\xi_0) + \nabla h(\xi_0)^T (x - \xi_0) + \frac{1}{2}(x - \xi_0)^T H(\xi_0)(x - \xi_0) + \ldots$. Easily, $h(\xi_0) = 0$, $\nabla h(\xi_0) = 0$, and the Hessian reads

$$H(\xi_0) = 2 \begin{bmatrix} [(a-d)_{x_1}]^2 + 4(b_{x_1})^2 & (a-d)_{x_1}(a-d)_{x_2} + 4b_{x_1}b_{x_2} \\ (a-d)_{x_1}(a-d)_{x_2} + 4b_{x_1}b_{x_2} & [(a-d)_{x_2}]^2 + 4(b_{x_2})^2 \end{bmatrix}_{\xi_0}.$$

Therefore, $\det H(\xi_0) = 16 \big[b_{x_1}(a-d)_{x_2} - b_{x_2}(a-d)_{x_1}\big]^2(\xi_0)$ and thus $\det H(\xi_0) = 0$ if and only if at ξ_0:

$$\begin{bmatrix}(a-d)_{x_1} & (a-d)_{x_2}\end{bmatrix} \begin{bmatrix} 0 & 1 \\ -1 & 0 \end{bmatrix} \begin{bmatrix} b_{x_1} \\ b_{x_2} \end{bmatrix} = 0,$$

that is $\nabla(a-d)(\xi_0)$ is parallel to $\nabla b(\xi_0)$, which we have excluded since the intersection is transversal. So $H(\xi_0)$ is positive definite, and the eigenvalues have the form $\lambda_\pm = \frac{(a+d)(x)}{2} \pm \frac{1}{2}\sqrt{\|z\|^2 + O(\|x - \xi_0\|^4)}$, where $z = H^{1/2}(\xi_0)(x - \xi_0)$: a double cone structure.

This is the reason why (generic) coalescing points of the eigenvalues of a symmetric function are also called **conical intersections**.[3] See the figure below, showing the

---

[3] They are also called "diabolical points", with reference to the popular toy "diablo".

generic coalescing point for the two surfaces of the eigenvalues $\lambda_\pm(x_1, x_2) = 1 \pm \sqrt{x_1^2 + x_2^2}$.

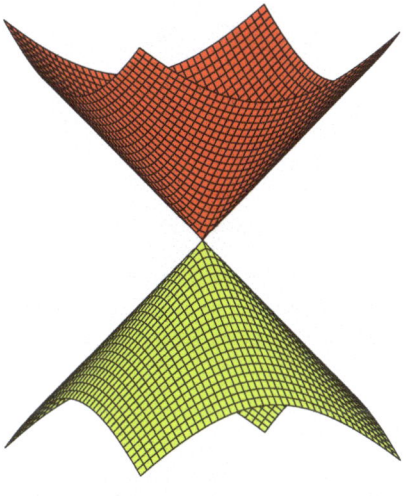

Conical Intersection

By virtue of Example 3.4, not only now we cannot expect global smoothness of the eigendecomposition, but in order to locate coalescing points, given their nature of being isolated conical intersections, we have very little hope to detect them by simply continuing curves of eigenvalues (or singular values) by freezing one parameter at the time. A new approach is needed, and it will be based upon the theoretical results of Sect. 3.2. Before looking at these results, let us summarize the differences between generic 1-parameter and 2-parameter cases; we do this relatively to the SVD, the case of the symmetric eigenproblem being much the same.

1. *One parameter.*

    - We expect a smooth matrix valued function to have distinct singular values (hence a smooth SVD).
    - When it occurs, coalescing of singular values will be broken under a generic small perturbation. In other words, singular values may come close to each other, but they are not expected to coalesce: *veering* phenomenon (see Example 3.5).

    What is particularly intriguing in the one-parameter case is precisely this possibility of "near coalescing" of singular values (eigenvalues); see figure below showing one such "avoided crossing".

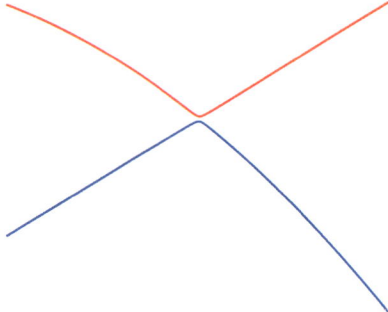

Veering or avoided crossing

What we are seeing here is actually a slice of a "double-cone" near a coalescing point. Indeed, this picture is ubiquitous whenever 1 parameter is varied, and it betrays the possible presence of a coalescing point for a (hidden) 2-parameter matrix function.

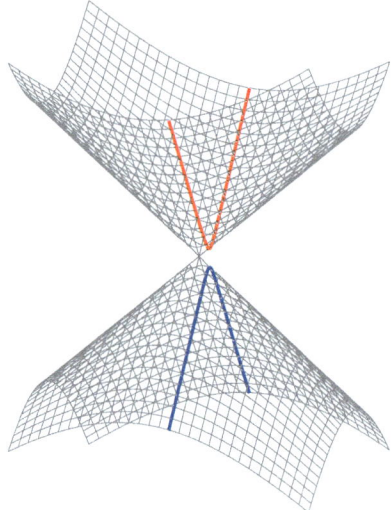

Slicing near conical intersections

2. *Two parameters.*

   - Coalescing of singular values has to be expected.
   - It occurs at isolated points and persists under small perturbations.
   - We should not see anything more degenerate than isolated coalescing of single pairs of singular values (transversal intersection).

It should be appreciated what these considerations say, and what they do not say. The gist is that for a two parameter function whose singular values coalesce along a curve, a generic perturbation will destroy the curve and only (some) isolated coalescing points will survive. For example, consider the (nongeneric) case

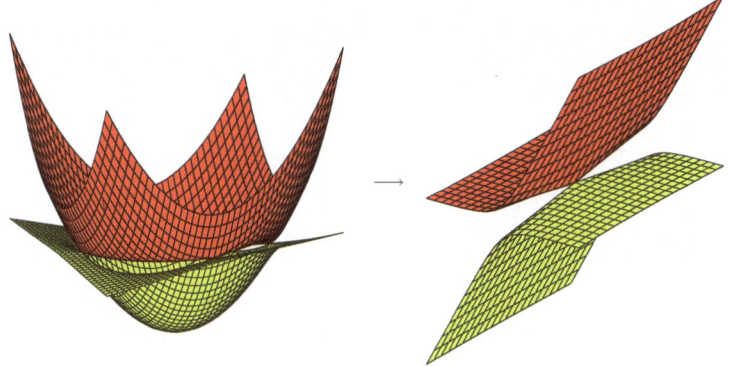

Fig. 13 Breaking of nongeneric coalescing and zoom-in on a surviving coalescing point

illustrated in the figure below. As soon as this nongeneric function is perturbed, the curve of coalescing points disappears and it is replaced by points of conical intersection. We illustrate this in Fig. 13, the second figure being an enlargement of a surviving coalescing point: indeed, a conical intersection of singular values.

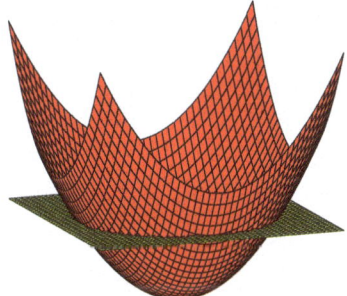

Nongeneric coalescing along a curve

We complete this section with an important example that clarifies in details the nature of the veering occurrence, and the associated mode exchange phenomenon (see the application on coupled oscillators, Sect. 1.3.3).

Example 3.5. Our purpose here is to constructively explain the veering phenomenon.

Suppose we have an analytic one-parameter function A, $A \in \mathscr{C}^\omega(\mathbb{R}, \mathbb{R}^{2\times 2})$. As we saw in Chap. 2, A admits an analytic SVD, regardless of whether or not the singular values coalesce (for the sake of precision, analyticity is not needed, any smooth A admitting a smooth SVD will do). Suppose that this function A has coalescing singular values at some (isolated) value of t. See the figure below, where we show the two singular values and (columnwise) the four components of the two left singular vectors.

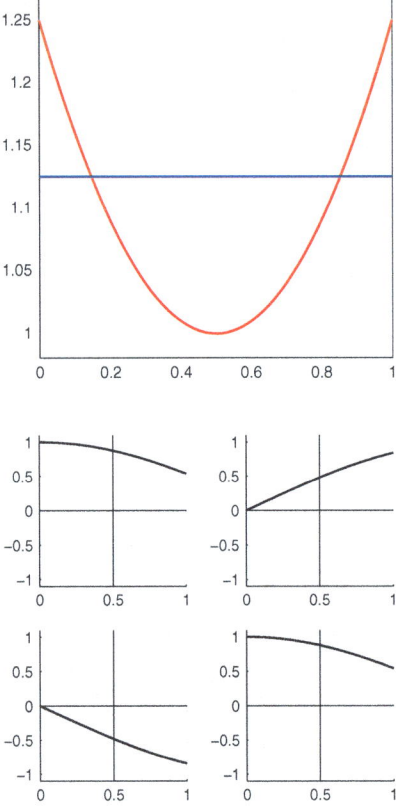

Unperturbed: Singular values and vectors

Next, we perturb (with a smooth, say analytic, perturbation) this function A, and let us assume that for the perturbed problem the singular values do not coalesce; hence, we surely have a smooth (analytic) SVD for this perturbed problem. Below, again we show the two singular values and (columnwise) the four components of the two left singular vectors. The most obvious facts are that the singular values veered, and the singular vectors appear to vary much more rapidly than before.

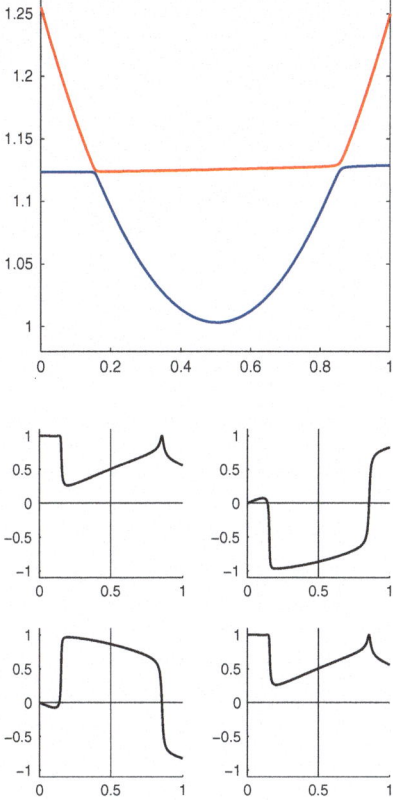

Perturbed: Singular values and vectors

Let us explain what we just observed, namely the jump in the singular vectors.

1. We have $A \in \mathscr{C}^\omega(\mathbb{R}, \mathbb{R}^{2\times 2})$, and all of $\sigma_1, \sigma_2, u_1, u_2, v_1, v_2$, are analytic. We also have $\sigma_1(t) = \sigma_2(t)$ at $t = \bar{t}$ (in the figure of the unperturbed case above, this occurred at two distinct points).
Let $E \in \mathscr{C}^k(\mathbb{R}, \mathbb{R}^{2\times 2})$ ($k \geq 1$, e.g., $k = \infty$) be such that $E(t) = 0$ for all $|t - \bar{t}| > \delta$, $\delta > 0$ small. Since it is expected to be the case, we will assume that the perturbation breaks down the coalescing, and we let $\tilde{\sigma}_1$ and $\tilde{\sigma}_2$ be the smooth singular values of the "perturbed" function $A + E$ and $\tilde{u}_1, \tilde{u}_2, \tilde{v}_1, \tilde{v}_2$, the corresponding smooth singular vectors. Now $\tilde{\sigma}_1$ and $\tilde{\sigma}_2$ veer away from each other at $\bar{t}$.

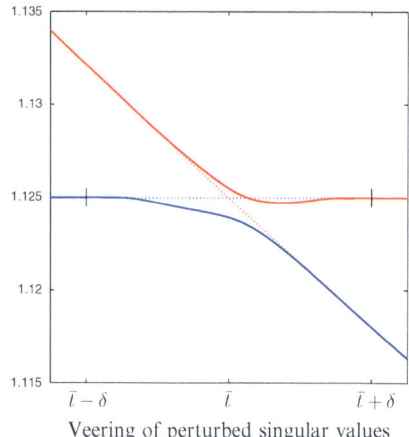
Veering of perturbed singular values

2. Since $A(t) = A(t) + E(t)$ for all $t < \bar{t}-\delta$, for these t we have that $\sigma_1(t) = \tilde{\sigma}_1(t)$ and $\sigma_2(t) = \tilde{\sigma}_2(t)$. Now, the degree of non-uniqueness of the singular vectors is given by their signs. So, we can assume that $u_1(t) = \tilde{u}_1(t)$, $u_2(t) = \tilde{u}_2(t)$, for $t < \bar{t} - \delta$, and the same thing for v_1, v_2 and $\tilde{v}_1, \tilde{v}_2$. For $t = \bar{t}$, $\sigma_1(t)$ and $\sigma_2(t)$ coalesce and cross each other, while $\tilde{\sigma}_1(t)$ and $\tilde{\sigma}_2(t)$ do not coalesce for any value of t see above figure.

3. We also have $A(t) = A(t) + E(t)$ for all $t > \bar{t} + \delta$. Hence, for these t, we must have $\sigma_1(t) = \tilde{\sigma}_2(t)$, $\sigma_2(t) = \tilde{\sigma}_1(t)$, $u_1(t) = \pm\tilde{u}_2(t)$, $u_2(t) = \pm\tilde{u}_1(t)$, for $t > \bar{t}+\delta$, and the same thing for the right singular vectors. In other words, since $[\tilde{u}_1\ \tilde{u}_2]$ is a smooth orthonormal frame for all $t \in \mathbb{R}$, the only (two) possibilities, for $t > \bar{t} + \delta$, are:

$$[\tilde{u}_1\ \tilde{u}_2] = [u_1\ u_2]\begin{bmatrix} 0 & \pm 1 \\ \mp 1 & 0 \end{bmatrix}.$$

4. Because of (3), the singular vector $\tilde{u}_1$, respectively $\tilde{u}_2$, which starts off aligned to u_1, respectively u_2, rotates toward the direction of u_2, respectively u_1, over the interval $[\bar{t} - \delta, \bar{t} + \delta]$. In essence, they undergo an exchange. This fully explain the previous observation on the jump in the singular vectors.

As a final observation of practical relevance, we stress that, in this particular example, it would be actually much easier to follow the SVD path for the unperturbed problem with coalescing singular values (since the singular values and vectors have very small variations) than for the perturbed problem with distinct singular values!

3.2 Theoretical Results: Two Parameters

First, let us consider the case of the eigendecomposition of a symmetric function, then we will give the relevant extensions for the SVD.

3.2.1 Symmetric Eigenproblem

We saw in Sect. 3.1 that if a given one-parameter symmetric problem arises from a two-parameter problem in the vicinity of a CI point, then the two eigenvalues that coalesced in the two parameter problem undergo a veering effect, and the two associated eigenvectors undergo what appears to be an exchange with one another, mode-exchange; see Example 3.5. But we do not know where CI points are, after all we are trying to locate them!, so attempting to locate coalescing points trying to see where veering phenomena occur is at best impractical, surely only valid locally (near a CI), and possibly even misleading.

To arrive at a much sounder way to locate coalescing points, we need to realize that a remarkable fact happens when we monitor the variation of the eigenvectors along a closed loop enclosing the CI point: in a nutshell (and we will qualify the statement more completely below), what happens is that the eigenvectors "flip over", i.e. they change orientation, as we complete one loop around the CI point. This remarkable insight first appeared in the work of Herzberg and Longuet-Higgins, [59], where they effectively considered the following example.

Example 3.6. Consider the following symmetric problem

$$A(x_1, x_2) = \begin{bmatrix} x_1 & x_2 \\ x_2 & -x_1 \end{bmatrix}, \quad x \in \mathbb{R}^2. \tag{11}$$

The eigenvalues are $\lambda_\pm = \pm\sqrt{x_1^2 + x_2^2}$ and obviously only coalesce at the origin, which is a CI point. If we restrict to consider the variation of the eigenvalues and eigenvectors on the circles (all of them enclosing the CI point at the origin) $\Gamma = \{x : x_1^2 + x_2^2 = \rho^2\}$ for $\rho > 0$, then the eigenvalues are simply $\pm\rho$ and the orthogonal eigenvectors as one moves along Γ are (aside from possible changes of signs of the columns) just the columns of Q below:

$$Q(\theta) = \begin{bmatrix} \cos\theta/2 & -\sin\theta/2 \\ \sin\theta/2 & \cos\theta/2 \end{bmatrix}, \quad \theta \in [0, 2\pi].$$

[Since A restricted to Γ is a reflection, the above form of Q had to be expected.] Now, obviously the restriction of A to Γ is 2π-periodic, however Q has period 4π, not 2π! As a consequence, as we cover a loop around the origin, starting with $Q(0) = \begin{bmatrix} 1 & 0 \\ 0 & 1 \end{bmatrix}$, we get that $Q(2\pi) = -I$, that is the eigenvectors after one loop have assumed the opposite orientation. We stress that what we just observed is true regardless of how small is the circle centered at the origin, in other words we are not witnessing a local phenomenon, but a global (topological) one.

In Example 3.6, we must appreciate that there is nothing unexpected about having a smooth eigendecomposition of a periodic function A with distinct eigenvalues in which the orthogonal factor has twice the period of A; we knew that this was possible, see Remark 2.30. Also, because of Problem 2.36, we know that there cannot be a continuous orthogonal factor of period 2π. It is natural to ask whether the doubling of the period is related to the presence of the CI inside the circle Γ.

Problem 3.7. Consider the function A in (11) restricted to circles $\hat{\Gamma}$ centered at the point $\hat{x} \neq 0$, $\hat{x} = (\hat{x}_1, \hat{x}_2)$. That is, in (11), let $x_1 = \hat{x}_1 + \rho \cos \theta$ and $x_2 = \hat{x}_2 + \rho \sin \theta$. Let $\hat{\rho}^2 = \hat{x}_1^2 + \hat{x}_2^2$. Show that when $\rho^2 < \hat{\rho}^2$, any continuous orthogonal matrix of eigenvectors of A along $\hat{\Gamma}$ has period 2π.

The results in Example 3.6 and Problem 3.7 are pointing us in the right direction and we are now ready for the general results.

Theorem 3.8 (Symmetric (2×2) case: One coalescing point). *Consider $P \in \mathscr{C}^k(\Omega, \mathbb{R}^{2\times 2})$, $k \geq 1$, symmetric:*

$$P(x) = \begin{bmatrix} a(x) & b(x) \\ b(x) & d(x) \end{bmatrix},$$

with eigenvalues $\lambda_1(x) \geq \lambda_2(x)$ for all x in Ω. Let there be a unique point $\xi_0 \in \Omega$ where the eigenvalues coincide: $\lambda_1(\xi_0) = \lambda_2(\xi_0)$. Consider the $\mathscr{C}^k$ function $F : \Omega \to \mathbb{R}^2$,

$$F(x) = \begin{bmatrix} a(x) - d(x) \\ b(x) \end{bmatrix},$$

and let 0 be a regular value for both $a - d$ and b. Consider the two $\mathscr{C}^k$ curves Γ_1 and Γ_2 through ξ_0: $\Gamma_1 = \{x \in \Omega : a(x) - d(x) = 0\}$, $\Gamma_2 = \{x \in \Omega : b(x) = 0\}$. Assume that Γ_1 and Γ_2 intersect transversally at ξ_0.

Let Γ be a simple closed curve enclosing the point ξ_0, and let it be parametrized as a $\mathscr{C}_1^p$ ($p \geq 0$) function γ in the variable t. Let $m = \min(k, p)$, and let P_γ be the $\mathscr{C}^m$ function $P(\gamma(t))$, $t \in \mathbb{R}$. Then, for all $t \in \mathbb{R}$,

$$P_\gamma(t) = Q_\gamma(t) \Lambda_\gamma(t) Q_\gamma^T(t), \quad \text{and}$$

(i) $\Lambda_\gamma \in \mathscr{C}_1^m(\mathbb{R}, \mathbb{R}^{2\times 2})$ *and diagonal:* $\Lambda_\gamma(t) = \begin{bmatrix} \lambda_1(\gamma(t)) & 0 \\ 0 & \lambda_2(\gamma(t)) \end{bmatrix}$;

(ii) $Q_\gamma \in \mathscr{C}_2^m(\mathbb{R}, \mathbb{R}^{2\times 2})$ *real orthogonal, and $Q_\gamma(t+1) = -Q_\gamma(t)$, for all t.*

Sketch of Proof. A complete proof of Theorem 3.8 is in [46], to which we refer for details. But, the idea of the proof proceeds as follows. First, consider a small circle C around ξ_0 and explicitly show that the first (respectively, second) component of the eigenvector associated to λ_1 changes sign exactly when C crosses the curve $b(x) = 0$ and $a(x) - d(x) < 0$ (respectively, $a(x) - d(x) > 0$). Then, using a homotopy technique, argue that the same holds true along Γ. See Fig. 14. □

Fig. 14 Generic coalescing point, transversal intersection at ξ_0

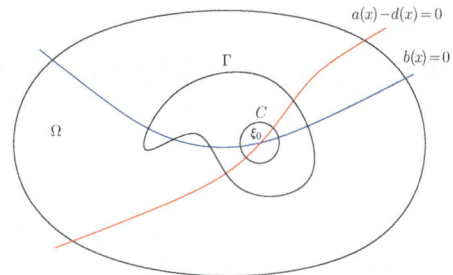

Remark 3.9. Theorem 3.8 states that it is a generic property when there is a conical intersection inside the loop Γ (as reflected in the hypothesis of transversal intersection of the curves $\Gamma_{1,2}$ at ξ_0) that a continuous function of orthogonal eigenvectors is 2-periodic along Γ, whereas the eigenvalues are 1-periodic. In other words, the simpler situation of Example 3.6 reflects what one should expect for any smooth function A. For simplicity, a coalescing point ξ_0 at which $\Gamma_{1,2}$ intersect transversally will be called a *generic coalescing point*.

With the aid of Theorem 3.1, it is possible to extend Theorem 3.8 to the case of a coalescing point for $(n \times n)$ symmetric functions. First, one isolates the coalescing point within a (2×2)-block, and then uses Theorem 3.8 relatively to this block; with this in mind, we will use the term generic coalescing point for a coalescing point of a $(n \times n)$ symmetric function if it is such for the (2×2)-block containing it, which has come about as a result of having used Theorem 3.1. There are nontrivial mathematical subtleties in following this course of action, and we refer to [46] for details, but here we simply state the relevant result.

Theorem 3.10 (Symmetric $(n \times n)$ case: One coalescing point). *Let $A \in \mathscr{C}^k(\Omega, \mathbb{R}^{n \times n})$ be symmetric. Let $\lambda_1(x), \ldots, \lambda_n(x)$, $x \in \Omega$, be its continuous eigenvalues. Suppose that*

$$\lambda_1(x) > \lambda_2(x) > \cdots > \lambda_j(x) \geq \lambda_{j+1}(x) > \cdots > \lambda_n(x), \; \forall x \in \Omega,$$

and

$$\lambda_j(x) = \lambda_{j+1}(x) \iff x = \xi_0 \in \Omega,$$

and ξ_0 be a generic coalescing point.

Let Γ be a simple closed curve in Ω enclosing the point ξ_0, and let it be parametrized as a $\mathscr{C}^p$ ($p \geq 0$) function γ in the variable t, so that the function $\gamma : t \in \mathbb{R} \to \Omega$ is $\mathscr{C}^p$ and 1-periodic. Let $m = \min(k, p)$, and let A_γ be the $\mathscr{C}^m$ function $A(\gamma(t))$, $t \in \mathbb{R}$.

Then, for all $t \in \mathbb{R}$, $A_\gamma(t)$ has the eigendecomposition

$$A_\gamma(t) = Q_\gamma(t) \Lambda_\gamma(t) Q_\gamma^T(t)$$

satisfying the following conditions:

(i) $\Lambda_\gamma \in \mathscr{C}_1^m(\mathbb{R}, \mathbb{R}^{n \times n})$ *and is diagonal:* $\Lambda_\gamma(t) = \mathrm{diag}(\lambda_1(\gamma(t)), \ldots, \lambda_n(\gamma(t)))$, $\forall t \in \mathbb{R}$;

(ii) $Q_\gamma \in \mathscr{C}_2^m(\mathbb{R}, \mathbb{R}^{n \times n})$ *is orthogonal, and for all* $t \in \mathbb{R}$

$$Q_\gamma(t+1) = Q_\gamma(t) D, \quad D = \begin{bmatrix} I_{j-1} & 0 & 0 \\ 0 & -I_2 & 0 \\ 0 & 0 & I_{n-j-1} \end{bmatrix}.$$

Remark 3.11. The noteworthy feature of Theorem 3.10 is that the eigenvectors associated to the CI point change sign over one period and are thus 2-periodic, but the eigenvectors relative to the other eigenvalues, noncoalescing inside Γ, maintain period 1.

The extension to two or more generic coalescing points is contained in the following theorem. The proof of the result is again in [46], to which we refer for details.

Theorem 3.12. *Let* $A \in \mathscr{C}^k(\Omega, \mathbb{R}^{n \times n})$, $k \geq 1$, *be symmetric and let* $\lambda_1(x) \geq \cdots \geq \lambda_n(x)$ *be its continuous eigenvalues. Suppose that, for every* $j = 1, \ldots, n-1$,

$$\lambda_j(x) = \lambda_{j+1}(x)$$

at d_j distinct generic coalescing points of eigenvalues in Ω, so that there are $\sum_{j=1}^{n-1} d_j$ such points. Let Γ be a simple closed curve enclosing all of these distinct generic coalescing points of eigenvalues. Let Γ be parametrized as a $\mathscr{C}^p$ ($p \geq 0$) function γ in the variable t, so that the function $\gamma : t \in \mathbb{R} \to \Omega$ is $\mathscr{C}^p$ and 1-periodic. Let $m = \min(k, p)$ and $A_\gamma \in \mathscr{C}^m(\mathbb{R}, \mathbb{R}^{n \times n})$ defined as $A_\gamma(t) = A(\gamma(t))$, for all $t \in \mathbb{R}$. Then, for all $t \in \mathbb{R}$,

$$A_\gamma(t) = Q_\gamma(t) \Lambda_\gamma(t) Q_\gamma^T(t)$$

such that:

(i) $\Lambda_\gamma \in \mathscr{C}_1^m(\mathbb{R}, \mathbb{R}^{n \times n})$ *is diagonal:* $\Lambda_\gamma(t) = \mathrm{diag}(\lambda_1(\gamma(t)), \ldots, \lambda_n(\gamma(t)))$, *for all* $t \in \mathbb{R}$;

(ii) $Q_\gamma \in \mathscr{C}^m(\mathbb{R}, \mathbb{R}^{n \times n})$ *is orthogonal, with*

$$Q_\gamma(t+1) = Q_\gamma(t) D, \quad \forall t \in \mathbb{R},$$

where D is a diagonal matrix of ± 1 given as follows:

$$D_{11} = (-1)^{d_1}, \quad D_{jj} = (-1)^{d_{j-1}+d_j} \text{ for } j = 2, \ldots, n-1, \quad D_{nn} = (-1)^{d_{n-1}}.$$

In particular, if $D = I_n$, then Q_γ is 1-periodic, otherwise Q_γ is 2-periodic. □

Remarks 3.13. (1) Notice that—relatively to the same pair of eigenvalues—an odd number of generic coalescing points (inside Γ) leads to negative sign(s) in D, and 2-periodic associated eigenvectors, while an even number returns 1-periodic eigenvectors.

(2) For the sake of completeness, we remark that a result like Theorem 3.12 has been further extended in [46] replacing the (generic) property of transversal intersection with that of curves crossing each other. So doing, one obtains a rigorous definition of order of contact of the eigenvalues, and the generalization of Theorem 3.12 will state that odd order of contact (counting multiplicity) of two eigenvalues would lead to a change of sign for the associated eigenvectors, whereas even order of contact would not.

3.2.2 SVD

The results of Sect. 3.2.1 are hereafter generalized to the case of the SVD, under some assumptions on the function A, specifically on the number and type of coalescing singular values.

First of all, consider the case when A is (2×2). Recalling Example 3.3, we can exclude consideration of the nongeneric case of a pair of zero singular values at any point $x \in \Omega$. In other words, at most one singular value will be allowed to vanish at any $x \in \Omega$; of course, having a vanishing singular value is a codimension 1 phenomenon, hence generic for functions depending on 2 parameters. Secondly, we will restrict consideration to matrices taking values in $\mathbb{R}^{n \times n}$, rather than in $\mathbb{R}^{m \times n}$; restriction to this square case allows us to avoid the difficulties caused by isolated loss of rank,[4] and it could be easily replaced by a condition of full rank of A. The following problem shows that this would not be a very restrictive request either.

Problem 3.14. You have to show the following facts.

(a) Consider functions $A \in \mathscr{C}^k(\Omega, \mathbb{R}^{m \times n})$, $m \geq n+2$, and $k \geq 1$. Show that it is a generic property that such functions A have full rank. [Hint: See the proof of Theorem 2.5.]
(b) Show that if $A \in \mathscr{C}^k(\Omega, \mathbb{R}^{m \times n})$, $m \geq n$, $k \geq 1$, is of full rank, then it has a $\mathscr{C}^k$ QR decomposition: $A = QR$, where $Q \in \mathscr{C}^k(\Omega, \mathbb{R}^{m \times n})$ is orthonormal ($Q^T Q = I$) and $R \in \mathscr{C}^k(\Omega, \mathbb{R}^{n \times n})$ is upper triangular. [Hint: Use the Gram-Schmidt process.]

So, we will restrict consideration to the SVD of a function $A \in \mathscr{C}^k(\Omega, \mathbb{R}^{n \times n})$, and see what happens when we take a loop encircling one—or several—coalescing point(s) for the singular values. We remark that we may have one singular value to go to 0 along the loop, hence (in order to maintain smoothness) we will need to consider a signed SVD along the loop.

---

[4] See Remark 2.8.

For the (2×2) case, we have the following result.

Theorem 3.15 (Signed SVD: 2×2 case). *Consider $A \in \mathscr{C}^k(\Omega, \mathbb{R}^{2\times 2})$, $k \geq 1$. For all $x \in \Omega$, write*

$$A(x) = \begin{bmatrix} a(x) & b(x) \\ c(x) & d(x) \end{bmatrix},$$

and let σ_1 and σ_2 be its two continuous singular values, labeled so that $\sigma_1(x) \geq \sigma_2(x) \geq 0$ for all x in Ω. Assume that there exists a unique point $\xi_0 \in \Omega$ where these singular values coincide, $\sigma_1(\xi_0) = \sigma_2(\xi_0)$. Consider the $\mathscr{C}^k$ functions $F, G : \Omega \to \mathbb{R}^2$ given by

$$F(x) = \begin{bmatrix} a^2(x) + c^2(x) - b^2(x) - d^2(x) \\ a(x)b(x) + c(x)d(x) \end{bmatrix}, G(x) = \begin{bmatrix} a^2(x) + b^2(x) - c^2(x) - d^2(x) \\ a(x)c(x) + b(x)d(x) \end{bmatrix}, \quad (12)$$

and assume that 0 is a regular value for the scalar valued functions given by the 1st and the 2nd components of F and G. Then, consider the $\mathscr{C}^k$ curves Γ_1, Γ_2, Γ_3, and Γ_4, given by the zero-set of the components of F and G: $\Gamma_1 = \{x \in \Omega : a^2(x) + c^2(x) - b^2(x) - d^2(x) = 0\}$, $\Gamma_2 = \{x \in \Omega : a(x)b(x) + c(x)d(x) = 0\}$, and $\Gamma_3 = \{x \in \Omega : a^2(x) + b^2(x) - c^2(x) - d^2(x) = 0\}$, $\Gamma_4 = \{x \in \Omega : a(x)c(x) + b(x)d(x) = 0\}$. Assume that, at ξ_0, the curves Γ_1 and Γ_2 intersect transversally.[5]

Let Γ be a simple closed curve enclosing the point ξ_0, and let it be parametrized as a $\mathscr{C}^p$ ($p \geq 0$) function γ in the variable t, so that the function $\gamma : t \in \mathbb{R} \to \Omega$ is $\mathscr{C}^p$ and 1-periodic. Let $m = \min(k, p)$, and let A_γ be the $\mathscr{C}_1^m$ function $A(\gamma(t))$, $t \in \mathbb{R}$. Then, for all $t \in \mathbb{R}$, $A_\gamma(t)$ has the signed singular value decomposition

$$A_\gamma(t) = U_\gamma(t) \Sigma_\gamma(t) V_\gamma^T(t)$$

such that:

(i) $\Sigma_\gamma \in \mathscr{C}_1^m(\mathbb{R}, \mathbb{R}^{2\times 2})$ *and diagonal,* $\Sigma_\gamma(t) = \begin{bmatrix} s_1(\gamma(t)) & 0 \\ 0 & s_2(\gamma(t)) \end{bmatrix}$ *with* $|s_i(\gamma(t))| = \sigma_i(\gamma(t))$, *for* $i = 1, 2$, *and for all* $t \in \mathbb{R}$;

(ii) $U_\gamma, V_\gamma \in \mathscr{C}_2^m(\mathbb{R}, \mathbb{R}^{2\times 2})$ *real orthogonal, and* $U_\gamma(t+1) = -U_\gamma(t)$, $V_\gamma(t+1) = -V_\gamma(t)$, *for all* $t \in \mathbb{R}$.

Remarks 3.16. (a) The functions F and G in the statement of Theorem 3.15 are just the functions arising from the symmetric problems $A^T A$ and $A A^T$, and the condition of transversality of the curves $\Gamma_{1,2}$ is of course generic.

(b) We observe that transversal intersection of Γ_1 and Γ_2 at ξ_0 rules out the possibility that the singular values be 0 there; indeed, if we had a pair of

---

[5]The same would be true if we assumed transversal intersection of the pair Γ_3 and Γ_4.

coalescing singular values equal to 0 then we would not have properly defined tangents at all, violating the assumption on 0 being a regular value.
(c) The decomposition of A_γ in Theorem 3.15 is essentially unique, within the class of $\mathscr{C}^m$ decompositions. The degree of non-uniqueness is given by the ordering of the diagonal and by joint (and global) choices of signs for the columns of U_γ and V_γ.

Next, we give a result relatively to the case of A taking values in $\mathbb{R}^{n \times n}$ and with several generic coalescing points inside a simple closed curve Γ. This result necessitates an appropriate definition of generic coalescing points relatively to the SVD, for which we need an SVD extension of the block-diagonalization result Theorem 3.1; these considerations are in [46], here we simply quote the result, whose flavor is quite similar to Theorem 3.12.

Theorem 3.17. *Let $A \in \mathscr{C}^k(\Omega, \mathbb{R}^{n \times n})$, $k \geq 1$. Let $\sigma_1(x) \geq \cdots \geq \sigma_n(x)$ be its continuous singular values and suppose that, for every $j = 1, \ldots, n-1$,*

$$\sigma_j(x) = \sigma_{j+1}(x)$$

at d_j distinct generic coalescing points of singular values in Ω. Let Γ be a simple closed curve enclosing all these coalescing points, parametrized by the 1-periodic function $\gamma \in \mathscr{C}_1^p(\mathbb{R}, \Omega)$. Let $m = \min(k, p)$ and $A_\gamma \in \mathscr{C}^m(\mathbb{R}, \mathbb{R}^{n \times n})$ defined as $A_\gamma(t) = A(\gamma(t))$, for all $t \in \mathbb{R}$. Then, for all $t \in \mathbb{R}$,

$$A_\gamma(t) = U_\gamma(t) S_\gamma(t) V_\gamma^T(t)$$

such that:

(i) *$S_\gamma \in \mathscr{C}^m(\mathbb{R}, \mathbb{R}^{n \times n})$ diagonal: $\Sigma_\gamma(t) = \text{diag}(s_1(\gamma(t)), \ldots, s_n(\gamma(t)))$, for all $t \in \mathbb{R}$, and $|s_i(\gamma(t))| = \sigma_i(\gamma(t))$, for all $i = 1, \ldots, n$, and all $t \in \mathbb{R}$;*
(ii) *$U_\gamma, V_\gamma \in \mathscr{C}^m(\mathbb{R}, \mathbb{R}^{n \times n})$ orthogonal, with*

$$U_\gamma(t+1) = U_\gamma(t) D, \quad V_\gamma(t+1) = V_\gamma(t) D, \quad \forall t \in \mathbb{R},$$

where D is as in Theorem 3.12.

3.2.3 From Periodicity to CIs

In Sects. 3.2.1 and 3.2.2 we witnessed what happens as we monitor the eigenvectors (singular vectors) as we move around a closed loop containing one, or several, coalescing point(s). In particular, we witnessed the impact that the presence of CIs has on the periodicity of the orthogonal factors. Here, we state a converse result stating that 2-periodic factors imply CIs.

Theorem 3.18. *Let $A \in \mathscr{C}^k(\Omega, \mathbb{R}^{n \times n})$, $k \geq 1$, be symmetric (respectively, a general function), with continuous and ordered eigenvalues (respectively,* continuous and ordered singular values). Let Γ be a simple closed curve in Ω, with no coalescing point for the eigenvalues (respectively, singular values) of A on it. Let γ be a $\mathscr{C}^p$, 1-periodic parametrization for Γ, let $q = \min(k, p)$, and let A_γ be the $\mathscr{C}^q$ restriction of A to Γ. Let Q_γ (respectively, U_γ and V_γ) be the smooth factor(s) of the eigendecomposition (respectively, SVD) of A_γ: $A_\gamma = Q_\gamma \Lambda_\gamma Q_\gamma^T$, $Q_\gamma \in \mathscr{C}^q(\mathbb{R}, \mathbb{R}^{n\times n})$ and orthogonal (respectively, $A_\gamma = U_\gamma \Sigma_\gamma V_\gamma^T$, $U_\gamma \in \mathscr{C}^q(\mathbb{R}, \mathbb{R}^{n\times n})$ and $V_\gamma \in \mathscr{C}^q(\mathbb{R}, \mathbb{R}^{n\times n})$, orthogonal).

Finally, let $Q_0 = Q_\gamma(0)$ and $Q_1 = Q_\gamma(1)$ (respectively, $U_0 = U_\gamma(0)$ and $U_1 = U_\gamma(1)$), and define $D = Q_0^T Q_1$ (respectively, $D = U_0^T U_1$). Next, let $2q$ be the (even) number of indices k_i, $k_1 < k_2 < \cdots < k_{2q}$, for which $D_{k_i k_i} = -1$. Let us group these indices in pairs $(k_1, k_2), \ldots, (k_{2q-1}, k_{2q})$. Then, λ_j and λ_{j+1} (respectively, σ_j and σ_{j+1}) coalesce at least once inside the region encircled by Γ, if $k_{2e-1} \leq j < k_{2e}$ for some $e = 1, \ldots, q$.

Remark 3.19. In Theorem 3.18, we claimed that the number of indices k_i for which $D_{k_i k_i} = -1$ is even. This is because of the following argument. With usual notation, relatively to the symmetric eigenproblem, we have $A_\gamma(t) = Q_\gamma(t) \Lambda_\gamma(t) Q_\gamma^T(t)$, with $Q_\gamma(t+1) = Q_\gamma(t) D$. Because of continuity, $\det(Q) = +1$ or $\det(Q) = -1$, for all t. Hence, since $D = Q_\gamma^T(0) Q_\gamma(1)$, then $\det(D) = 1$, which justifies the claim.

Remarks 3.20. (i) To illustrate the previous theorem for the SVD, suppose we have $n \geq 4$, and

$$D = \begin{bmatrix} -1 & & & & \\ & 1 & & & \\ & & 1 & & \\ & & & -1 & \\ & & & & \ddots \end{bmatrix}$$

Then, we expect that inside the region encircled by Γ, the pairs (σ_1, σ_2), (σ_2, σ_3), and (σ_3, σ_4), have coalesced at least once (and, in general, if not once, an odd number of times).

(ii) If a pair of singular values coalesce at an even number of (generic) points inside Γ, the corresponding singular vectors do not change sign.

(iii) In principle, we cannot rule out the possibility that a pair of eigenvalues (singular values) coalesce infinitely many times, though this is certainly not a generic situation. Further, this situation may, or may not, be detected.

(iv) Also, in principle, inside Γ we may have a combination of generic and non-generic coalescing points. For example, take the function

$$A(x) = \begin{bmatrix} (x_1 - 1)(x_1^2 + x_2^2) & x_2 \\ x_2 & 0 \end{bmatrix}.$$

The eigenvalues coalesce at the origin and at the point $(1, 0)$. Taking Γ to be a sufficiently large circle around the origin, one obtains $D = -I$, and can thus infer the existence of a point of intersection. In this case, the point $(1,0)$ has been responsible for the values of -1 in D, whereas the origin is a degenerate coalescing point and has no impact on the matrix D. Notice that the value 0 is not a regular value for the function $(x_1 - 1)(x_1^2 + x_2^2)$.

Large Problems

For large dimensional problems we may want to monitor only a small number of eigenvalues (singular values), say just the $q \ll n$ most dominant eigenpairs (singular triplets) of A. The extensions required to the previous results are in [47].

To illustrate, consider the case of the SVD. We now end up working with the reduced SVD along γ:

$$\left(U^{(q)}(t)\right)^T A_\gamma(t) V^{(q)}(t) = \Sigma^{(q)}(t),$$

where $U^{(q)}$ and $V^{(q)}$ each comprise of just q columns. The main difference with respect to the previous case is that we may have an odd number of -1 on the diagonal of

$$D^{(q)} = \left(U^{(q)}(0)\right)^T U^{(q)}(1).$$

Example 3.21. For example, consider the case of $q = 4$, and suppose we find that

$$D^{(q)} = \begin{bmatrix} 1 & & & \\ & 1 & & \\ & & -1 & \\ & & & 1 \end{bmatrix}.$$

Then, what we can conclude is that (σ_3, σ_4) and (σ_4, σ_5) coincide at an odd number of points inside Γ, even though we have not been monitoring σ_5 (which itself may have further coalesced with σ_6 etc., but this we cannot infer based upon just looking at $D^{(q)}$).

3.3 Localization Algorithm for Conical Intersections

The results of the previous section lend naturally to algorithmic techniques which find regions of Ω inside which some singular values (or eigenvalues) of A coalesce.

The basic idea is simple and it is akin to the "Intermediate Value Theorem" in Calculus: "Change of sign of a continuous function implies a root". For us, it will be:

Change of sign in the matrices of singular vectors along a loop Γ

coalescing pair of singular values inside Γ.

In other words, period 2 implies coalescing. [Detection of CI.]

We observe that Γ only needs to be continuous and does not have to be a curve close to the CI (in other words, we do not rely on a local result).

Of course, to implement the above idea one needs to make a lot of choices. Namely, (1) Γ has to be chosen, (2) integration along Γ has to be performed, and (3) the parameter values where the coalescings occur must be accurately located. The way that these tasks are carried out will impact performance of the localization algorithm, and we now proceed to clarify the choices we made.

3.3.1 Choosing Γ

We will work with Ω being a rectangular region, $\Omega = \{x = (x_1, x_2) \in \mathbb{R}^2 : a \leq x_1 \leq b, \ c \leq x_2 \leq d \}$. [The case when Ω is a simply connected region with piecewise linear boundary are handled with no major modifications, as is the case where Ω is a disk; hence, through appropriate subdivisions, and reparametrizations, one can tackle more complicated situations.]

Subdivide Ω in a cartesian grid of $N \times M$ boxes, not necessarily all of the same size. That is, we let $\Omega = \cup_{i=1:N, j=1:M} B_{ij}$, where $B_{ij} = \{(x_1, x_2) : x_1^{(i)} \leq x_1 \leq x_1^{(i+1)}, \ x_2^{(j)} \leq x_2 \leq x_2^{(j+1)} \}, i = 1, \ldots, N, j = 1, \ldots, M$. Let us call $x_{i,j} = (x_1^{(i)}, x_2^{(j)})$, and similarly $x_{i+1,j}$, $x_{i,j+1}$ and $x_{i+1,j+1}$, the vertices of the box B_{ij}. Relatively to a box, Γ is its contour. Now, rather than finding just one SVD along the loop Γ, it is more convenient to find two SVD and compare them at the same arrival point: starting from the same point (the South-West corner of the square), one SVD will move along the contour East and then North, and the other will cover the portion of the contour moving North and then East. See the figure below, and we note that the matrix D we form is just the matrix D of Theorem 3.18:

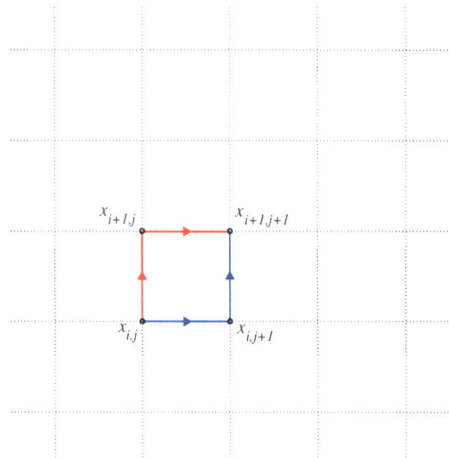

1. given
$$A(x_{i,j}) = U_0 \Sigma_0 V_0^T$$
compute continuous SVD of A along the two (red and blue) paths;
2. save orthogonal factors at $x_{i+1,j+1}$:

U_1, V_1, U_2, V_2;
3. compare U_1 and U_2:

$D = (U_1)^T U_2$.

Remark 3.22. We observe that the parametrization of the contour Γ is irrelevant insofar as the information retrieved from the matrix D.

3.3.2 Integration Along Γ

Naturally, a 1-d solver for computing (smooth) SVD (or eigendecomposition) paths is required. This 1-d solver is the workhorse of the algorithm, and it must balance efficiency and robustness. Algorithmically, we dealt with this problem in Sect. 2.3 and our algorithm follows the one outlined therein. Now as then, we must stress that continuation of an SVD path is not trivial, and it is all the more challenging the closer we are to a coalescing point for the singular values, because of the veering phenomenon; see Example 3.5, specifically the last sentence there. It is thus mandatory that the 1-d solver fulfills these two key features:

1. Proceed with adaptive stepsizes;
2. Be robust and able to handle the case when singular values coalesce and distinguish it from the case of near-coalescing.

3.3.3 Refinement

The basic algorithm consists of going through the grid by covering one box at the time, which of course we do so to transverse each edge of any box only once and never compute twice the SVD at any given point, in order to isolate coalescing points within a box. However, the grid resolution is obviously very important, as the next example makes clear.

Example 3.23. Let $\Omega = \{(x_1, x_2) \in \mathbb{R}^2 : -1 \leq x_1 \leq 1, -1 \leq x_2 \leq 1\}$,

$$B(x_1, x_2) = \mathrm{diag}(x_1^2 + x_2^2, .81, .36),$$

$$C = \begin{bmatrix} -0.179\ldots & -0.294\ldots & -0.722\ldots \\ 0.787\ldots & 0.626\ldots & -0.594\ldots \\ -0.884\ldots & -0.980\ldots & -0.602\ldots \end{bmatrix},$$

$$E(x_1, x_2) = .5(x_1 + x_2)(x_1 + 1/3) C,$$

and $A(x_1, x_2) = B(x_1, x_2) + E(x_1, x_2)$, for all $(x_1, x_2) \in \Omega$.

Below, there are two figures. The first shows the result of the localization module subdividing Ω into a 2×2 grid. Coalescing points for the pair σ_2-σ_3 are detected in boxes B_{11} and B_{21}. The second one is a figure relative to a subdivision of Ω into a 10×10 grid. Now coalescing points are detected in 10 boxes. Eight coalescing points had gone undetected through the 2×2 grid due to the fact that they occurred in pair over each box of that grid. The legend shows: "×": $\sigma_1 = \sigma_2$, "+": $\sigma_2 = \sigma_3$.

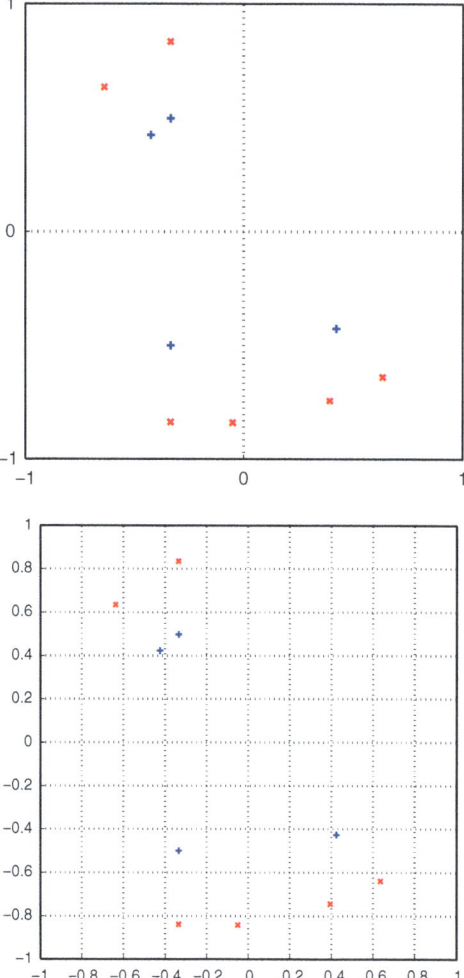

Clearly, the above localization procedure can be implemented so to refine just a box where there is a coalescing point, progressively halving each vertical/horizontal side ("bisection" approach). So doing, in principle we are able to locate a coalescing point with arbitrary degree of accuracy, see figure below.

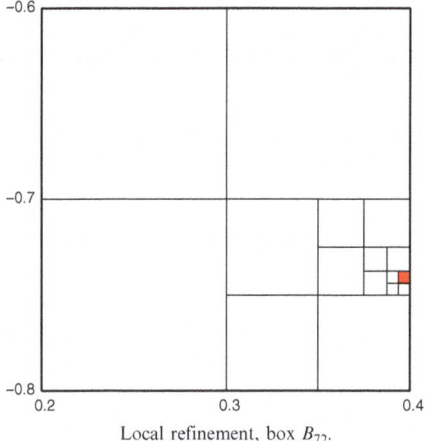

Local refinement, box B_{72}.

Obviously, this refinement process will converge only linearly and it is bound to be expensive, a fact which is greatly alleviated by a parallel implementation of the above refinement technique.

Zoom-In. As an alternative, once we have a region where the i-th and $(i+1)$-st singular values coalesce, we have used a (modified) Newton method to find a root of the gradient of

$$f(x) = \bigl(\sigma_i(x) - \sigma_{i+1}(x)\bigr)^2$$

(see Remark below for justification of why f is a smooth function). By combining the local refinement technique, and Newton's method to accurately locate the coalescing point, the workload of the method reflects very well the location of the coalescing points, see figure below.

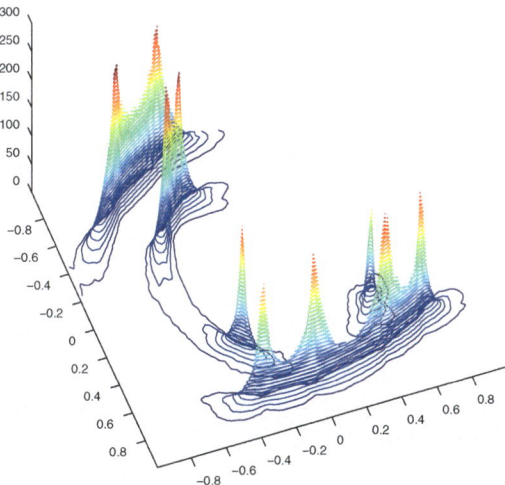

Distribution of the workload over the region Ω.

Table 1 Experiments, $m = 3,200$

n	Average # of steps	Min-max # steps	Rejected steps
200	132	75–679	1,635
400	129	72–503	1,558
800	135	71–642	1,808
1,600	137	77–702	1,682
3,200	151	71–549	1,819

Remark 3.24. We claim that the above function $f(x) = \bigl(\sigma_i(x) - \sigma_{i+1}(x)\bigr)^2$ is as smooth as the original function A, in a neighborhood of the coalescing point ξ, as long as the pair of singular values σ_i and σ_{i+1} is the only pair of singular values coalescing at ξ and no sum of two singular values is 0 at ξ (these are the generic assumptions under which we have been working). To validate this claim, we argue as follows.

(i) Consider the symmetric function $S = A^T A$. It is well known the eigenvalues of S are nothing but the squares of the singular values of A. Hence, with obvious labeling, we have $\lambda_j = \sigma_j^2$, $j = 1, \ldots, n$, and the λ_j's are distinct in a neighborhood of ξ, if $j \neq i, i+1$. Since they are distinct, the eigenvalues λ_j's, $j \neq i, i+1$, are smooth in a neighborhood of ξ; a fortiori, so are the singular values σ_j's, $j \neq i, i+1$.

(ii) Consider the discriminant of S, $d(S)$, that is the discriminant of the characteristic polynomial of S. It is well known that $d(S)$ is a homogeneous polynomial of degree $2(n-1)$ in the entries of S, and moreover it can be expressed (within a multiplicative constant) in terms of the roots of the characteristic polynomial (the eigenvalues of S) as

$$d(S) = \prod_{k<j}(\lambda_k - \lambda_j)^2 .$$

(iii) Using the expression in (ii), that $\lambda_j = \sigma_j^2$, factoring $\sigma_k^2 - \sigma_j^2 = (\sigma_k - \sigma_j)(\sigma_k + \sigma_j)$, and isolating $\sigma_i - \sigma_{i+1}$, shows that

$$(\sigma_i - \sigma_{i+1})^2 = d(S)/g(\sigma_1, \ldots, \sigma_n)$$

where numerator and denominator are as smooth as A.

Example 3.25. For completeness, in Table 1, we report on results for a large dimensional problem, where $A(x_1, x_2) \in \mathbb{R}^{m \times n}$, $\Omega = [-1,1] \times [-1,1]$, grid 10×10, and we follow just $q = 6$ dominant singular values. In this case, pairs of singular values coalesce at ~ 20 points. A noteworthy aspect is that the number of steps is largely independent of the number of columns of A, and it is related just to the number of conical intersection inside Ω.

3.3.4 Curves of CI Points

Finally, what happens when the function A depends on 3 (or more) parameters? Consider the case of $A = A(x_1, x_2, x_3)$. Now, it is a generic property that singular values coalesce along curves and the goal is to compute these curves of coalescing points (in the Physical Chemistry literature, these are called "seams" of CIs).

The process by which we do this is standard practice in continuation techniques, and we recall it here below:

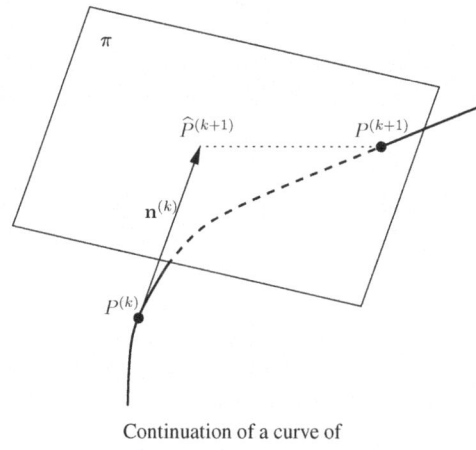

Continuation of a curve of
coalescing points: $\sigma_i(x) = \sigma_{i+1}(x)$.

1. prediction along an approximate tangent $\mathbf{n}^{(k)}$ (quadratic interpolant);
2. restriction of A to the plane π through $\hat{P}^{(k+1)}$ perpendicular to $\mathbf{n}^{(k)}$;
3. correction via the zoom-in technique.

3.4 Persisting Double Eigenvalue

Although not a generic phenomenon, there are situations (accidental, or due to special structure/symmetries in the problem) where we have a double eigenvalue of a symmetric function of two parameters which persists along a curve in parameter space. Obviously, this means that all points along this curve where the two eigenvalues coalesce are not generic conical intersection points.

This nongeneric situation is bound to occur (and be of relevance) in applications where extensive use of physically motivated symmetries renders a nongeneric eigenproblem. Surely, a generic perturbation of this nongeneric problem will destroy persistence of the double eigenvalue curve, but this is not going to be true if the perturbation has to remain within the same symmetry class from which the original model was derived.

The prototypical example is that of two uncoupled oscillators (see (1) with $k_2 = 0$), where we saw that the eigenfrequencies (eigenvalues) coalesced varying just one parameter; generic perturbations will either couple the oscillators or introduce friction, or both, and the eigenfrequencies will no longer coalesce varying one parameter, but any restricted perturbation which does not introduce coupling or friction will not eliminate the non-generic coalescing occurrences. Indeed, any block diagonal system of the type $A(x, y) = \begin{bmatrix} E & 0 \\ 0 & F \end{bmatrix}$ is a candidate for this persistence phenomenon, as we will soon clarify.

Below are the questions we want to address here.

- Given a symmetric matrix valued function A of two parameters (x, y) such that A has a double eigenvalue λ_0 at the point $\xi_0 \equiv (x_0, y_0)$. When/how does this eigenvalue persist—as double eigenvalue—along a curve passing through ξ_0?
- Given symmetric and positive definite matrix valued functions M and K of two parameters (x, y) such that at the point $\xi_0 \equiv (x_0, y_0)$ the pencil (K, M) has a double eigenvalue λ_0. When/how does this eigenvalue persist—as double eigenvalue—along a curve passing through ξ_0?

In Remark 2.18, we actually already saw what must happen for the eigenvalues to split along a curve, so we know that—necessarily—the condition we gave there must be violated in order for the eigenvalues not to split. However, in order to use the condition of Remark 2.18 we would need to select a certain curve along which the eigenvalue does not split; here below we will rephrase the result in a way that is independent of the curve and only requires partial derivative information at a point where the double eigenvalue persists.

To simplify the exposition, we will presently assume that the functions depend analytically on the parameters.

Theorem 3.26 (Double Eigenvalue: Symmetric Eigenproblem). *Let $A \in \mathscr{C}^\omega(\Omega, \mathbb{R}^{n \times n})$ be symmetric and let it have a double eigenvalue, call it λ_0, at a point ξ_0. Let $U = \begin{bmatrix} u_1 & u_2 \end{bmatrix}$ be a matrix whose columns are two orthonormal eigenvectors associated to λ_0 for $A_0 \equiv A(\xi_0)$. Then, λ_0 persists as a double eigenvalue of A—along some curve—only if $BC = CB$, with $B = U^T A_x(\xi_0)U$, $C = U^T A_y(\xi_0)U$.*

Remark 3.27. The matrix U is not unique, of course, but the end result is independent of the choice of U. Also, we note that the curve along which the eigenvalue persists is not known, nor needs to be known.

Proof. Suppose we have a curve $\gamma(t) = \begin{bmatrix} x(t) \\ y(t) \end{bmatrix}$, $t \in \mathbb{R}$, depending analytically on t, such $\gamma(0) = \xi_0$, and let $d = \begin{bmatrix} d_1 \\ d_2 \end{bmatrix} = \frac{d\gamma}{dt}|_{t=0}$. Call $A_\gamma(t), t \geq 0$, the analytic function $A(\gamma(t))$. Because of analyticity, A_γ admits an analytic eigendecomposition through $t = 0$. Let $\lambda(t)$ and $u(t)$ correspond to an eigenpair passing through λ_0. Everything can be expanded as (all matrices are evaluated at $t = 0$, that is at ξ_0)

$$A(t) = A_0 + tA_1 + \cdots, \quad A_1 = A_x d_1 + A_y d_2, \cdots,$$
$$\lambda(t) = \lambda_0 + t\lambda_1 + \cdots, \quad u(t) = v_0 + tv_1 + \cdots.$$

[Note that v_0 is the limit as $t \to 0$ of $u(t)$, and it is not known ahead of time. All we can say is that v_0 will need to be a combination of u_1 and u_2.]

From the eigenvalue relation $Au = \lambda u$ we get

$$A_0 v_0 = \lambda_0 v_0 \quad \text{and so} \quad v_0 = c_1 u_1 + c_2 u_2,$$
$$A_0 v_1 + A_1 v_0 = \lambda_0 v_1 + \lambda_1 v_0, \tag{13}$$

where $c_1^2 + c_2^2 = 1$. Now, left multiply the second relation in (13) by u_1^T, use $A_0 u_1 = \lambda_0 u_1$ and the form of v_0:

$$(u_1^T A_1 u_1) c_1 + (u_1^T A_1 u_2) c_2 = \lambda_1 c_1.$$

Similarly, left-multiplying the second relation in (13) by u_2^T:

$$(u_2^T A_1 u_1) c_1 + (u_2^T A_1 u_2) c_2 = \lambda_1 c_2.$$

Therefore, we must have

$$M \begin{bmatrix} c_1 \\ c_2 \end{bmatrix} = \lambda_1 \begin{bmatrix} c_1 \\ c_2 \end{bmatrix}, \quad M = \begin{bmatrix} u_1^T A_1 u_1 & u_1^T A_1 u_2 \\ u_2^T A_1 u_1 & u_2^T A_1 u_2 \end{bmatrix}.$$

So, for λ_0 to persist as double eigenvalue in some direction d (recall that $A_1 = A_x d_1 + A_y d_2$), we must have that λ_1 is a double eigenvalue of M for that d.

Since M is symmetric, this means $M_{11} = M_{22}$ and $M_{12} = 0$. This translates into

$$\begin{cases} (u_1^T A_x u_1) d_1 + (u_1^T A_y u_1) d_2 = (u_2^T A_x u_2) d_1 + (u_2^T A_y u_2) d_2 \\ (u_1^T A_x u_2) d_1 + (u_1^T A_y u_2) d_2 = 0 \end{cases}.$$

The last system can be further rewritten as

$$N \begin{bmatrix} d_1 \\ d_2 \end{bmatrix} = 0, \quad N = \begin{bmatrix} (u_1^T A_x u_1) - (u_2^T A_x u_2) & (u_1^T A_y u_1) - (u_2^T A_y u_2) \\ u_1^T A_x u_2 & u_1^T A_y u_2 \end{bmatrix}. \tag{14}$$

So, persistence of λ_0 as double eigenvalue requires $\det(N) = 0$.

Let us rewrite N as follows. Let

$$B = \begin{bmatrix} u_1^T A_x u_1 & u_1^T A_x u_2 \\ u_1^T A_x u_2 & u_2^T A_x u_2 \end{bmatrix} = U^T A_x(\xi_0) U, \quad C = \begin{bmatrix} u_1^T A_y u_1 & u_1^T A_y u_2 \\ u_1^T A_y u_2 & u_2^T A_y u_2 \end{bmatrix} = U^T A_y(\xi_0) U,$$

so that $N = \begin{bmatrix} b_{11} - b_{22} & c_{11} - c_{22} \\ b_{12} & c_{12} \end{bmatrix}$ and so $\det(N) = 0$ is the same as the requirement $b_{11}c_{12} - b_{22}c_{12} - b_{12}c_{11} + b_{12}c_{22} = 0$. But this last requirement is the same as requiring $BC = CB$. □

Remarks 3.28. (i) We expect that (a generic property), if N in (14) is singular then it is of rank 1, that is there is only one curve along which the eigenvalue through λ_0 stays double. If N has rank 0, then the double eigenvalue would persist along any direction (and in this case, trivially $B = bI$ and $C = cI$).

(ii) Observe that if λ_0 persists as a double eigenvalue, then λ_1 is a double eigenvalue of M. Therefore, $u_1^T A_1 u_1 = u_2^T A_1 u_2$ and $u_1^T A_1 u_2 = 0$, $\lambda_1 = u_1^T A_1 u_1$, and $\begin{bmatrix} c_1 \\ c_2 \end{bmatrix}$ is any unit vector. This means that the limiting value v_0 of $v(t)$ is not determined: any unit vector in the plane spanned by u_1, u_2, would be a possible limit!

- This should be contrasted to the case when λ_0 does not persist as double eigenvalue. Then, there are two distinct eigenvalues of M, and two independent associated eigenvectors: Each of these would give a (unique, up to sign) pair of unit vectors $\begin{bmatrix} c_1 \\ c_2 \end{bmatrix}$ and a well defined limit, in general distinct from u_1 or u_2. In other words, if the double eigenvalue splits, then there are well defined eigenvectors paths. If the eigenvalue stays double, any eigenvector path in the plane spanned by u_1 and u_2 could be retrieved.

Example 3.29. Consider a block diagonal system of the type $A(x, y) = \begin{bmatrix} E & 0 \\ 0 & F \end{bmatrix}$ where E and F are square, and symmetric, functions, but not necessarily of the same dimension. Suppose that A has a double eigenvalue λ_0, which is a simple eigenvalue of both E and F at ξ_0. Notice that, if E and F are 2-parameter functions, then ξ_0 is not a conical intersection point.

Now the matrix of the associated eigenvectors of A has the form $U = [u_1 \ u_2] = \begin{bmatrix} v_1 & 0 \\ 0 & v_2 \end{bmatrix}$, and in this case it is trivial to verify that $BC = CB$ with $B = U^T A_x(\xi_0) U$, $C = U^T A_y(\xi_0) U$, hence the double eigenvalue will persist. Geometrically, this is quite to be expected: when the two surfaces of the eigenvalues of E and F intersect, they do so along a curve!

For the generalized eigenproblem $(K - \lambda M)u = 0$, with K and M symmetric positive definite, the situation is quite similar and the verification of the following result is left to the reader.

Theorem 3.30 (Double Eigenvalue: Generalized Eigenproblem). *Let $K, M \in \mathscr{C}^\omega(\Omega, \mathbb{R}^{n \times n})$ be symmetric positive definite and let the pencil (K, M) have a double eigenvalue λ_0 at a point ξ_0. Let $U = [u_1 \ u_2]$ be a matrix whose columns are two orthonormal eigenvectors associated to λ_0. Then, λ_0 persists as a double* *eigenvalue—along some curve—only if* $BC = CB$, *with* $B = U^T(K_x - \lambda_0 M_x)_{\xi_0} U$, $C = U^T(K_y - \lambda_0 M_y)_{\xi_0} U$.

4 Three Parameters: Coalescing Eigenvalues of Hermitian Functions

In this chapter, we consider complex valued, Hermitian, matrix valued functions depending on three parameters, $A \in \mathscr{C}^k(\Omega, \mathbb{C}^{n \times n})$, $A^* = A, k \geq 1$ (or also $k = \omega$). Recall that Ω is a region of $\mathbb{R}^3$ diffeomorphic to the unit ball. Similarly to what we did in Chap. 3, and for a reason which will be clarified below, our specific interest will be in finding values $x \in \Omega$ where the eigenvalues of A coalesce.

The presentation in this chapter is based on our works [48, 49]. But of course, there are several precursors. For example, there are several works in mathematical physics ultimately related to coalescing of eigenvalues of three-parameter Hermitian matrices; notably, the pioneering work of Stone, [56], and of Berry, [57, 58], with the name of Berry indivisibly attached to the fundamental concept of *geometric phase*, but also more recent work in *random matrices*, e.g. [61–63]; further, there are works in structural engineering, e.g. see [65] which in the end are concerned with Hermitian matrices depending on three (real) parameters.

In Sect. 4.1, we will review the fundamental *geometric phase* and clarify its role in detecting eigenvalues coalescing. In Sect. 4.2, we will review the relevant theoretical results on the impact of having one or several coalescing points inside a given surface. Section 4.3 contains algorithmic details and illustrative examples.

Several considerations are helpful in our analysis of the present complex Hermitian case.

(i) Theorem 3.1 continues to hold—with the appropriate, and obvious, modifications—also in the present case. Given its relevance, we give without proof the following result, which combines the (extensions of) Theorem 3.1 and Remarks 3.2.

Theorem 4.1. *Let R be a closed pluri-rectangular region in $\mathbb{R}^3$, and assume that the continuous eigenvalues of $A \in \mathscr{C}^k(R, \mathbb{C}^{n \times n})$, $k \geq 0$, $A^* = A$, can be labeled so to belong to p disjoint sets for all $x \in R$: $\lambda_1(x), \ldots, \lambda_{n_1}(x)$ in $\Lambda_1(x)$, $\lambda_{n_1+1}(x), \ldots, \lambda_{n_1+n_2}(x)$ in $\Lambda_2(x)$, and so forth. Then, there exists unitary $U \in \mathscr{C}^k(R, \mathbb{C}^{n \times n})$, such that $U^* A U$ is block diagonal:*

$$U^*(x) A(x) U(x) =: S = \begin{bmatrix} S_1(x) & \cdots & 0 \\ \vdots & \ddots & \vdots \\ 0 & \cdots & S_p(x) \end{bmatrix}, \forall x \in R,$$

where for all $j = 1, \ldots, p$, $S_j \in \mathscr{C}^k(R, \mathbb{C}^{n_j \times n_j})$, $S_j^* = S_j$, and the eigenvalues of $S_j(x)$ are those in $\Lambda_j(x)$, for all $x \in R$. In particular, if all eigenvalues of A are distinct, then A admits a $\mathscr{C}^k$ unitary eigendecomposition.

(ii) Secondly, we recall that even when the eigenvalues are distinct, the unitary Schur factor (the function of eigenvectors) is always unique up to a smooth phase matrix.

(iii) Finally, we remark that coalescing of eigenvalues of Hermitian matrices is a *real codimension 3* phenomenon. As usual, the codimension tells us how many parameters we need in order to expect the phenomenon to take place. The codimension count is easy to appreciate as follows. Take a 2×2 complex Hermitian function:

$$A = \begin{bmatrix} a & b + ic \\ b - ic & d \end{bmatrix}$$

where a, b, c are real valued, depending on (x_1, x_2, x_3). To have coalescing eigenvalues A must be of the form

$$\begin{bmatrix} a & 0 \\ 0 & a \end{bmatrix}$$

that is, we must satisfy the system of three (real valued) equations:

$$F = 0, \quad F \equiv \begin{bmatrix} \frac{a-d}{2} \\ b \\ c \end{bmatrix}.$$

Much like we did in Chap. 3, we will say that a coalescing point ξ_0 is a *generic coalescing point* for A if $F(\xi_0) = 0$ and $DF(\xi_0)$ is invertible (transversal intersection of surfaces). Naturally, it is a generic property—for a smooth Hermitian family of matrices—that if eigenvalues coalesce they do so at a generic coalescing point. Moreover, proceeding exactly like we did in Example 3.4, it is easy to see that a **generic coalescing** point is a **conical intersection**.

As a consequence of the above, we have that:

- Coalescing of eigenvalues has to be expected when the Hermitian function depends on 3 real parameters;
- it occurs at isolated points;
- it persists under small perturbations;
- nothing more degenerate should occur.

Example 4.2. In the Table below, we summarize the key differences between eigendecompositions of generic (real) symmetric and (complex) Hermitian functions. In particular, we note the key difference when considering a 1-parameter eigendecomposition for a given ordering of the eigenvalues.

	Real symmetric	Complex Hermitian
1-parameter smooth continuation on a curve with no coalescing point (ordered eigenvalues)	"Uniquely" defined (eigenvectors defined up to sign change)	Not uniquely defined (eigenvectors defined up to arbitrary smooth phase factor $e^{i\theta_j(t)}$)
Codimension of coalescing	2	3

So, if A is Hermitian, it is natural to work with 3 parameters. But, new ideas and tools are needed in order to detect coalescing points.

To understand which ideas are useful, let consider the implication of the above points (i) and (ii), when applied to the case of A restricted to a simple closed loop $\Gamma \subset \Omega$.

So, we have $A \in \mathscr{C}_1^k(\mathbb{R}, \mathbb{C}^{n \times n})$, $k \geq 1$, Hermitian, with distinct eigenvalues. Thus, we know from Theorem 2.6 that A has a $\mathscr{C}^k$ unitary decomposition for all $t \in \mathbb{R}$:

$$U^*(t)A(t)U(t) = \Lambda(t), \quad \Lambda(t) = \text{diag}(\lambda_1(t), \ldots, \lambda_n(t)), \quad \lambda_1(t) > \cdots > \lambda_n(t).$$

Further, from the proof of Theorem 2.6, we know that this eigendecomposition can be found as solution of

$$\dot{U} = UH(A, U),$$
$$H^* = -H, \quad H_{hj} = (U^*\dot{A}U)_{hj}/(\lambda_j - \lambda_h), \quad h < j,$$
$$\dot{\lambda}_j = (U^*\dot{A}U)_{jj}, \tag{15}$$

where, as we know, the diagonal of H is not uniquely defined. One way to arrive at a uniquely defined H is to impose the condition

$$H_{jj} = u_j^*(t)\dot{u}_j(t) = 0, \quad \forall t \in \mathbb{R}, \quad j = 1, \ldots, n, \tag{16}$$

where we are partitioning U columnwise: $U(t) = [u_1(t), \ldots, u_n(t)]$, for all t.

Remark 4.3. If $A \in \mathscr{C}_1^k(\mathbb{R}, \mathbb{C}^{n \times n})$, and we look at the decomposition over one period, then—in general—the above construction using (16) does not render a periodic factor U. However, we know (see Remark 2.30) that there is a smooth 1-periodic decomposition! The present "innocent" sounding remark is actually of practical relevance, since it is precisely the lack of periodicity resulting from the choice (16) which will be exploited.

Remark 4.4. Before proceeding, we want to remark that at least three different ways have been proposed in order to resolve the lack of uniqueness in a smooth eigendecomposition path, namely in resolving the ambiguity in the values of H_{jj} above in (15).

(i) Stone [56] proposed to take U so that

$$\mathrm{Im}\bigl(u_j^*(t+dt)u_j(t)\bigr) = \mathcal{O}(dt^2)$$

for all t and j.

(ii) Bunse-Gerstner et al. [24] proposed to select U so that (on the relevant interval $[0, 1]$)

$$U = \mathrm{argmin}\Bigl(\int_0^1 \|\dot{U}\|_F \, dt\Bigr),$$

and called the resulting choice the *Minimum Variation Decomposition*, or MVD for short.

(iii) Dieci and Eirola [26] proposed to use the condition (16).

In [48, Theorem 2.6], it was recently proven that

The three choices (i)–(ii)–(iii) all lead to the same unitary Schur factor.

As a consequence, we will unambiguously refer to an MVD for the decomposition obtained by enforcing (16) while finding U.

4.1 Geometric Phase, the Minimum Variation Decomposition, and Phase Rotating/Preserving Surfaces

Now consider a closed curve Γ in parameter space, and let it be parametrized as

$$\gamma : t \in [0, 1] \mapsto \gamma(t) \in \mathbb{R}^3,$$

smooth, with $\gamma(0) = \gamma(1)$. Suppose that all eigenvalues of $A(\gamma(t))$ are distinct for all t, and let

$$A(\gamma(t)) = U(t)\Lambda(t)U^*(t)$$

be an MVD of $A(\gamma(\cdot))$, with ordered eigenvalues.

How are $U(0)$ and $U(1)$ related? Obviously, they must be related by a phase matrix, since both $U(0)$ and $U(1)$ comprise unitary eigenvectors associated to the same, distinct, eigenvalues. So, we must have:

$$U(1) = U(0) \begin{bmatrix} e^{i\alpha_1} & & \\ & \ddots & \\ & & e^{i\alpha_n} \end{bmatrix}.$$

Now, for each $j = 1, \ldots, n$, we will demand that $\alpha_j \in (-\pi, \pi]$, in which case α_j is known as the *geometric phase* (or *Berry phase*) associated to λ_j.

Remark 4.5. Note that the geometric phase is related to having used the MVD along Γ. Also, note that if A happened to be real symmetric, then $\alpha_j = 0$ or π, so that $e^{i\alpha_j} = \pm 1$, as it should; in this case, we recall that having the value -1 was associated to having a coalescing point inside the loop Γ.

The geometric phase turns out to be the right object to consider. This is because of the following result.

Theorem 4.6 ([48, Theorem 2.4]). *Two different parametrizations of the same curve will lead to the same unitary factors at all points of the curve if and only if they are both MVDs.*

As a consequence, the MVD (and only the MVD) allows for a definition of phase along a loop that is independent of how the loop is parametrized. Thus, in conclusion, we have the key property that

The geometric phase is a property of the loop Γ and not of a parametrization γ of Γ.

That the geometric phase was going to reveal something important about coalescing eigenvalues was first realized by Stone, [56]. He had the remarkable insight that we do not have to just look at the value of the geometric phase along a loop, but we must look at the variation of the geometric phase as a family of loops covers a sphere-like surface.

To fix ideas, consider $\mathbb{S}^2$ (the unit sphere) and on it consider the following normalized geographical coordinates:

$$\begin{cases} x(s,t) = \sin(\pi s) \cos(2\pi t) \\ y(s,t) = \sin(\pi s) \sin(2\pi t) \\ z(s,t) = \cos(\pi s) \end{cases} \quad (17)$$

with $s, t \in [0, 1]$ representing, respectively, latitude and longitude. So, we can think of the family of parallels $\{\gamma_s\}_{s \in [0,1]}$ that covers $\mathbb{S}^2$, where γ_0, γ_1 represent (respectively) South pole and North pole, and we can think of s as a "normalized longitude" (meridian):

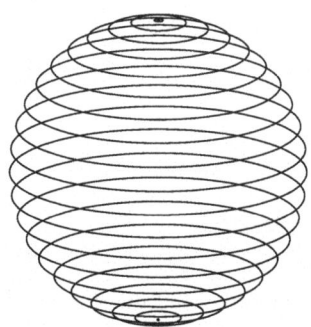

Accordingly, we can define n smooth maps:

$$\alpha_j : s \in [0,1] \mapsto \alpha_j(s),$$

where $\alpha_j(s)$ is now the (smooth) Berry phase associated to λ_j along γ_s.

Remarks 4.7. (i) Note that now $\alpha_j(s)$, $j = 1, \ldots, n$, must be allowed to possibly go beyond $(-\pi, \pi]$ (to retain smoothness). This is simply a reflection of the fact that the logarithm is a multivalued function, hence we must allow to move across adjacent Riemann sheets to maintain smoothness of the argument α_j.

(ii) The parallels γ_0 and γ_1 are just points; therefore, for all $j = 1, \ldots, n$, $\alpha_j(0)$ can (and will) be chosen to be 0 and $\alpha_j(1)$ has to be a multiple of 2π (because A is constant along $\gamma_{0,1}$ and so $U_0(1) = U_0(0)$ and $U_1(1) = U_1(0)$).

There is nothing exceptional about $\mathbb{S}^2$ nor about a covering made out of parallels. The same considerations apply to more general surfaces and loops' coverings. We call S a *2-sphere* if it is the embedded image of $\mathbb{S}^2$ in $\mathbb{R}^3$ under a smooth map f.

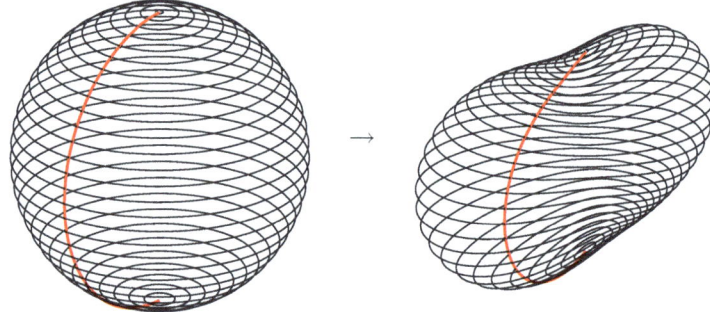

Covering inherited from the sphere.

Finally, we will make use of the following definition (see [56]).

Definition 4.8. A 2-sphere S is called ***phase rotating*** for A if $\alpha_j(1) \neq 0$ for at least one j. Otherwise, S is called ***phase preserving*** for A.

In [48, Theorem 2.12], it is further proven that *Definition 4.8 does not depend on the choice of loops' covering*.

And, with this, we are ready to make precise the original insight of Stone.

Theorem 4.9. *If a 2-sphere S is phase rotating, then there is a coalescing point for the eigenvalues of A inside S.*

4.2 Some Theoretical Results

As insightful as it is, Theorem 4.9 does not tell us which eigenvalues may have coalesced, nor it is helpful in understanding the reverse implication. So, our present challenge is to determine **if** and **when** a coalescing point is revealed by having a phase rotating surface.

Guided by what we learned in Chap. 3, we guess that—for a certain eigenvalue—an odd number of generic coalescing points inside a 2-sphere should be detected by monitoring the variation in the Berry phases as the 2-sphere is swept, but not otherwise. This guess is nearly exact.

First, we will consider the $(2,2)$-case.

Example 4.10. Consider the following problem

$$A = \begin{bmatrix} x & y+iz \\ y-iz & -x \end{bmatrix},$$

for $(x, y, z) \in B_\rho$, a ball of radius ρ centered at the origin. The origin is the only coalescing point inside B_ρ, and it is a generic coalescing point. We want to look at the eigendecomposition of A along a covering of the sphere S_0 of radius ρ. Normalized geographical coordinates (17) give the following rewriting of A:

$$A = \rho \begin{bmatrix} \sin(\pi s)\cos(2\pi t) & \sin(\pi s)\sin(2\pi t) + i\cos(\pi s) \\ \sin(\pi s)\sin(2\pi t) - i\cos(\pi s) & -\sin(\pi s)\cos(2\pi t) \end{bmatrix},$$

with $s, t \in [0, 1]$. The eigenvalues of A are constant for all s, t: $\lambda_\pm = \pm\rho$. We can choose an eigenvector associated to λ_+, call it q^+, to be

$$q_s^+(t) = \frac{1}{\sqrt{2 + 2\sin(\pi s)\cos(2\pi t)}} \begin{bmatrix} 1 + \sin(\pi s)\cos(2\pi t) \\ \sin(\pi s)\sin(2\pi t) - i\cos(\pi s) \end{bmatrix},$$

for all $(s, t) \in [0, 1] \times [0, 1] \setminus (1/2, 1/2)$. Note that each $q_s^+(\cdot)$, $s \in (0, 1) \setminus \{1/2\}$, has unit length and is 1-periodic, and does not satisfy (16). In order to extract the geometric phase, we know that we need the eigenvectors $u_s^+(\cdot)$ satisfying (16). Since clearly $u_s^+(t) = e^{i\eta_s^+(t)} q_s^+(t)$, where $\eta_s^+(\cdot)$ is a real-valued smooth function with $\eta_s^+(0) = 0 \pmod{2\pi}$, then we need to find η_s^+.

A simple computation shows that

$$\dot{\eta}_s^+(t) = i(q_s^+(t))^* \partial_t q_s^+(t),$$

from which it follows that:

$$\eta_s^+(1) - \eta_s^+(0) = i \int_0^1 (q_s^+(t))^* \partial_t q_s^+(t)\, dt,$$

where $\eta_s^+(1)$ is, modulo 2π, the principle phase associated to λ_+, that is the geometric phase, for each s.

A direct computation gives

$$i \int_0^1 \left(q_s^+(t)\right)^* \partial_t q_s^+(t)\, dt = \begin{cases} \pi - \pi\cos(\pi s) & \text{if } s \in [0, 1/2) \\ -\pi - \pi\cos(\pi s) & \text{if } s \in (1/2, 1] \end{cases}.$$

Therefore, invoking the continuity of $\alpha^+(\cdot)$ over $[0, 1]$, we obtain the Berry phase function:

$$\alpha^+(s) = \pi\left(1 - \cos(\pi s)\right), \text{ for all } s \in [0, 1].$$

Finally, observe that α^+ computed above satisfies $\alpha^+(0) = 0$, $\alpha^+(1) = 2\pi$. A similar argument shows that $\alpha^-(0) = 0$, $\alpha^-(1) = -2\pi$. Hence, by Definition 4.8, S_0 is phase-rotating.

The situation we just saw in Example 4.10 is true in general.

Theorem 4.11. *Let $A \in \mathscr{C}^1(\Omega, \mathbb{C}^{2\times 2})$ be Hermitian. Let $\lambda_1(\xi) \geq \lambda_2(\xi)$ be its continuous eigenvalues, for all $\xi \in \Omega$. Let $\xi_0 \in \Omega$ be a generic coalescing point for A in Ω, and suppose that ξ_0 is the only coalescing point in Ω. Let $S \subset \Omega$ be a 2-sphere. If the interior part of S contains ξ_0, then S is phase-rotating, with $\alpha_{1,2}(1) = \pm 2\pi$.*

Sketch of Proof. The main steps of the proof are the following (filling in the details is a doable exercise, or see [48]).

(i) Without loss of generality, restrict consideration to the case of $A = \begin{bmatrix} a & b+ic \\ b-ic & -a \end{bmatrix}$.

(ii) Use the implicit function theorem and genericity of the coalescing point to perform a change of variables, so that (locally) the problem is reduced to the one of Example 4.10.

(iii) Argue as in Example 4.10 to obtain that a small sphere $\mathbb{S}^2$ centered at the origin is phase rotating.

(iv) Finally, through a homotopy argument, extend validity of the result to S. (For this step, a powerful topological result, the "Annulus theorem", is needed.)

To tackle the general (n, n) case, and the possibility of having several (generic) coalescing points in Ω, the general strategy is similar to what we did in Chap. 3: Theorem 4.1 is repeatedly used to isolate (locally) a coalescing pair of eigenvalues, Theorem 4.11 is then used on the smaller $(2, 2)$ blocks, and finally the results are patched together. We caution that there are several, and nontrivial, mathematical subtleties in pursuing this course of action, and we refer once more to [48] for the relevant details. However, for our present purposes, it suffices to give the general result.

Theorem 4.12 (Hermitian $n \times n$ case). Let $A \in \mathscr{C}^1(\Omega, \mathbb{C}^{n \times n})$ be Hermitian. Let $\lambda_1(\xi), \ldots, \lambda_n(\xi)$ be its continuous eigenvalues, labeled in descending order. Suppose that, for any $j = 1, \ldots, n-1$, we have:

$$\lambda_j = \lambda_{j+1}$$

solely at d_j distinct generic coalescing points in Ω. Let $S \subset \Omega$ be a 2-sphere whose interior part contains all those coalescing points. Then we have that:

(i) $\alpha_1(1) = 0$ (resp. 2π) (mod 4π) if d_1 is even (resp. odd)
(ii) $\alpha_j(1) = 0$ (resp. 2π) (mod 4π) if $d_{j-1} + d_j$ is even (odd) and $j = 2, \ldots, n-1$
(iii) $\alpha_n(1) = 0$ (resp. 2π) (mod 4π) if d_{n-1} is even (resp. odd)

with $|\alpha_1(1)| \leq 2d_1\pi$, $|\alpha_j(1)| \leq 2(d_{j-1}+d_j)\pi$ for $j = 2, \ldots, n-1$, and $|\alpha_n(1)| \leq 2d_{n-1}\pi$.

Remarks 4.13. (i) Theorem 4.12 clarifies that the Berry phase functions associated to non-coalescing eigenvalues do not undergo any increment as we cover S. In particular (see Theorem 4.9), if all eigenvalues of A are distinct on Ω, then S is phase-preserving.

(ii) The geometric phases add up nicely, though not fully predictably. Nevertheless, an interesting quantity is always conserved (recall that all coalescing points are inside S):

$$\sum_{j=1}^{n} \alpha_j(s) = 0, \text{ for all } s \in [0, 1]. \tag{18}$$

(iii) An *even number* of generic coalescing points involving the same pair of eigenvalues may fail to contribute to whether or not S is phase rotating.

Example 4.14. This example is used to illustrate that nongeneric coalescing points may go unnoticed. Consider $\mathbb{S}^2$ and

$$A(\xi) = \begin{bmatrix} x^p & y+iz \\ y-iz & -x^p \end{bmatrix}, p = 1, 2, 3, \ldots.$$

Here, there is a unique coalescing point in the interior of $\mathbb{S}^2$, the origin. We are looking for solutions of $F(x, y, z) = 0$: $\begin{bmatrix} x^p \\ y \\ z \end{bmatrix} = 0$, and DF is clearly singular at the origin for $p \geq 2$. Still, a trivial verification shows that $\mathbb{S}^2$ is phase-preserving (respectively, phase-rotating) whenever p is even (respectively, odd). □

Rather than degenerate cases as in Example 4.14, it is more interesting to understand what kind of inferences we are able to make using a result like Theorem 4.12. The juice of the story is to realize that Theorem 4.12 validates the following claim (with A, Ω, and S as usual):

If there exists an index k, $1 \leq k \leq n$, for which $\alpha_k(1) \neq 0$, then λ_k must have coalesced at least once inside Ω with λ_{k-1} or λ_{k+1}.

We illustrate in the example below.

Example 4.15. Suppose we have a function A with distinct eigenvalues on the unit sphere, giving us the following different scenarios for the geometric phases.

Here we have the figure below. Observe that we have $\alpha_1(1) = -2\pi$, $\alpha_3(1) = 2\pi$, and $\alpha_2(1) = 0$. Therefore, we can conclude that, inside $\mathbb{S}^2$, λ_1 and λ_2 coalesce (at least once), and that so do λ_2 and λ_3.

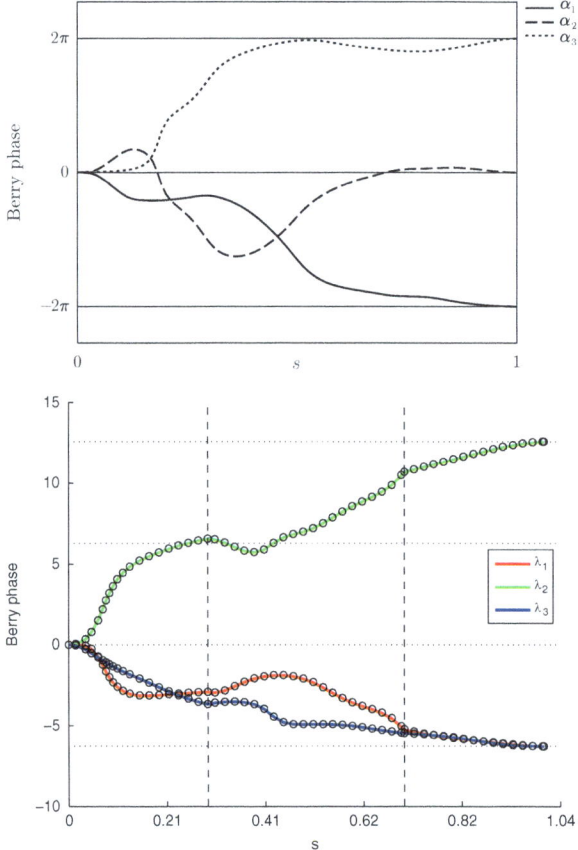

Similarly, in the case above we observe $\alpha_1(1) = -2\pi$, $\alpha_3(1) = -2\pi$, and $\alpha_2(1) = 4\pi$. Again, we can conclude that, inside $\mathbb{S}^2$, λ_1 and λ_2 coalesce, and so do λ_2 and λ_3.

Next, suppose we observe the situation below. Now we have $\alpha_1(1) = \alpha_5(1) = 2\pi, \alpha_4(1) = \alpha_6(1) = -2\pi, \alpha_2(1) = -4\pi$, and $\alpha_3(1) = 4\pi$. What we can conclude is that, inside $\mathbb{S}^2$, there is at least one coalescing point for each pair of consecutive eigenvalue.

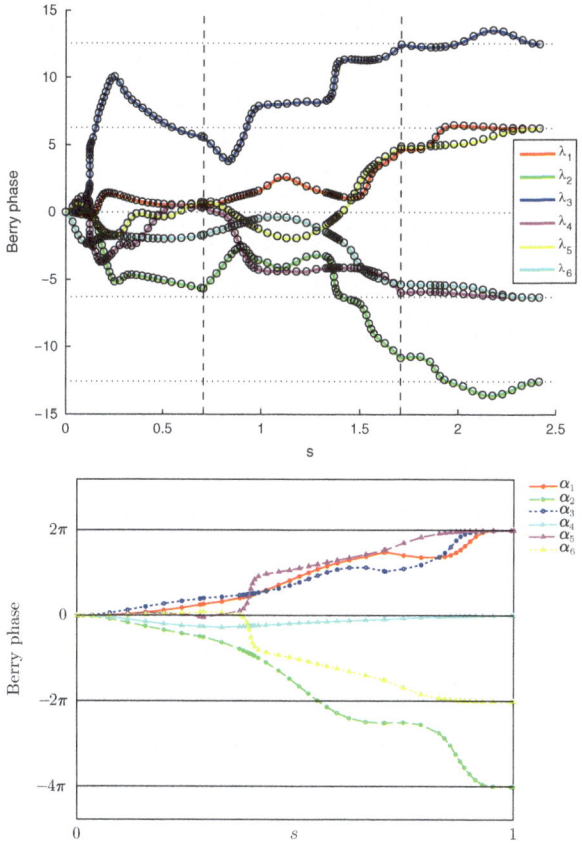

Finally, for the figure above we have $\alpha_1(1) = \alpha_3(1) = \alpha_5(1) = 2\pi, \alpha_4(1) = 0$, $\alpha_2(1) = -4\pi$, and $\alpha_6(1) = -2\pi$. All we can conclude is that, inside $\mathbb{S}^2$, λ_1 coalesces with λ_2, that λ_2 coalesces with λ_3, and then that λ_5 coalesces with λ_6.

4.3 Algorithm and Examples

Based upon Theorem 4.12, we now outline an algorithm which locates (generic) coalescing points. Of course, there are several choices which need to be made, in particular how to cover Ω, how to perform integration along parallels/meridians, and how to accurately locate the parameter values where the coalescings occur. These concerns are addressed in detail in [49], here we recall some of the key features.

First of all, we will work with Ω being a cube, which will be further refined in a number of smaller cubes (see below). [The cube is not a smooth embedding of $\mathbb{S}^2$, but this has no practical consequence, since we will step exactly at the corners, and thus will work along piecewise-smooth paths.]

To "cover" a basic cube we proceed from the South Pole (barycenter of the bottom square) to the North Pole, by integrating along squares (the parallels) on the three different regions of the cube: Bottom, Lateral, Top; see the figure below. We notice that the parametrization along the parallels is irrelevant insofar as the information retrieved about the geometric phases.

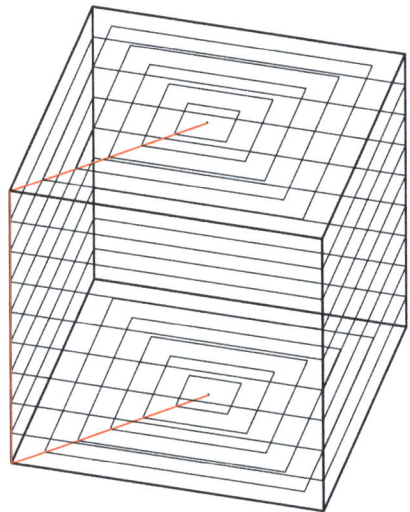

Of course, the original cube is typically first subdivided in a number of smaller cubes, the geometric phases are computed along each of these cubes, and—based on these—coalescing eigenvalues are detected and the location of the coalescing points is further accurately approximated:

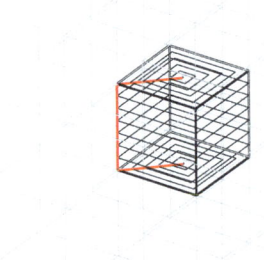

- subdivide region in cubes;
- cover each cube with square-shaped loops and compute Berry phases along each loop;
- so doing, we have obtained n functions $\alpha_j(s)$, where s parametrizes the (red) C-shaped curve;
- if a cube is phase rotating, local refinement and/or a zoom-in module are triggered in order to locate the coalescing point accurately.

Also, for reasons of efficiency, the original cube is typically first subdivided in "columns" (parallelepipeds). Since each column can be monitored independently (i.e., in parallel), we implemented our algorithm in a distributed computing environment. Further, in a given column, information from the top face of a cube is reused for the bottom face of the cube lying on top of it.

When a cube is declared phase rotating, and a coalescing pair has been detected, a **zoom-in** procedure to accurately locate the coalescing point is triggered, to minimize $(\lambda_j(x) - \lambda_{j+1}(x))^2$. In the present context, it has been quite beneficial to do a coordinate search to find a good initial guess, rather than unduly refining the cube (which is quite expensive). Still, in case of zoom-in failure (or if multiple coalescing points are detected) **local refinement** is performed (the cube is progressively divided in 8 sub-cubes). This combination of local refinement and zoom-in has proved to be very robust in locating coalescing points:

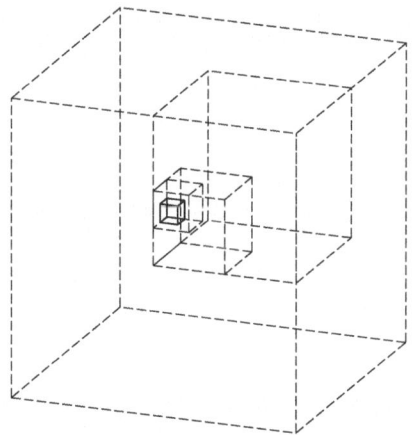

A key component of the algorithm is the adaptive integration along loops (the parallels), as well as adaptively choosing the meridian values where the loops are located. The integration module will be discussed below, since it is not the same as the basic integration scheme we had discussed in Chaps. 2 and 3.

4.3.1 A Predictor–Corrector 1-d Continuation

Here we describe the scheme we adopted to integrate along a parallel. For a given value s of the meridian, let $B(t)$, $t \in [0, 1]$, be the restriction of the function A to the parallel corresponding to the value s of the meridian. We will assume that there are no coalescing points along the parallel.[6]

Suppose we have a partition of the interval $[0, 1]$ (in practice, the mesh is found adaptively):

$$0 = t_0 < t_1 < \cdots < t_N = 1 , \; h_k = t_k - t_{k-1} , \; k = 0, 1, \ldots .$$

Our goal is a method that:

(i) at leading order of expense, costs as much as one eigendecomposition per step,
(ii) it is guaranteed to have 2nd order of accuracy globally, also for the Berry phases, and
(iii) it requires no explicit integration of the DE for the factors (in other words, we do not want to form the vector field in (15)).

Now, recall that we seek $U(t)$, $\Lambda(t)$: $U^*(t)B(t)U(t) = \Lambda(t)$. We can assume that the eigenvalues are ordered for all t: $\lambda_1(t) > \cdots > \lambda_n(t)$.

Let's call U_k the computed approximations to $U(t_k)$.

1-d Continuation Algorithm

- (*Initialization*). At $t_0 = 0$, let $B(0) = U_0 \Lambda(0) U_0^*$.
  For $k = 1, \ldots, N$:
- (*Predictor phase*). First we find **an** exact eigendecomposition at the points t_k: $B(t_k) = Q_k \Lambda(t_k) Q_k^*$, where Q_k is unitary and $\Lambda(t_k)$ is the real diagonal matrix of the ordered eigenvalues of $B(t_k)$.
- (*Corrector phase*). Then, we adjust the factor Q_k by bringing it as close as possible (in the Frobenius norm) to the previous approximation U_{k-1}, while preserving a Schur decomposition of $B(t_k)$. That is, we seek a phase matrix Φ_k such that $\|U_{k-1} - Q_k \Phi_k\|_F$ is minimized. Finally, we set $U_k = Q_k \Phi_k$.
  We note that the problem $\min_{\Phi_k} \|U_{k-1} - Q_k \Phi_k\|_F$, with Φ_k a phase matrix, has the unique solution $\Phi_k = \exp(i \operatorname{Arg}(D_k))$, where $D_k = \operatorname{diag}(Q_k^* U_{k-1})$, and Arg is the principal argument of a complex number.

In the above continuation algorithm, we perform only one eigendecomposition per step, and do not explicitly integrate (15). The following result (see [49, Theorem 2.7] for a proof) shows that the technique is 2nd order accurate, as desired.

Theorem 4.16. *Given* $B \in \mathscr{C}_1^k([0, 1], \mathbb{C}^{n \times n})$, $k \geq 3$, *Hermitian and 1-periodic (in t) with no coalescing eigenvalues. Let a mesh be given, with $h_k \leq h$ sufficiently*

---

[6]Recall that the occurrence of a coalescing point along a curve is non generic. However, in case it happens, then it is actually detected by our algorithm (since the stepsize is pushed below the minimum allowed stepsize). This is actually a somewhat pleasant occurrence, since after all we are trying to detect coalescing points!

small, and let U_k be the approximation to $U(t_k)$ found from the above predictor-corrector method, for $k = 1, 2, \ldots, N$.

Then, the factor U_N is a second order (i.e., $\mathcal{O}(h^2)$) accurate approximation to the exact factor $U(1)$ and therefore the approximate phase matrix $U_N^* U_0 = U_N^* U(0)$ (which is diagonal) is also a second order accurate approximation to the exact phase matrix $U(1)^* U(0)$. Finally, the computed Berry phases $\alpha_j(s)$, $j = 1, \ldots, n$, are also second order accurate, for any $s \in [0, 1]$.

Remark 4.17. The fact that the global error is of second order accuracy appears surprising. To appreciate this statement, consider what we do at the first mesh-point t_1. Clearly, $U(t_1) = U_0 + \mathcal{O}(h_1)$, and so U_0 (which is what we use in order to correct some Schur factorization Q_1) is only a first order approximation to the exact factor at t_1, and as a consequence a standard error analysis would give us a 0-th order method. However, our algorithm behaves like a 2-nd order method used to integrate the differential equation, in spite of the fact that no integration of the differential equation is practically performed. Indeed, locally, our trivial predictor turns out to be $\mathcal{O}(h^3)$ accurate in enforcing minimum variation: a nice surprise.

Adaptive Stepsize Choice

- (*Along a parallel*). Along a parallel of length L, the stepsize selection is based on a standard error per step control. We let

$$\rho_U = \|U_k - U_{k-1}\|_1, \quad \rho_\Lambda = \max_j \frac{|\lambda_j(t_k) - \lambda_j(t_{k-1})|}{1 + |\lambda_j(t_k)|},$$

and then—for a given error tolerance `tolls`—we set $\rho = \frac{\max(\rho_U, \rho_\Lambda)}{\text{tolls}}$. Given a value for the maximum stepsize $h_{\max}$, we adjust the stepsize according to the criterion $h_{\text{new}} = \min(h_{\max}, h_{k-1}/\rho)$. If $\rho \leq 1.2$, the step is accepted, otherwise it is not. In either case, the new stepsize is given by h_{new}, and the computation is continued until either $h_{\text{new}} < h_{\min}$ or we reached the last point.[7]

- (*Along the meridian, the C-curve*). Choosing the stepsize to move along the meridian is delicate. The key concern, and the reason to use adaptive stepsizes, is to ensure that the Berry phases are accurately approximated, in particular not to miss a rapid variation of the same (we need to avoid jumping to the wrong branch of the logarithm).

Below, let

$$s_0 = 0 < s_1 < s_2 \cdots < s_K = 1$$

---

[7]Typical values used are `tolls` $= 10^{-1}$, $h_{\max} = L/10$, $h_{\min} = 10^{-14}$.

be the mesh (which will be found adaptively) along the meridian, and $\Delta_k = s_{k+1} - s_k$, $k = 0, 1, \ldots, K - 1$. Also, let α be the vector of the (approximate) Berry phases, $\alpha : [0, 1] \to \mathbb{R}^n$, and finally let tollp be a fixed tolerance value to monitor variation in the phases. We adopted the following strategy.[8]

(a) Compute $\delta\alpha = \|\alpha(s_k + \Delta_k) - \alpha(s_k)\|_\infty / \text{tollp}$;
(b) Let $\Delta_{\text{new}} = \min(2\Delta_k, \Delta_{\max}, \Delta_k/\delta\alpha)$;
(c) If $\delta\alpha \le 1.5$, the step is accepted, otherwise it is not. In either case, the new stepsize is given by Δ_{new}, and computation is continued until either the step is successful or $\Delta_{\text{new}} < \Delta_{\min}$.

Finally, we monitor that (18) is not violated: we reject a step, and halve it, if $|\sum_j (\alpha_j(s_{k+1}) - \alpha_j(s_k))| > \text{tolls}$.

We conclude this section with two illustrative examples. The first is chosen to highlight that it is computationally demanding to integrate *near* a coalescing point. The second is chosen to illustrate what can happen in non-generic cases.

Example 4.18. Take $A(x, y, z) = \begin{bmatrix} x & y + iz \\ y - iz & -x \end{bmatrix}$, and $\Omega = [-10^{-k}, 2 - 10^{-k}] \times [-1, 1]^2$, $k = 0, 1, 2, 3$. Obviously, there is only one coalescing point, at the origin. For all k, we get $\alpha_1(1) = 2\pi$, $\alpha_2(1) - 2\pi$ and the coalescing point is detected.

Now, the case $k = 0$ is trivial: the origin is well inside the cube, and integration along the faces of the cube is rather simple. But when, say, $k = 3$, then the integration is demanding with the stepsize becoming of order 10^{-4} along the meridian and the parallels.

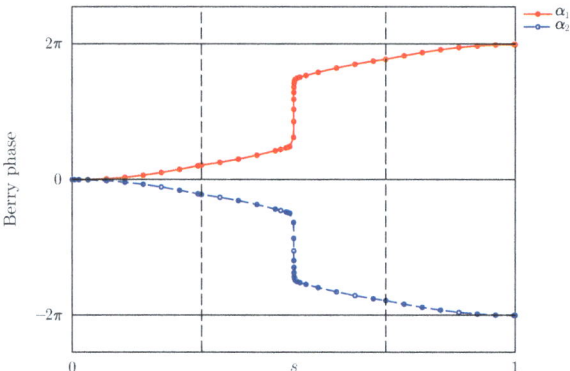

Example 4.19. The algorithm typically fails when used to find non-generic coalescing points. The region here is the cube $\Omega = [-1, 1]^3$. Take

---

[8] Typical values used are tollp $= \pi/6$, which corresponds to restricting a phase variation to be at most $\pi/4$, $\Delta_{\max}$ to be $1/10$, $\Delta_{\min} = 10^{-14}$, $\Delta_0 = \Delta_{\max}$.

Continuous Decompositions and Coalescing Eigenvalues for Matrices... 259

$$A(x,y,z) = \begin{bmatrix} x^2 & y+iz \\ y-iz & -x^2 \end{bmatrix}, \quad \text{and}$$

$$B(x,y,z) = \begin{bmatrix} x^2+y^2-r^2 & iz \\ -iz & -(x^2+y^2-r^2) \end{bmatrix}, r = \frac{1}{2}.$$

The function A has one coalescing point at the origin, and it is a non-generic coalescing point; the eigenvalues of the function B, instead, coalesce along the circle of radius $\frac{1}{2}$ centered at the origin in the xy-plane; the entire circle is made up of non-generic coalescing points.

In both cases, the algorithm returns $\alpha_1(1) = \alpha_2(1) = 0$, and no coalescing point is detected for either A or B.

Interestingly, if we use x^3 instead of x^2 in A, then the algorithm detects the coalescing point at the origin. Probably, our technique is capable of locating coalescing points of "odd multiplicity", but the theory is lacking.

4.3.2 Application: Density of Degeneracies

An interesting application of locating coalescing points for Hermitian matrices depending on three real parameters arises in the context of random matrix theory. Below, motivated by work in (computational) mathematical physics of Wilkinson, Walker et alia (see [61–63]) we want to study the spatial distribution of coalescing points for random matrix models.

Following these cited works, we consider $A \in \mathscr{C}^\omega([0, 2\pi]^3, \mathbb{C}^{n \times n})$, periodic in each component of x, of the following form:

$$A(x) = \sum_{j=1}^{3} [A_{2j-1} \cos(x_j) + A_{2j} \sin(x_j)], \quad x \in [0, 2\pi]^3, \quad (19)$$

where $A_1, \ldots, A_6$, are random matrices selected by independent samples from the *Gaussian unitary ensemble* (GUE). In other words, each A_j is a Hermitian matrix with entries being independently distributed elements from the Gaussian distribution with mean 0 and variance 1, and thus $A_j = B_j + iC_j$, with $B_j = B_j^T$ and $C_j = -C_j^T$ with entries being independently Gaussian distributed, $j = 1, \ldots, 6$.

Our goal is to find the spatial distribution, i.e. the density, of generic coalescing points, $M(n)$, for the function A, in function of the dimension n. In the physics literature, these generic coalescing points are called *degeneracies*.

The idea is simple: We will **count** the number of coalescing points $M(n)$ in the cube $[0, 2\pi]^3$; of course, because of periodicity, we will work with the half-cube $[\pi, 2\pi] \times [0, 2\pi]^2$. But, how large is $M(n)$?

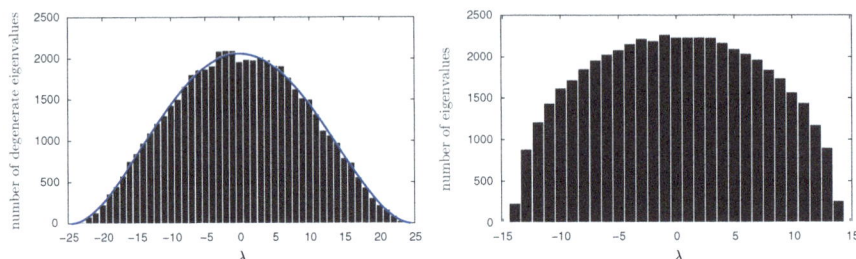

Fig. 15 Distribution of degeneracies for a realization with $n = 50$, and of the eigenvalues of 1,000 random Hermitian matrices, $n = 50$

Table 2 Experiments, $n = 20$

No. of cubes	Average # of CIs
14×14×7	992
28×28×14	1,072
56×56×28	1,097
112×112×56	1,102

In [62], Walker and Wilkinson, using Wigner's semi-circle law on the distribution of eigenvalues of random Hermitian matrices, obtained the following remarkable formula valid in the asymptotic regime (as $n \to \infty$):

$$M(n) \stackrel{asy}{\sim} \frac{512 n^{5/2}}{135\sqrt{3\pi}}. \tag{20}$$

From (20), clearly there are a lot of degeneracies already for moderate values of n, and so the computations will become quickly very expensive as n grows. Here, exploiting the parallel implementation of our algorithm has been absolutely mandatory.

Based upon our experiments, we make a couple of observations.

1. The distribution of coalescing points in the cube appears to be uniform. This means that a refinement strategy will effectively reduce to globally refining the cube in equal sub-cubes.
2. There is clear support to the fact that the coalescing eigenvalues are distributed according to the power law $c(\rho(E))_+^4$, where $\rho(E)$ indicates the density of states and it is given by $\rho(E) = \frac{1}{\pi\sqrt{3}}(n - E^2/12)^{1/2}$. [Here, we are using the notation $z_+ = \max(z, 0)$.] This should be compared with Wigner's semicircle law $a(n - x^2/4)_+^{1/2}$, a law giving the distribution of the eigenvalues of Hermitian matrices (Fig. 15).

Finally, we conclude with Table 2, showing the values $M(n)$, for $n = 20$, computed by our algorithm. These values are averages over 5 experiments, each experiment being a different sixtuplet of random Hermitian matrices.

From (20), the expected number of degeneracies from the asymptotic estimate is 1,105. Comparison with our results shows excellent agreement with the asymptotic value; e.g., the error on the finest grid is just 0.27 %. We also observe that refining the subdivision (i.e., increasing the number of cubes) CI's there were undetected at lower resolution become detected; this is because these new CI's are associated to pairs of eigenvalues that coalesced an even number of times inside the larger (sub)-cube.

References

1. R.A. Horn, C.R. Johnson, *Matrix Analysis* (Cambridge University Press, New York, 1985)
2. C. Adams, R. Franzosa, *Introduction to Topology Pure and Applied* (Pearson Prentice Hall, Upper Saddle River, 2008)
3. M.W. Hirsch, *Differential Topology* (Springer, New York, 1976)
4. V.I. Arnold, *Geometrical Methods in the Theory of Ordinary Differential Equations*, 2nd edn. (Springer, New York, 1988)
5. J.K. Hale, *Ordinary Differential Equations* (Krieger Publishing Co, Malabar, 1980)
6. H.B. Keller, *Lectures on Numerical Methods in Bifurcation Problems* (Springer/Tata Institute of Fundamental Research, New York/Bombay, 1987)
7. Y.A. Kuznetsov, *Elements of Applied Bifurcation Theory* (Springer, New York, 1995)
8. E. Allgower, K. Georg, *Numerical Continuation Methods* (Springer, New York, 1990)
9. W.C. Rheinboldt, *Numerical Analysis of Parameterized Nonlinear Equations* (Wiley, New York, 1986)
10. G.H. Golub, C.F. Van Loan, *Matrix Computations*, 2nd edn. (The Johns Hopkins University Press, Baltimore, 1989)
11. T. Kato, *A Short Introduction to Perturbation Theory for Linear Operators* (Springer, New York, 1982) [Kato shows that analytic and Hermitian, then it has analytic eigendecomposition]
12. P. Lax, *Linear Algebra and Its Applications*, 2nd edn. (Wiley, New York, 2007)
13. T. Kato, *Perturbation Theory for Linear Operators*, 2nd edn. (Springer, Berlin, 1976)
14. F. Rellich, *Perturbation Theory of Eigenvalue Problems* (Gordon and Breach, New York, 1969) [Shows that smooth Hermitian may have non-smooth eigenspaces]
15. M. Baer, *Beyond Born-Oppenheimer: Electronic Nonadiabatic Coupling Terms and Conical Intersections* (Wiley, Hoboken, 2006)
16. A.P. Seyranian, A.A. Mailybaev, *Multiparameter Stability Theory with Mechanical Applications* (World Scientific, Singapore, 2003) [Very thorough account, includes local analysis at coalescing eigenvalues]
17. E. Allgower, K. Georg, Continuation and path following. Acta Numerica **2**, 1–64 (1993)
18. L.T. Watson, M. Sosonkina, R.C. Melville, A.P. Morgan, H.F., Walker, Algorithm 777: HOMPACK 90: A suite of Fortran 90 codes for globally convergent homotopy algorithms. ACM Trans. Math. Software **23**, 514–549 (1997)
19. S. Campbell, Numerical solution of higher index linear time varying singular systems of DAEs. SIAM J. Sci. Stat. Comp. **6**, 334–348 (1988)
20. P. Kunkel, V. Mehrmann, Canonical forms for linear DAEs with variable coefficients. J. Comp. Appl. Math. **56**, 225–251 (1994)
21. E.J. Doedel, A.R. Champneys, T.F. Fairgrieve, Yu.A. Kuznetsov, B. Sandstede, X.J. Wang, AUTO97: Continuation and bifurcation software for ordinary differential equations (with HomCont). Available at ftp.cs.concordia.ca, Concordia University (1997)

22. Y. Sibuya, Some global properties of matrices of functions of one variable. Math. Anal. **161**, 67–77 (1965) [Smooth diagonalization, considerations on periodicity of eigendecompositions]
23. V.A. Eremenko, Some properties of periodic matrices. Ukrainian Math. J. **32**, 19–26 (1980) [Somewhat extends periodicity results of Sibuya]
24. A. Bunse-Gerstner, R. Byers, V. Mehrmann, N.K. Nichols, Numerical computation of an analytic singular value decomposition by a matrix valued function. Numer. Math. **60**, 1–40 (1991) [Analytic SVD for analytic function]
25. H. Gingold, P.F. Hsieh, Globally analytic triangularization of a matrix function. Lin. Algebra Appl. **169**, 75–101 (1992) [Extends Kato's results to Schur form when eigenvalues are aligned]
26. L. Dieci, T. Eirola, On smooth orthonormal factorizations of matrices. SIAM J. Matrix Anal. Appl. **20**, 800–819 (1999) [Analyzes Schur, SVD, block-analogs, as well as impact of coalescing and genericity]
27. J.L. Chern, L. Dieci, Smoothness and periodicity of some matrix decompositions. SIAM Matrix Anal. Appl. **22**, 772–792 (2000) [Constant rank, periodicity and impact of coalescing on it]
28. K.A. O'Neil, Critical points of the singular value decomposition. SIAM J. Matrix Anal. **27**, 459–473 (2005)
29. L. Dieci, M.G. Gasparo, A. Papini, Smoothness of Hessenberg and bidiagonal forms. Med. J. Math. **5**(1), 21–31 (2008)
30. L. Dieci, T. Eirola, Applications of smooth orthogonal factorizations of matrices, in *IMA Volumes in Mathematics and Its Applications*, ed. by E. Doedel, L. Tuckermann, **119** (Springer-Verlag, New York, 1999), pp. 141–162
31. M. Baumann, U. Helmke, Diagonalization of time varying symmetric matrices, Proc. Inter. Conference on Computational Sci. **3**, pp. 419–428 (2002)
32. L. Dieci, A. Papini, Continuation of Eigendecompositions. Future Generat. Comput. Syst. **19**(7), 363–373 (2003)
33. J. Demmel, L. Dieci, M. Friedman, Computing connecting orbits via an improved algorithm for continuing invariant subspaces. SIAM J. Sci. Comput. **22**, 81–94 (2000)
34. L. Dieci, M. Friedman, Continuation of invariant subspaces. Appl. Numer. Lin. Algebra **8**, 317–327 (2001)
35. W.J. Beyn, W. Kles, V. Thümmler, Continuation of low-dimensional invariant subspaces in dynamical systems of large dimension, in *Ergodic Theory, Analysis and Efficient Simulation of Dynamical Systems*, ed. by B. Fiedler (Springer, Berlin, 2001), pp. 47–72
36. L. Dieci, J. Rebaza, Point-to-periodic and periodic-to-periodic connections. BIT **44**, 41–62 (2004). Erratum of the same in BIT **44**, 617–618 (2004)
37. D. Bindel, J. Demmel, M. Friedman, Continuation of invariant subspaces for large bifurcation problems, in *SIAM Conference on Applied Linear Algebra* (The College of William and Mary, Williamsburg, 2003)
38. K. Wright, Differential equations for the analytical singular value decomposition of a matrix. Numer. Math. **63**, 283–295 (1992)
39. V. Mehrmann, W. Rath, Numerical methods for the computation of analytic singular value decompositions. Electron. Trans. Numer. Anal. **1**, 72–88 (1993)
40. L. Dieci, M.G. Gasparo, A. Papini, Continuation of singular value decompositions. Med. J. Math. **2**, 179–203 (2005)
41. S. Chow, Y. Shen, Bifurcations via singular value decompositions. Appl. Math. Comput. **28**, 231–245 (1988)
42. L. Dieci, M.G. Gasparo, A. Papini, *Path Following by SVD*. Lecture Notes in Computer Science, vol. 3994 (Springer, New York, 2006), pp. 677–684
43. O. Koch, C. Lubich, Dynamical low-rank approximation. SIAM J. Matrix Anal. Appl. **29**, 434–454 (2007)
44. A. Nonnenmacher, C. Lubich, Dynamical low rank approximation: applications and numerical experiments. Math. Comp. Simulat. **79**, 1346–1357 (2008)
45. L. Dieci, A. Pugliese, Singular values of two-parameter matrices: An algorithm to accurately find their intersections. Math. Comp. Simulat. **79**(4), 1255–1269 (2008)

46. L. Dieci, A. Pugliese, Two-parameter SVD: Coalescing singular values and periodicity. SIAM J. Matrix Anal. **31**(2), 375–403 (2009)
47. L. Dieci, M.G. Gasparo, A. Papini, A. Pugliese, Locating coalescing singular values of large two-parameter matrices. Math. Comp. Simul. **81**(5), 996–1005 (2011)
48. L. Dieci, A. Pugliese, Hermitian matrices depending on three parameters: Coalescing eigenvalues. Lin. Algebra Appl. **436**, 4120–4142 (2012)
49. L. Dieci, A. Papini, A. Pugliese, Approximating coalescing points for eigenvalues of Hermitian matrices of three parameters. SIAM J. Matrix Anal. **34**(2), 519–541 (2013)
50. H. Gingold, A method of global block diagonalization for matrix-valued functions. SIAM J. Math. Anal. **9**(6), 1076–1082 (1978) [On plurirectangular regions, disjoint spectra]
51. P.F. Hsieh, Y. Sibuya, A global analysis of matrices of functions of several variables. J. Math. Anal. Appl. **14**, 332–340 (1966)
52. M.E. Henderson, Multiple parameter continuation: Computing implicitly defined k-manifolds. Int. J. Bifur. Chaos Appl. Sci. Eng. **12**, 451–476 (2002)
53. W.C. Rheinboldt, J.V. Burkardt, A locally parameterized continuation process. ACM Trans. Math. Software **9**, 215–235 (1983)
54. W. Rheinboldt, On the computation of multi-dimensional solution manifolds of parametrized equations. Numer. Math. **53**, 165–181 (1988)
55. J. von Neumann, E. Wigner, Eigenwerte bei adiabatischen prozessen. Physik Zeitschrift **30**, 467–470 (1929) [First realization of codimension of phenomenon of coalescing eigenvalues]
56. A.J. Stone, Spin-orbit coupling and the intersection of potential energy surfaces in polyatomic molecules. Proc. Roy. Soc. Lond. **A351**, 141–150 (1976)
57. M.V. Berry, Quantal phase factors accompanying adiabatic changes. Proc. Roy. Soc. Lond. **A392**, 45–57 (1984)
58. M.V. Berry, Geometric phase memories. Nat. Phys. **6**, 148–150 (2010)
59. G. Herzberg, H.C. Longuet-Higgins, Intersection of potential energy surfaces in polyatomic molecules. Discuss. Faraday Soc. **35**, 77–82 (1963) [Study of model problem with coalescing eigenvalues on loop around origin]
60. D.R. Yarkony, Conical intersections: The new conventional wisdom. J. Phys. Chem. A **105**, 6277–6293 (2001)
61. P.N. Walker, M.J. Sanchez, M. Wilkinson, Singularities in the spectra of random matrices. J. Math. Phys. **37**(10), 5019–5032 (1996)
62. P.N. Walker, M. Wilkinson, Universal fluctuations of Chern integers. Phys. Rev. Lett. **74**(20), 4055–4058 (1995)
63. M. Wilkinson, E.J. Austin, Densities of degeneracies and near-degeneracies. Phys. Rev. A **47**(4), 2601–2609 (1993)
64. N.C. Perkins, C.D. Mote Jr., Comments on curve veering in eigenvalue problems. J. Sound Vib. **106**, 451–463 (1986)
65. A. Srikantha Phani, J. Woodhouse, N.A. Fleck, Wave propagation in two-dimensional periodic lattices, J. Acoust. Soc. Am. **119**, 1995–2005 (2006)
66. C. Pierre, Mode localization and eigenvalue loci veering phenomena in disordered structures. J. Sound Vib. **126**, 485–502 (1988)
67. A. Gallina, L. Pichler, T. Uhl, Enhanced meta-modelling technique for analysis of mode crossing, mode veering and mode coalescence in structural dynamics. Mech. Syst. Signal Process. **25**, 2297–2312 (2011)
68. L. Dieci, J. Lorenz, Computation of invariant tori by the method of characteristics. SIAM J. Numer. Anal. **32**, 1436–1474 (1995)
69. L. Dieci, J. Lorenz, Lyapunov–type numbers and torus breakdown: Numerical aspects and a case study. Numer. Algorithms **14**, 79–102 (1997)
70. T.F. Coleman, D.C. Sorensen, A note on the computation of and orthonormal basis for the null space of a matrix. Math. Program. **29**, 234–242 (1984)
71. W. Beyn, On well-posed problems for connecting orbits in dynamical systems. Cont. Math. **172**, 131–168 (1994)

72. T. Eirola, J. von Pfaler, Numerical Taylor expansions for invariant manifolds. Numer. Math. **99**, 25–46 (2004)
73. Y.W. Kwon, H. Bang, The finite element method using MATLAB, 2nd edn. (CRC Press, Boca Raton, 2000)
74. W. Wasow, On the spectrum of Hermitian matrix valued functions. Resultate der Mathematik **2**, 206–214 (1979)
75. M.V. Berry, M. Wilkinson, Diabolical points in the spectra of triangles. Proc. Roy. Soc. Lond. Ser. A **392**, 15–43 (1984)

Stability of Linear Problems: Joint Spectral Radius of Sets of Matrices

Nicola Guglielmi and Marino Zennaro

Abstract It is well known that the stability analysis of step-by-step numerical methods for differential equations often reduces to the analysis of linear difference equations with variable coefficients. This class of difference equations leads to a family $\mathscr{F}$ of matrices depending on some parameters and the behaviour of the solutions depends on the convergence properties of the products of the matrices of $\mathscr{F}$. To date, the techniques mainly used in the literature are confined to the search for a suitable norm and for conditions on the parameters such that the matrices of $\mathscr{F}$ are contractive in that norm. In general, the resulting conditions are more restrictive than necessary. An alternative and more effective approach is based on the concept of *joint spectral radius* of the family $\mathscr{F}$, $\rho(\mathscr{F})$. It is known that all the products of matrices of $\mathscr{F}$ asymptotically vanish if and only if $\rho(\mathscr{F}) < 1$. The aim of this chapter is that to discuss the main theoretical and computational aspects involved in the analysis of the joint spectral radius and in applying this tool to the stability analysis of the discretizations of differential equations as well as to other stability problems. In particular, in the last section, we present some recent heuristic techniques for the search of optimal products in finite families, which constitutes a fundamental step in the algorithms which we discuss. The material we present in the final section is part of an original research which is in progress and is still unpublished.

N. Guglielmi (✉)
Dipartimento di Matematica Pura ed Applicata, Università degli Studi di L'Aquila,
Via Vetoio - Loc. Coppito, 67100 L'Aquila, Italy
e-mail: guglielm@univaq.it

M. Zennaro
Dipartimento di Matematica e Geoscienze, Università degli Studi di Trieste,
Via A. Valerio 12/1, 34100 Trieste, Italy
e-mail: zennaro@units.it

1 Introduction

The aim of this chapter is to describe some methods to compute the so-called *joint spectral radius* of a family of matrices.

Simply speaking, the joint spectral radius of a set of matrices measures the highest possible rate of growth of the products generated by the set. For a single matrix A this quantity is clearly identified by the spectral radius $\rho(A)$, but for a set of matrices such a simple characterization is not available and one has to analyze the product semigroup, where the products have to be considered with no ordering restriction and allowing repetitions of the matrices.

The joint spectral radius appears to be an important measure in many applications including discrete switched systems (see, e.g., Mason and Shorten [40]), convergence analysis of subdivision schemes, refinement equations and wavelets (see, e.g., Daubechies and Lagarias [15], Hechler et al. [31], Villemoes [53]), numerical stability analysis of ordinary differential equations (see, e.g., Guglielmi and Zennaro [21]), as well as functional equations, coding theory (see, e.g., Moision et al. [41]) and combinatorics (see, e.g., Berstel [5], Jungers et al. [35]). For an extensive discussion of some applications and a complete list of references, see the recent survey monography by Jungers [33].

We remark that other interesting *joint spectral characteristics* of a set of matrices have been investigated. We quote, for example, the *lower spectral radius*, which was defined by Gurvits [29] and measures the lowest possible rate of growth of products of matrices from a set. However, we do not discuss any of them in these notes.

In this article we do not intend to make an exhaustive survey of all the known theoretical results regarding the joint spectral radius, for which we recommend all the papers cited in the References and, in particular, the survey [33]. We want instead to stress more some computational aspects and also propose a heuristic procedure for the actual computation of the joint spectral radius.

The article is organized as follows.

In Sect. 2 we describe a natural framework for a stability analysis through the joint spectral radius, that is discrete switched linear systems, and give the key definition of uniform asymptotic stability providing an illustrative example. Then we introduce the joint spectral radius by giving several equivalent definitions and characterizations, emphasizing the fundamental infimum (and in some cases minimum) property over the set of operator norms which leads to the concept of *extremal norm*.

In Sect. 3 we define and discuss some fundamental issues for the subsequent theory, that is *defectivity* of a family of matrices, *trajectories* of vectors under the action of the product semigroup and the related extremal norms. We give the definition of *spectrum maximizing product* and, moreover, we introduce the notion of *balanced complex polytope*, which generalizes to the complex case the well-known notion of centrally symmetric polyhedron. Such geometrical objects constitute the basic bricks for the exact computation of the joint spectral radius in a finite time, since they turn out to define, under suitable assumptions, the unit balls of the above mentioned extremal norms.

To this aim we describe how to compute the *polytope norm* of a vector in several cases (the general complex case, the real case and the nonnegative case). In the same section we treat the important class of families of matrices sharing an invariant cone, which includes the case of nonnegative families, a subject recently studied in the literature.

In Sect. 4 we deal with finiteness properties of families of matrices and with the possible finite construction of an extremal polytope norm. Here we give the main theoretical results and discuss the sharpness of the assumptions necessary to obtain a finiteness result.

In Sect. 5 we describe the basic algorithms and provide several illustrative examples, belonging to different classes of problems and including a concrete application in interpolatory subdivision.

Eventually, in Sect. 6 we describe a fast heuristic procedure for the search of a spectrum maximizing product, which is a very important issue for the overall efficiency of the proposed methods. This section provides the results of an unpublished very recent original investigation.

2 Discrete Linear Dynamical Systems and Joint Spectral Radius

Many mathematical and real-life problems lead to the study of the *stability* features of a *discrete linear dynamical system (DLDS)* such as

$$x(k) = A_{i_k} x(k-1), \quad k = 1, 2, \ldots, \tag{1}$$

where $x(0) \in \mathbf{C}^n$ and $A_{i_k} \in \mathbf{C}^{n \times n}$ are elements of a given *bounded* family of matrices

$$\mathscr{F} = \{A_i\}_{i \in \mathscr{I}}, \quad \mathscr{I} \text{ set of indices.} \tag{2}$$

In the literature (1) is also refereed as a switched discrete linear system.

For a recent review concerning stability criteria for switched (discrete and continuous) linear systems we refer the reader to [47].

In particular, it is important to find necessary and sufficient conditions on the matrix family $\mathscr{F}$ in order that the DLDS (1) be *uniformly asymptotically stable (u.a.s.)* in the following sense.

Definition 1. The DLDS (1) is said to be uniformly asymptotically stable (u.a.s.) if it holds that

$$\lim_{k \to \infty} x(k) = 0 \tag{3}$$

for any initial condition $x(0) \in \mathbf{C}^n$ and for any possible sequence of matrices $\{A_{i_k}\}_{k \geq 1}$ chosen in the family $\mathscr{F}$.

As we shall see in Sect. 2.4, the uniform asymptotic stability of the DLDS (1) is strictly related to the so called *joint spectral radius*, or simply *spectral radius*, of the associated matrix family (2), which will be introduced in Sect. 2.2.

2.1 Some Introductory Examples

We introduce two simple examples of DLDS associated to families of two matrices which depend on a parameter. In both cases the goal is to find values of the parameter that make the DLDS u.a.s. In this section, we only give the final results and leave to the next sections the justification as an application of the theory we shall present.

Example 1 (see Willson [54]). The following DLDS is pertinent to the design of a simple digital frequency-shift-keying (FSK) oscillator:

$$x(k) = A_{i_k} x(k-1), \quad i_k \in \{1,2\}, \quad k \geq 1,$$

where

$$A_1 = \begin{bmatrix} 0 & 1 \\ -\frac{1}{2} & 0 \end{bmatrix}, \qquad A_2 = \begin{bmatrix} \alpha & 1 \\ -\frac{1}{2} & 0 \end{bmatrix}. \tag{4}$$

The goal is to find the largest open interval $(\bar{\alpha}_1, \bar{\alpha}_2)$ such that the system is u.a.s. for all $\alpha \in (\bar{\alpha}_1, \bar{\alpha}_2)$. ◇

Example 2. We consider the following DLDS, which is not related to any real-life problem:

$$x(k) = \beta A_{i_k} x(k-1), \quad i_k \in \{1,2\}, \quad k \geq 1,$$

where

$$A_1 = \begin{bmatrix} -1 & 1 & -1 \\ -1 & -1 & 1 \\ 0 & 1 & 1 \end{bmatrix}, \qquad A_2 = \begin{bmatrix} -1 & 1 & -1 \\ -1 & -1 & 0 \\ 1 & 1 & 1 \end{bmatrix}. \tag{5}$$

The goal is find the greatest $\bar{\beta} > 0$ such that the system is u.a.s. for all β with $|\beta| < \bar{\beta}$. ◇

2.2 The Joint Spectral Radius

We consider a bounded family $\mathscr{F} = \{A_i\}_{i \in \mathscr{I}}$ of complex $n \times n$-matrices such as in (2). Various different quantities have been associated to $\mathscr{F}$ in the literature that

Stability of Linear Problems: Joint Spectral Radius of Sets of Matrices

turn out to be equal to one another, that define its *joint spectral radius*, or simply *spectral radius*.

Let $\|\cdot\|$ be a given norm on $\mathbf{C}^n$ and let the same symbol $\|\cdot\|$ denote also the corresponding induced $n \times n$-matrix norm defined by

$$\|A\| = \max_{\|x\|=1} \|Ax\|.$$

Then, for each $k = 1, 2, \ldots$, consider the set

$$\Sigma_k(\mathscr{F}) = \{A_{i_k} \cdots A_{i_1} \mid i_1, \ldots, i_k \in \mathscr{I}\} \tag{6}$$

of all products of *length* (or *degree*) k and the number

$$\hat{\rho}_k(\mathscr{F}) = \sup_{P \in \Sigma_k(\mathscr{F})} \|P\|^{1/k}. \tag{7}$$

Definition 2 (Joint spectral radius (see Rota and Strang [48, 49])). The number

$$\hat{\rho}(\mathscr{F}) = \limsup_{k \to \infty} \hat{\rho}_k(\mathscr{F}) \tag{8}$$

is called the joint spectral radius (j.s.r.) of the family $\mathscr{F}$.

Note that the numbers $\hat{\rho}_k(\mathscr{F})$ depend on the particular norm $\|\cdot\|$ used in (7) whereas, by the equivalence of all the norms in finite dimensional spaces, it turns out that $\hat{\rho}(\mathscr{F})$ is independent of it.

Analogously, let $\rho(\cdot)$ denote the spectral radius of an $n \times n$-matrix and then, for each $k = 1, 2, \ldots$, consider the number

$$\bar{\rho}_k(\mathscr{F}) = \sup_{P \in \Sigma_k(\mathscr{F})} \rho(P)^{1/k}. \tag{9}$$

Definition 3 (Generalized spectral radius (see Daubechies and Lagarias [15])). The number

$$\bar{\rho}(\mathscr{F}) = \limsup_{k \to \infty} \bar{\rho}_k(\mathscr{F}) \tag{10}$$

is called the generalized spectral radius (g.s.r.) of the family $\mathscr{F}$.

In their paper [15], Daubechies and Lagarias also proved that

$$\bar{\rho}_k(\mathscr{F}) \leq \bar{\rho}(\mathscr{F}) \leq \hat{\rho}(\mathscr{F}) \leq \hat{\rho}_k(\mathscr{F}) \quad \text{for all } k \geq 1, \tag{11}$$

from which it easily follows that

$$\hat{\rho}(\mathscr{F}) = \inf_{k \geq 1} \hat{\rho}_k(\mathscr{F}) = \lim_{k \to \infty} \hat{\rho}_k(\mathscr{F}) \tag{12}$$

and

$$\bar{\rho}(\mathscr{F}) = \sup_{k\geq 1} \bar{\rho}_k(\mathscr{F}). \tag{13}$$

The fact that

$$\hat{\rho}(\mathscr{F}) \leq \bar{\rho}(\mathscr{F}) \tag{14}$$

is much more difficult to prove, but a few proofs of different type were given later by Berger and Wang [3], Elsner [17], Shih et al. [46] and Shih [45].

In the light of inequalities (11) and (14), we conclude that the j.s.r. and the g.s.r. of $\mathscr{F}$ are the same number, which can be simply called the *spectral radius* of the (bounded) family of matrices $\mathscr{F}$ and denoted by $\rho(\mathscr{F})$.

The above definitions and results are generalizations of the well-known analogues for *single* families, i.e. $\mathscr{F} = \{A\}$.

In particular, the equality $\hat{\rho}(\mathscr{F}) = \bar{\rho}(\mathscr{F})$ is the generalization of the so called *Gelfand limit*

$$\rho(A) = \lim_{k\to\infty} \|A^k\|^{1/k}.$$

Nevertheless, unlike the case of single families, in which the possible problems are just a matter of computational complexity and/or ill-conditioning, in general the actual computation of $\rho(\mathscr{F})$ is not an easy task at all. Substantially, one of the main difficulties comes out from the fact that the spectral radius of matrices is not submultiplicative and that, more in general, from the simple knowledge of $\rho(A_1)$ and $\rho(A_2)$, nothing can be said for $\rho(A_1 A_2)$.

Some proposals of algorithms for the approximation and the computation of $\rho(\mathscr{F})$ will be the subject of the next Sects. 5 and 6.

We have seen that the sequence $\{\hat{\rho}_k(\mathscr{F})\}_{k\geq 1}$ always admits a limit (see (12)). On the contrary, in general this is not the case for the sequence $\{\bar{\rho}_k(\mathscr{F})\}_{k\geq 1}$.

Example 3 (see Jungers [33]). A simple example is given by the 2×2 matrix family $\mathscr{F} = \{A_1, A_2\}$ with

$$A_1 = \begin{bmatrix} 0 & 1 \\ 0 & 0 \end{bmatrix}, \quad A_2 = \begin{bmatrix} 0 & 0 \\ 1 & 0 \end{bmatrix},$$

for which it is immediately seen that $\rho(\mathscr{F}) = 1$ and

$$\bar{\rho}_k(\mathscr{F}) = \begin{cases} 0 & \text{if } k \text{ is odd} \\ 1 & \text{if } k \text{ is even} \end{cases}$$

◇

Therefore, it makes sense to introduce the following definition.

Definition 4 (Asymptotic regularity (see Guglielmi and Zennaro [26])). A bounded family $\mathscr{F}$ of complex $n \times n$ matrices is said to be asymptotically regular if it holds that

$$\rho(\mathscr{F}) = \lim_{k \to \infty} \bar{\rho}_k(\mathscr{F}). \tag{15}$$

Guglielmi and Zennaro [26] find sufficient conditions in order that a bounded family $\mathscr{F}$ be asymptotically regular.

Remark that any single family is asymptotically regular since $\hat{\rho}_k(A) = \rho(A)$ for all $k \geq 1$.

2.3 The Common Spectral Radius and Extremal Norms

An important characterization of the spectral radius $\rho(\mathscr{F})$ of a matrix family is the generalization of the following formula, which is well-known for single families:

$$\rho(A) = \inf_{\|\cdot\| \in \mathscr{N}} \|A\|,$$

where $\mathscr{N}$ denotes the set of all possible induced $n \times n$-matrix norms.

In order to state this characterization, we define the *norm of the family* $\mathscr{F} = \{A_i\}_{i \in \mathscr{I}}$ as

$$\|\mathscr{F}\| = \hat{\rho}_1(\mathscr{F}) = \sup_{i \in \mathscr{I}} \|A_i\|.$$

Proposition 1 (see [17,48]). *The spectral radius of a bounded family $\mathscr{F}$ of complex $n \times n$-matrices is characterized by the equality*

$$\rho(\mathscr{F}) = \inf_{\|\cdot\| \in \mathscr{N}} \|\mathscr{F}\|. \tag{16}$$

We remark that the right-hand side of (16) is often referred to as the *common spectral radius (c.s.r.)* of the family $\mathscr{F}$.

The foregoing result is an immediate consequence of the fact that, given any norm $\|\cdot\|$ on $x \in \mathbf{C}^n$ and any number $\epsilon > 0$, the norm

$$\|x\|_{*,\epsilon} = \max\left\{\|x\|, \sup_{k \geq 1} \sup_{P \in \Sigma_k(\mathscr{F})} \frac{\|Px\|}{(\rho(\mathscr{F}) + \epsilon)^k}\right\} \tag{17}$$

satisfies the inequality

$$\|\mathscr{F}\|_{*,\epsilon} \leq \rho(\mathscr{F}) + \epsilon. \tag{18}$$

Given a family $\mathscr{F}$, an important question is whether the c.s.r. in (16) is a minimum or not. In this respect, we have the following definition.

Definition 5 (Extremal norm). We say that a norm $\|\cdot\|_*$ satisfying the condition

$$\|\mathscr{F}\|_* = \rho(\mathscr{F}) \tag{19}$$

is extremal for the family $\mathscr{F}$.

Already for single families, in general extremal norms are not assured to exist. Necessary and sufficient conditions for their existence is given in Sect. 3.

Here we give two simple examples in which the extremal norm exists and is quite easy to compute.

Example 4. Consider the family $\mathscr{F}$ of the previous Example 3. We have already observed that $\rho(\mathscr{F}) = 1$. On the other hand, we also clearly have

$$\|A_1\|_1 = \|A_2\|_1 = \|A_1\|_2 = \|A_2\|_2 = \|A_1\|_\infty = \|A_2\|_\infty = 1,$$

which implies that the norms $\|\cdot\|_1$, $\|\cdot\|_2$ and $\|\cdot\|_\infty$ are extremal, where, for given $x \in \mathbf{C}^n$,

$$\|x\|_1 = \sum_{i=1}^n |x_i|, \quad \|x\|_2 = \sqrt{\sum_{i=1}^n |x_i|^2}, \quad \|x\|_\infty = \max_{1 \leq i \leq n} |x_i|.$$

$\diamond$

Example 5. Consider a family $\mathscr{F} = \{A_i\}_{i \in \mathscr{I}}$, where A_i is *hermitian* for all $i \in \mathscr{I}$. Then the norm $\|\cdot\|_2$ is extremal. In fact, since $\|A_i\|_2 = \rho(A_i)$ for all $i \in \mathscr{I}$, by using (11) we obtain

$$\|\mathscr{F}\|_2 = \bar{\rho}_1(\mathscr{F}) \leq \rho(\mathscr{F}) \leq \hat{\rho}_1(\mathscr{F}) = \|\mathscr{F}\|_2.$$

$\diamond$

2.4 The Stability Theorem

Going back to the DLSD (1), we observe that its uniform asymptotic stability is equivalent to the fact that the sets $\Sigma_k(\mathscr{F})$ in (6) tend to $\{0\}$ as $k \to \infty$, where 0 stands for the zero matrix. Therefore we have the following result.

Theorem 1 (see [3]). *The DLDS (1) is u.a.s. if and only if*

$$\rho(\mathscr{F}) < 1.$$

Once again, we have a natural generalization of the well-known result that holds for a single family $\mathscr{F} = \{A\}$, the corresponding DLDS of which has constant coefficients and is u.a.s. if and only if $\rho(A) < 1$.

Example 6. We consider the DLDS of Example 1, whose associated family is $\mathscr{F}_\alpha = \{A_1, A_2\}$ given by (4).

In view of Theorem 1, the goal is to find the largest open interval $(\bar{\alpha}_1, \bar{\alpha}_2)$ such that $\rho(\mathscr{F}_\alpha) < 1$ for all $\alpha \in (\bar{\alpha}_1, \bar{\alpha}_2)$.

Since

$$\rho(A_1) = \frac{\sqrt{2}}{2} \quad \text{and} \quad \rho(A_2) = \begin{cases} \frac{\sqrt{2}}{2} & \text{if } |\alpha| \leq \sqrt{2} \\ \frac{1}{2}\left(|\alpha| + \sqrt{\alpha^2 - 2}\right) & \text{if } |\alpha| \geq \sqrt{2} \end{cases},$$

it is evident that

$$\rho(\mathscr{F}_\alpha) < 1 \implies \rho(A_2) < 1 \iff |\alpha| < \frac{3}{2}.$$

Indeed, by using the forthcoming theory, it is possible to prove that

$$\bar{\alpha}_1 = -\frac{9}{8} \text{ (the unique } \alpha < 0 \text{ such that } \rho\left(A_1 A_2^2\right) = 1\text{)},$$

$$\bar{\alpha}_2 = \frac{9}{8} \text{ (the unique } \alpha > 0 \text{ such that } \rho\left(A_1 A_2^2\right) = 1\text{)}.$$

◇

Example 7. We consider the DLDS of Example 2, whose associated family is $\beta\mathscr{F}$ with $\mathscr{F} = \{A_1, A_2\}$ given by (5).

In view of Theorem 1, the goal is to find $\bar{\beta} = 1/\rho(\mathscr{F})$.

Using the forthcoming theory, we shall prove that

$$\bar{\beta} = \rho\left(\left((A_1^2 A_2)^2 A_1 A_2\right)^3 A_2 A_1 A_2\right)^{-\frac{1}{27}} = 0.559\ldots$$

◇

In the next sections we shall see that the relationships between the DLDS (1) and the theory of the joint spectral radius do not end with Theorem 1, but also involve other important aspects. In particular, in Sect. 3.3 we shall see the strict connections between the *trajectories* of (1) and the unit balls of some particular extremal norms of the associated family $\mathscr{F}$.

3 Defectivity, Extremal Norms and Trajectories

It is well-known that, for a single family $\mathscr{F} = \{A\}$, the existence of an extremal norm is equivalent to the fact that the matrix A is *nondefective*, i.e. all of the blocks relevant to the eigenvalues of maximum modulus are diagonal in its Jordan canonical form.

Whenever $\rho(A) > 0$, another equivalent property is that, with $\hat{A} = \rho(A)^{-1}A$, the power set $\Sigma(\hat{A}) = \{\hat{A}^k \mid k \geq 1\}$ is bounded. This second result generalizes to a bounded family $\mathscr{F}$ as follows.

3.1 Defectivity and Reducibility

Given a bounded family $\mathscr{F} = \{A_i\}_{i \in \mathscr{I}}$ of complex $n \times n$-matrices with $\rho(\mathscr{F}) > 0$, let us consider the *normalized* family

$$\hat{\mathscr{F}} = \{\rho(\mathscr{F})^{-1} A_i\}_{i \in \mathscr{I}}, \qquad (20)$$

whose spectral radius is $\rho(\hat{\mathscr{F}}) = 1$. Then consider the matrix semigroup generated by $\hat{\mathscr{F}}$, i.e.

$$\Sigma(\hat{\mathscr{F}}) = \bigcup_{k \geq 1} \Sigma_k(\hat{\mathscr{F}}).$$

Definition 6 (Defectivity). A bounded family $\mathscr{F}$ of complex $n \times n$-matrices is said to be *defective* if the corresponding normalized family $\hat{\mathscr{F}}$ is such that the semigroup $\Sigma(\hat{\mathscr{F}})$ is an unbounded set of matrices. Otherwise, if $\Sigma(\hat{\mathscr{F}})$ is bounded, then the family $\mathscr{F}$ is said to be *nondefective*.

Note that the above definition of defective family does not involve directly the spectral properties of its elements.

We also recall the following definition.

Definition 7 (Common reducibility). A bounded family $\mathscr{F} = \{A_i\}_{i \in \mathscr{I}}$ of complex $n \times n$-matrices is said to be commonly reducible, or simply reducible, if there exist a nonsingular $n \times n$-matrix M and two integers $n_1, n_2 \geq 1$, $n_1 + n_2 = n$, such that, for all $i \in \mathscr{I}$, it holds that

$$M^{-1} A_i M = \begin{bmatrix} B_i & C_i \\ 0 & D_i \end{bmatrix},$$

where the blocks B_i, C_i, D_i are $n_1 \times n_1$-, $n_1 \times n_2$- and $n_2 \times n_2$-matrices, respectively. If a family $\mathscr{F}$ is not reducible, then it is said to be commonly irreducible, or simply irreducible.

Remark 1. Reducibility of a family $\mathscr{F}$ means the existence of a common nontrivial linear invariant subspace for all the matrices of $\mathscr{F}$.

Then we are in a position to state a first important result concerning defectivity.

Theorem 2 (see Elsner [17]). *A bounded family $\mathscr{F}$ of complex $n \times n$-matrices which is defective is also commonly reducible.*

We remark that, whereas a defective family is always reducible, the opposite implication is not necessarily true. For example, for $n \geq 2$ all single families $\mathscr{F} = \{A\}$ are clearly reducible, but not necessarily defective.

Moreover, we remark that the concept of reducibility introduced by Definition 7 is different from the classical concept of reducibility of matrices, in which M is required to be a permutation matrix. In order to avoid confusion between the two concepts, Guglielmi and Zennaro [20] used the term *decomposable* instead of (commonly) reducible.

3.2 Existence of Extremal Norms

The following characterization of the existence of extremal norms can be found, for example, in Berger and Wang [3] (see also [2]).

Theorem 3. *A bounded family $\mathscr{F}$ of complex $n \times n$-matrices admits an extremal norm $\|\cdot\|_*$ if and only if it is nondefective.*

Indeed, if a family $\mathscr{F}$ is nondefective, any given norm $\|\cdot\|$ on $x \in \mathbf{C}^n$ determines the extremal norm

$$\|x\|_* = \max\left\{\|x\|, \sup_{k\geq 1} \sup_{P\in \Sigma_k(\mathscr{F})} \frac{\|Px\|}{\rho(\mathscr{F})^k}\right\}. \tag{21}$$

Moreover, it is obvious that the norm $\|\cdot\|_*$ satisfies

$$\hat{\rho}_k(\mathscr{F}) = \rho(\mathscr{F}) \quad \text{for all } k \geq 1. \tag{22}$$

3.3 Trajectories and Extremal Norms

The definitions and results reported in this section may be found in Guglielmi et al. [27].

Definition 8 (Trajectory). Given a bounded family $\mathscr{F}$ of complex $n \times n$-matrices and a vector $x \in \mathbf{C}^n$, we say that the set

$$\mathscr{T}[\mathscr{F}, x] = \{x\} \cup \{Px \mid P \in \Sigma(\mathscr{F})\},$$

is the trajectory of $\mathscr{F}$ exiting from x.

Observe that the trajectory is the set obtained by applying all the products P of matrices of $\mathscr{F}$ to the vector x. In other words, the trajectory $\mathscr{T}[\mathscr{F}, x]$ is the set of the solutions with initial value $x(0) = x$ of the DLDS (1) for any possible sequence of matrices $\{A_{i_k}\}_{k \geq 1}$ chosen in the family $\mathscr{F}$.

Therefore, in short, we also say that $\mathscr{T}[\mathscr{F}, x]$ is the trajectory with initial value $x(0) = x$ of the DLDS (1) associated to the family $\mathscr{F}$.

We have the following characterization.

Proposition 2. *Let $\mathscr{F}$ be a bounded family of complex $n \times n$-matrices and let $x \in \mathbf{C}^n$. Then* $\mathrm{span}\left(\mathscr{T}[\mathscr{F}, x]\right)$ *is the smallest linear subspace V of $\mathbf{C}^n$ containing x such that $\mathscr{F}(V) \subseteq V$.*

In the light of Remark 1, the following result is straightforward.

Corollary 1. *Let $\mathscr{F}$ be an irreducible bounded family of complex $n \times n$-matrices and let $x \in \mathbf{C}^n$, $x \neq 0$. Then*

$$\mathrm{span}\left(\mathscr{T}[\mathscr{F}, x]\right) = \mathbf{C}^n. \tag{23}$$

In order to establish a relationship between the trajectories of a DLDS (1) and some extremal norms of the associated family $\mathscr{F}$, we recall the following well-known concepts.

Definition 9. We shall say that a set $\mathscr{X} \subset \mathbf{C}^n$ is absolutely convex if, for all $x', x'' \in \mathscr{X}$ and $\lambda', \lambda'' \in \mathbf{C}$ such that $|\lambda'| + |\lambda''| \leq 1$, it holds that $\lambda' x' + \lambda'' x'' \in \mathscr{X}$.

Definition 10. Let $\mathscr{X} \subset \mathbf{C}^n$. Then the intersection of all absolutely convex sets containing $\mathscr{X}$ will be called the absolutely convex hull of $\mathscr{X}$ and will be denoted by $\mathrm{absco}(\mathscr{X})$.

Now, given a matrix family $\mathscr{F} = \{A_i\}_{i \in \mathscr{I}}$, we consider its normalization $\hat{\mathscr{F}}$ (see (20)) and, for $x \in \mathbf{C}^n$, we define the set

$$\mathscr{S}[\hat{\mathscr{F}}, x] = \mathrm{clos}\left(\mathrm{absco}\left(\mathscr{T}[\hat{\mathscr{F}}, x]\right)\right), \tag{24}$$

that is, the closure of the absolutely convex hull of the *normalized trajectory* $\mathscr{T}[\hat{\mathscr{F}}, x]$.

Proposition 3. *Let $\mathscr{F}$ be a nondefective bounded family of complex $n \times n$-matrices and, given a vector $x \in \mathbf{C}^n$, let (23) hold. Then the set $\mathscr{S}[\hat{\mathscr{F}}, x]$ is the unit ball of an extremal norm for $\mathscr{F}$.*

Example 8. As an example, in Fig. 1 we show the set $\mathscr{S}[\hat{\mathscr{F}}, x]$ for the family $\hat{\mathscr{F}} = \{\hat{A}_1, \hat{A}_2\}$, where

$$\hat{A}_1 = \vartheta \begin{bmatrix} 1 & 1 \\ 0 & 1 \end{bmatrix}, \quad \hat{A}_2 = \vartheta \begin{bmatrix} 1 & 0 \\ 1 & 1 \end{bmatrix}, \quad \vartheta = 2/(1 + \sqrt{5}), \tag{25}$$

Fig. 1 The trajectory and its absolutely convex hull for the family (25)

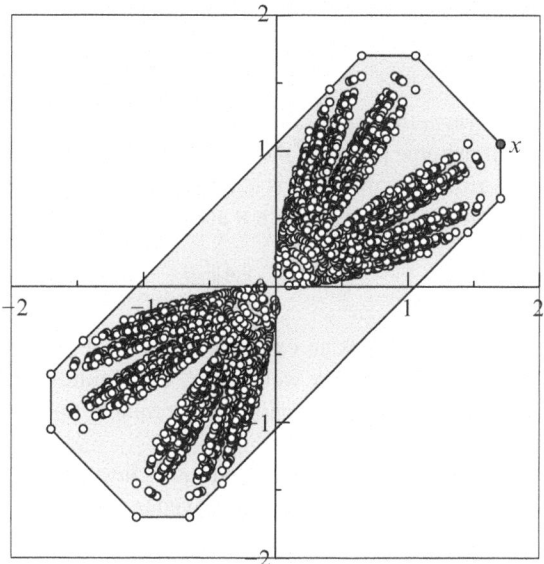

and $x = [1, (1 + \sqrt{5})/2]^T$. It is interesting to observe that, in this case, the set $\mathscr{S}[\hat{\mathscr{F}}, x]$ is a centrally symmetric real polytope. ◇

We also observe that the family (25) is nonnegative, so that it has the invariant cone $\mathbf{R}^2_+$. In this special case we can look for a more specific class of extremal norms.

3.4 Families with an Invariant Cone: Monotone Norms

In this section we consider families of matrices $\mathscr{F}$ that share a common invariant cone $K \subset \mathbf{R}^n$ like, for example, families of *nonnegative matrices*, in which case $K = \mathbf{R}^n_+ = \{x \in \mathbf{R}^n \mid x_i \geq 0, \ i = 1, \ldots, n\}$.

For a discussion of properties of matrices with an invariant cone we refer the reader to Vandergraft [52] (see also [11]).

Recall that a function g is *monotone* with respect to a cone K if $g(x) \geq g(y)$, whenever $x - y \in K$.

If g is a monotone norm defined on the cone K, then it is extended onto $\mathbf{R}^n$ in a standard way: the unit ball of that norm is

$$\{x \in \mathbf{R}^n \mid \|x\| \leq 1\} = \mathrm{co}_s\left(\{x \in K \mid g(x) \leq 1\}\right), \qquad (26)$$

where, given a set $\mathscr{X}$,

$$\text{co}_s(\mathscr{X}) = \text{co}\left(\mathscr{X} \cup -\mathscr{X}\right)$$

is the *symmetric convex hull* of $\mathscr{X}$.

All extreme points of the ball defined by (26) are in the cones K and $-K$. Since the norm of any matrix A is attained at an extreme point of the unit ball, we see that, if A leaves K invariant, then it attains its norm in the cone K. Thus,

$$\|A\| = \max_{x \in K, \, g(x) \le 1} g(Ax).$$

In particular, if g is an extremal norm for a family $\mathscr{F}$, then its extension defined by (26) is extremal as well. Thus, for families with a common invariant cone it suffices to construct a monotone extremal norm g on that cone (see Guglielmi and Protasov [19]).

Recall that a face F of a cone K is invariant for a matrix A if $Ax \subset F$ for all $x \in F$. The following result is proved in [19].

Proposition 4. *If a finite family of matrices $\mathscr{F} = \{A_i\}_{i \in \mathscr{I}}$ has a common invariant cone K and does not have any common invariant face of that cone, then $\mathscr{F}$ has a monotone extremal norm on K.*

If $K = \mathbf{R}_+^n$, then the faces of K are included in coordinate planes, i.e., they are sets of the type $F_{i_1 \ldots i_d} = \{x \in \mathbf{R}_+^n \mid x_{i_1} = \cdots x_{i_d} = 0\}$, $d = 1, \ldots, n-1$.

Definition 11. A family of nonnegative matrices $\mathscr{F} = \{A_i\}_{i \in \mathscr{I}}$ is called *positively irreducible* if it does not have any common invariant face of the cone $K = \mathbf{R}_+^n$.

Applying Proposition 4 to the case $K = \mathbf{R}_+^n$, we obtain the following result.

Corollary 2. *A finite positively irreducible family $\mathscr{F} = \{A_i\}_{i \in \mathscr{I}}$ of nonnegative matrices has a monotone extremal norm on $\mathbf{R}_+^n$.*

3.5 Balanced Complex Polytopes and Related Norms

A special class of norms that plays an important role in the theory of the j.s.r. of families of matrices is the class of *complex polytope norms*, that is norms whose unit ball is a *balance complex polytope (b.c.p.)* (see Guglielmi and Zennaro [23]), which is a generalization of a real symmetric polytope to the complex vector space $\mathbf{C}^n$.

In order to define the b.c.p.s and the related norms, we observe that, if $\mathscr{X} = \{x^{(i)}\}_{1 \le i \le m}$ is a finite set of vectors, then

$$\text{absco}(\mathscr{X}) = \left\{x \in \mathbf{C}^n \;\middle|\; x = \sum_{i=1}^m \lambda_i x^{(i)} \text{ with } \lambda_i \in \mathbf{C} \text{ and } \sum_{i=1}^m |\lambda_i| \le 1\right\} \quad (27)$$

and, in this case, it is a closed subset of $\mathbf{C}^n$.

In what follows, if $\mathscr{X}'$ and $\mathscr{X}''$ are two subsets of $\mathbf{C}^n$, we shall write $\mathscr{X}' \subset \mathscr{X}''$ ($\mathscr{X}' \supset \mathscr{X}''$) to denote proper inclusions, i.e., $\mathscr{X}' \subseteq \mathscr{X}''$ ($\mathscr{X}' \supseteq \mathscr{X}''$) and $\mathscr{X}' \neq \mathscr{X}''$.

Definition 12 (Balanced complex polytope). We shall say that a bounded set $\mathscr{P} \subset \mathbf{C}^n$ is a balanced complex polytope (b.c.p.) if there exists a finite set $\mathscr{X} = \{x^{(i)}\}_{1 \leq i \leq m}$ of vectors such that $\mathrm{span}(\mathscr{X}) = \mathbf{C}^n$ and

$$\mathscr{P} = \mathrm{absco}(\mathscr{X}). \tag{28}$$

Moreover, if $\mathrm{absco}(\mathscr{X}') \subset \mathrm{absco}(\mathscr{X})$ for all $\mathscr{X}' \subset \mathscr{X}$, then $\mathscr{X}$ will be called an essential system of vertices (e.s.v.) for $\mathscr{P}$, whereas any vector $ux^{(i)}$ with $u \in \mathbf{C}$, $|u| = 1$, will be called a vertex of $\mathscr{P}$.

Remark that, geometrically speaking, a b.c.p. $\mathscr{P}$ is not a classical polytope. In fact, if we identify the complex space $\mathbf{C}^n$ with the real space $\mathbf{R}^{2n}$, we can easily see that $\mathscr{P}$ is not bounded by hyperplanes. In general, even the intersection $\mathscr{P} \cap \mathbf{R}^n$ is not a polytope in $\mathbf{R}^n$.

Example 9. Consider the b.c.p.

$$\mathscr{P} = \mathrm{absco}\left(\{[1,\ \mathrm{i}]^\mathrm{T}, [1,\ -\mathrm{i}]^\mathrm{T}\}\right).$$

It can be easily verified that $\mathscr{P} \cap \mathbf{R}^2$ is the unit circle

$$\mathscr{C} = \left\{[x_1,\ x_2]^\mathrm{T} \in \mathbf{R}^2 \mid x_1^2 + x_2^2 = 1\right\}, \tag{29}$$

that is not a polytope in $\mathbf{R}^2$. ◇

On the other hand, it is easy to see that, if the b.c.p. $\mathscr{P}$ admits an essential system of real vertices, then $\mathscr{P} \cap \mathbf{R}^n$ is a polytope in $\mathbf{R}^n$.

Now we extend the concept of polytope norm to the complex case in a straightforward way.

Lemma 1. *Any b.c.p. $\mathscr{P}$ is the unit ball of the norm $\|\cdot\|_{\mathscr{P}}$ on $\mathbf{C}^n$ given by the Minkowski functional associated to $\mathscr{P}$, that is, for all $z \in \mathbf{C}^n$,*

$$\|z\|_{\mathscr{P}} = \inf\{\rho > 0 \mid z \in \rho \mathscr{P}\}. \tag{30}$$

Definition 13. We shall call complex polytope norm any norm $\|\cdot\|_{\mathscr{P}}$ whose unit ball is a b.c.p. $\mathscr{P}$.

Taking Definition 12 into account, the Minkowski functional (30) may be rewritten as

$$\|z\|_{\mathscr{P}} = \min\left\{\sum_{i=1}^{m} |\lambda_i| \;\Big|\; z = \sum_{i=1}^{m} \lambda_i x^{(i)}\right\}, \tag{31}$$

where $\mathscr{X} = \{x^{(i)}\}_{1 \leq i \leq m}$ is an e.s.v. for $\mathscr{P}$.

Observe that formula (31) suggests a way to actually compute $\|z\|_{\mathscr{P}}$.

Also note that (31) can be written as a real nonlinear semidefinite optimization problem. In fact, setting $\lambda_i = \alpha_i + i\beta_i$ we can write it as

$$\min \ f = \sum_{i=1}^{m} \tau_i$$

$$\text{subject to } \sum_{i=1}^{m} \left(\alpha_i \, \Re(x^{(i)}) - \beta_i \, \Im(x^{(i)})\right) = \Re(z)$$

$$\sum_{i=1}^{m} \left(\alpha_i \, \Im(x^{(i)}) + \beta_i \, \Re(x^{(i)})\right) = \Im(z)$$

$$\text{and } \sqrt{\alpha_i^2 + \beta_i^2} \leq \tau_i, \qquad \tau_i \geq 0, \quad i = 1, \ldots, m.$$

Indeed, in [23] the following more efficient formula is proved.

Proposition 5. *If $m \geq 2n$, then it holds that*

$$\|z\|_{\mathscr{P}} = \min\left\{ \sum_{j=1}^{2n-1} |\lambda_{i_j}| \ \Big| \ z = \sum_{j=1}^{2n-1} \lambda_{i_j} x^{(i_j)} \ \text{and} \right.$$

$$\left. \{i_1, \ldots, i_{2n-1}\} \subset \{1, \ldots, m\} \right\}. \tag{32}$$

An exhaustive study of the geometry of b.c.p.s of dimension $n = 2$ has been done by Vagnoni and Zennaro [51], who also proposed a first efficient procedure for the computation of the related norm.

The next theorem shows that the set of the complex polytope norms is dense in the set of all norms defined on $\mathbf{C}^n$ and that, consequently, the corresponding set of induced matrix complex polytope norms is dense in the set $\mathscr{N}$ of all induced $n \times n$-matrix norms.

Theorem 4 (Density theorem). *Let $\|\cdot\|$ be a norm on $\mathbf{C}^n$. Then for any $\epsilon > 0$ there exists a b.c.p. $\mathscr{P}_\epsilon$ whose corresponding complex polytope norm $\|\cdot\|_\epsilon$ satisfies the inequalities*

$$\|x\| \leq \|x\|_\epsilon \leq (1+\epsilon)\|x\| \quad \text{for all } x \in \mathbf{C}^n.$$

Moreover, denoting by $\|\cdot\|$ and $\|\cdot\|_\epsilon$ also the corresponding induced matrix norms, it holds that

$$(1+\epsilon)^{-1}\|A\| \leq \|A\|_\epsilon \leq (1+\epsilon)\|A\| \quad \text{for all } A \in \mathbf{C}^{n \times n}.$$

The foregoing theorem immediately implies the following refinement of Proposition 1, which makes the class of complex polytope norms quite suitable for the approximation of the j.s.r.

Proposition 6. *The spectral radius of a bounded family $\mathscr{F}$ of complex $n \times n$-matrices is characterized by the equality*

$$\rho(\mathscr{F}) = \inf_{\|\cdot\| \in \mathscr{N}_{pol}} \|\mathscr{F}\|, \tag{33}$$

where $\mathscr{N}_{pol}$ denotes the set of all possible induced $n \times n$-matrix complex polytope norms.

We remark that, on the contrary, in general the class of induced inner product norms does not enjoy the same infimum property.

3.6 Special Cases

Before concluding this section we mention two important special cases. The first is the real case, where, using the fact that $\mathscr{P} = \text{absco}(\mathscr{X}) = \text{co}_s(\mathscr{X})$ is a real centrally symmetric polytope, computing the norm $\|z\|_{\mathscr{P}}$ is a standard linear programming problem of the form

$$\min f = \sum_{i=1}^{m} (\lambda_i + \mu_i)$$

$$\text{subject to } \sum_{i=1}^{m} \left(\lambda_i x^{(i)} + \mu_i(-x^{(i)})\right) = z$$

$$\text{and } \lambda_i \geq 0, \ \mu_i \geq 0, \quad i = 1, \ldots, m.$$

As a second special case, we define a monotone polytope norm, which can be used in the construction of a monotone extremal norm for a nonnegative family of matrices.

To this aim, we consider the cone $K = \mathbf{R}_+^n$ and, for a given set $Q \subset K$, we define

$$\text{co}_-(Q) = \left(\text{co}(Q) - K\right) \cap K = \left\{x \in K \mid x = y - t, \ y \in \text{co}(Q), t \in K\right\}.$$

Then, given a set of vertices $\mathscr{X} = \{x^{(i)}\}_{1 \leq i \leq m}$, $x^{(i)} \in K$, $i = 1, \ldots, m$, we define the *positive polytope*

$$\mathscr{P} = \text{co}_-(\mathscr{X}).$$

Therefore, in order to compute $\|z\|_{\mathscr{P}}$ for a given $z \in K$, we are reduced to solve the linear programming problem

$$\min \ f = \sum_{i=1}^{m} \lambda_i$$

$$\text{subject to} \ \sum_{i=1}^{m} \lambda_i x^{(i)} \geq z \text{ (componentwise)}$$

$$\text{and} \ \lambda_i \geq 0, \quad i = 1, \ldots, m.$$

4 The Finiteness Property and Extremal Complex Polytope Norms

In this section we introduce the so-called *finiteness property* for families of matrices and show the mutual relationships between such a property and the existence of an extremal norm.

More precisely, we investigate conditions under which a nondefective family $\mathscr{F}$ admits an extremal complex polytope norm, that is, under which the infimum in (41) is indeed a minimum.

Since the unit balls of these norms are determined by a finite essential system of vertices (see Definition 12), they are potentially suitable for the design of efficient algorithms aimed at the actual computation and approximation of the j.s.r. via the detection of an extremal norm.

While in this section we illustrate the theoretical background, in Sect. 5 we shall present the practical algorithmic proposals.

4.1 The Finiteness Conjecture and the Normed Finiteness Conjecture

From now on we mainly consider finite families $\mathscr{F} = \{A_i\}_{1 \leq i \leq d}$ of complex $n \times n$-matrices.

Definition 14 (Finiteness property). A bounded family $\mathscr{F}$ of complex $n \times n$-matrices is said to have the finiteness property if there exist $k^* \geq 1$ and a product $\bar{P} \in \Sigma_{k^*}(\mathscr{F})$ such that

$$\rho(\bar{P})^{1/k^*} = \bar{\rho}_{k^*}(\mathscr{F}) = \rho(\mathscr{F}). \tag{34}$$

Definition 15 (Spectrum-maximizing product). If $\mathscr{F}$ is a bounded family of complex $n \times n$-matrices, any matrix $\bar{P} \in \Sigma_{k^*}(\mathscr{F})$ satisfying (34) for some $k^* \geq 1$ is called a spectrum-maximizing product (s.m.p.) for $\mathscr{F}$.

Definition 16 (Minimal s.m.p.). An s.m.p. $\bar{P}$ of a bounded family of complex $n \times n$-matrices $\mathscr{F}$ is called minimal if is not a power of another s.m.p. of $\mathscr{F}$.

Observe that the finiteness property means the existence of at least one (minimal) s.m.p. $\bar{P}$.

It is evident that, if a finite family $\mathscr{F}$ has the finiteness property, then it is possible to compute $\rho(\mathscr{F})$ by means of a finite number of evaluations of spectral radii of matrices, even if this number might be very large and, in most cases, unknown a priori.

Example 10 (see Lagarias and Wang [37]). Let $\mathscr{F} = \{A_1, A_2\}$, where

$$A_1 = \alpha^k \begin{bmatrix} 0 & 0 \\ 0 & 1 \end{bmatrix}, \quad A_2 = \alpha^{-1} \begin{bmatrix} \cos(\beta_k) & \sin(\beta_k) \\ -\sin(\beta_k) & \cos(\beta_k) \end{bmatrix},$$

with $\beta_k = \frac{2\pi}{k}$ and $1 < \alpha < (\cos(\beta_k))^{-1}$.

It turns out that $\bar{\rho}(\mathscr{F}) = \hat{\rho}(\mathscr{F}) = 1$ and

$$\bar{\rho}_j(\mathscr{F}) < 1 \text{ for } j \leq k \quad \text{and} \quad \bar{\rho}_{k+1}(\mathscr{F}) = 1,$$

which, even for an apparently very simple family, indicates that we may have to compute an arbitrarily large number of spectral radii in order to find the joint spectral radius. ◇

For nondefective families, the finiteness property also may be read as a necessary and sufficient condition for the existence of *maximal growth solutions* of *periodic* character to the DLDS (1).

To prove the sufficiency, we consider a nondefective family $\mathscr{F}$ and, in order to simplify the discussion, we assume without any substantial restriction that a leading eigenvalue λ_{max} of the s.m.p. $\bar{P}$ is real and positive, i.e.,

$$\lambda_{max} = \rho(\mathscr{F})^{k^*}.$$

Then, if $\bar{P} = A_{i_{k^*}} \ldots A_{i_1}$, we choose the particular distance of (1) in which the sequence $\{A_{i_k}\}_{k \geq 1}$ is the *periodic* repetition of the ordered k^*-tuple of matrices $\{A_{i_1}, \ldots, A_{i_{k^*}}\}$. Moreover, we choose the initial value $x(0) = x$, where $x \neq 0$ is a leading eigenvector of $\bar{P}$ related to λ_{max}.

By setting

$$\hat{x}(0) = x \quad \text{and} \quad \hat{x}(j) = \rho(\mathscr{F})^{-j} A_{i_j} \ldots A_{i_1} x, \ 1 \leq j \leq k^* - 1,$$

it easily turns out that the solution is given by the sequence of vectors

$$x(k) = \rho(\mathscr{F})^k \hat{x}(j_k), \tag{35}$$

where $k = sk^* + j_k$ for $s \geq 1$ and $0 \leq j_k \leq k^* - 1$.

Now we observe that, since $\mathscr{F}$ is nondefective, any solution of (1) must satisfy the bound

$$\|x(k)\|_* \leq \rho(\mathscr{F})^k \|x(0)\|_*,$$

where $\|\cdot\|_*$ is an extremal norm.

Therefore, we can conclude that (35) is a solution of maximal growth. Finally, we observe that, apart from the growth factor $\rho(\mathscr{F})^k$, such a solution is periodic.

To prove the necessity, we observe that, if the DLDS (1) has a solution of periodic character of maximal growth like (35) corresponding to a certain sequence $\{A_{i_k}\}_{k\geq 1}$, then it is evident that the product $\bar{P} = A_{i_{k^*}} \ldots A_{i_1}$ is an s.m.p. of the associated family $\mathscr{F}$ and that $x(0)$ is a leading eigenvector of it.

For the sake of completeness, we just mention that, if the family $\mathscr{F}$ is defective, then the solution (35) is not of maximal growth and that any solution of maximal growth, if any, has not a periodic character. For theoretical results related to this topic, we refer the reader to Guglielmi and Zennaro [20, 22].

Starting from previous work of Daubechies and Lagarias [15], Lagarias and Wang [37] formulated the following conjecture.

Conjecture 1 (Finiteness conjecture). All finite families of matrices have the finiteness property.

At least in principle, the validity of the finiteness conjecture would imply that, in general, the j.s.r. is computable exactly in a finite time.

Unfortunately, the conjecture was disproved by Bousch and Mairesse [10] and, later, by Blondel et al. [7]. The following theorem gives an elementary counterexample.

Theorem 5 (see [7]). *There are uncountably many values of the parameter* $\alpha \in [0, 1]$ *such that the family* $\mathscr{F} = \{A_1, A_2\}$, *with*

$$A_1 = \begin{bmatrix} 1 & 1 \\ 0 & 1 \end{bmatrix}, \qquad A_2 = \alpha \begin{bmatrix} 1 & 0 \\ 1 & 1 \end{bmatrix}, \tag{36}$$

does not satisfy the finiteness conjecture.

Very recently, Hare et al. [30] improved the statement of Theorem 5 by finding an explicit value of α such that the family $\mathscr{F}$ does not have the finiteness property.

However, in many cases the finiteness property holds. Indeed, Lagarias and Wang [37] were able to give sufficient conditions in terms of existence of extremal norms guaranteeing that the finiteness property holds.

They formulated the following weaker version of the finiteness conjecture.

Conjecture 2 (Normed finiteness conjecture). Let $\|\cdot\|$ be a given induced matrix norm. Then every finite nondefective family $\mathscr{F}$ for which $\|\cdot\|$ is an extremal norm has the finiteness property.

They also proved the following result.

Theorem 6. *The finiteness conjecture is true if and only if the normed finiteness conjecture is true for all induced matrix norms.*

As a consequence of the counterexamples in [7, 10] to the finiteness conjecture, the normed finiteness conjecture cannot be true for all induced matrix norms. However, Lagarias and Wang [37] proved it for families of real matrices for the class of *piecewise analytic norms* and, for the particular subclasses of *piecewise algebraic norms* and *polytope norms*, they also gave an upper bound to the degree k^* of a minimal s.m.p. $\bar{P}$. Moreover, as a corollary, Guglielmi et al. [27] got the following extension to complex polytope norms.

Theorem 7. *The normed finiteness conjecture is true for complex polytope norms for nondefective finite families of complex $n \times n$-matrices.*

Furthermore, it has been recently conjectured that the finiteness property holds for special classes of matrices. In particular, we recall the following conjecture by Blondel et al. [9] (see also Jungers and Blondel [32]).

Conjecture 3. Every pair of $n \times n$ sign-matrices with entries in $\{-1, 0, +1\}$ has the finiteness property.

It has been proven that this would be equivalent to the finiteness property of finite families of matrices with rational entries. Cicone et al. [12] proved that Conjecture 3 holds true for $n = 2$, but the proof for the general case is still missing.

4.2 Limit Spectrum-Maximizing Products and Leading Eigenvectors

Now we recall the following definition from Guglielmi and Zennaro [20].

Definition 17 (Limit spectrum-maximizing product). Assume that $\hat{\mathscr{F}}$ is a normalized bounded family of complex $n \times n$-matrices (i.e., $\rho(\hat{\mathscr{F}}) = 1$) and that there exists a sequence of products $P_k \in \Sigma_{d_k}(\hat{\mathscr{F}})$, d_k nondecreasing integers, such that

$$\lim_{k \to \infty} P_k = \tilde{P}, \qquad (37)$$

where $\tilde{P} \in \overline{\Sigma(\hat{\mathscr{F}})}$ and $\rho(\tilde{P}) = 1$. Then $\tilde{P}$ will be called a limit spectrum-maximizing product (l.s.m.p.) for $\hat{\mathscr{F}}$.

Example 11. Consider the family $\mathscr{F} = \{A_1, A_2\}$ defined by (36) for $\alpha = \frac{4}{5}$ and let $\hat{\mathscr{F}} = \{\hat{A}_1, \hat{A}_2\}$ be the normalized family.

We can prove that $\rho(\mathscr{F}) = 1 + \frac{1}{\sqrt{5}}$ and that there exists a unique spectrum maximizing product $P = A_1 A_2$ (see Guglielmi and Zennaro [24]). However, the sequence of products $\tilde{Q}_k = \hat{A}_1 \left(\hat{A}_1 \hat{A}_2 \right)^k$ is convergent to the limit

$$\tilde{Q}_\infty = \lim_{k \to \infty} \tilde{Q}_k = \begin{pmatrix} \frac{\sqrt{5}+1}{4} & \frac{1}{2} \\ \frac{\sqrt{5}-1}{4} & \frac{3-\sqrt{5}}{4} \end{pmatrix},$$

which satisfies $\rho(\tilde{Q}_\infty) = 1$ and, hence, is an l.s.m.p. for the normalized family $\hat{\mathscr{F}}$.
◊

Note that, for a normalized family $\hat{\mathscr{F}}$, an s.m.p. $\bar{P}$ is an l.s.m.p., too. To see this, just put $P_k = \bar{P}$ for all $k \geq 1$. Moreover, if the family $\hat{\mathscr{F}}$ is nondefective, another possibility is to consider the power sequence $\{\bar{P}^k\}_{k\geq 1}$ and, since $\Sigma(\hat{\mathscr{F}})$ is bounded, to extract a subsequence $\{\bar{P}^{k_s}\}_{s\geq 1}$ converging to some $\tilde{P} \in \overline{\Sigma(\hat{\mathscr{F}})}$, which obviously satisfies $\rho(\tilde{P}) = 1$. For the sake of brevity, we shall say that such a limit point of the sequence $\{\bar{P}^k\}_{k\geq 1}$ is an *infinite power* of the matrix $\bar{P}$. For nondefective families, Guglielmi and Zennaro [22] proved the following result.

Theorem 8. *Let $\mathscr{F}$ be a (possibly infinite) nondefective bounded family of complex $n \times n$-matrices. Then there exists an l.s.m.p. $\tilde{P}$ for the normalized family $\hat{\mathscr{F}}$.*

On the contrary, for defective families they gave some counterexamples to the existence of l.s.m.p.'s whenever the dimension of the matrices is $n \geq 4$.

The following notion of *leading eigenvector* for a nondefective family $\mathscr{F}$ will be fundamental for the forthcoming extremality result in Sect. 4.3.

Definition 18. An eigenvector $x \neq 0$ of a matrix P related to an eigenvalue λ with $|\lambda| = \rho(P)$ is called a leading eigenvector of P.

Definition 19 (Leading eigenvector). Let $\mathscr{F}$ be a nondefective bounded family of complex $n \times n$-matrices. A leading eigenvector $x \neq 0$ of either an s.m.p. $\bar{P}$ of $\mathscr{F}$ or of an l.s.m.p. $\tilde{P}$ of the normalized family $\hat{\mathscr{F}}$ is called leading eigenvector of $\mathscr{F}$ (and of $\hat{\mathscr{F}}$ too).

Remark 2. Because of Theorem 8, any nondefective bounded family $\mathscr{F}$ has at least one leading eigenvector.

4.3 Existence of Extremal Complex Polytope Norms

Due to the essential finiteness of information needed to determine the unit ball of a polytope or a complex polytope norm, for computational purposes it would be nice to be able to reverse Theorem 7. As we shall see later in this section, Guglielmi et al. [27] were able to do it under some restrictive assumptions on the nondefective finite family $\mathscr{F}$. Moreover, due to the lack of counterexamples, they also formulated the following general conjecture.

Conjecture 4 (Complex polytope extremality (CPE) conjecture). Every finite nondefective family $\mathscr{F}$ that has the finiteness property admits an extremal complex polytope norm.

However, also this conjecture was disproved by Jungers and Protasov [34] as follows.

Theorem 9. *Let $0 \leq \alpha_i \leq 2\pi$, where α_i is rationally independent of π, for $i = 1, 2$. Let A_i the rotation matrix in $\mathbf{R}^3$ by the angle α_i with respect to an axis a_i and assume the axes a_1 and a_2 are orthogonal. Then the family $\mathscr{F} = \{A_1, A_2\}$ does not have an extremal complex polytope norm.*

Now we report the existence results proved in [27], which involve the normalized trajectories $\mathscr{T}[\hat{\mathscr{F}}, x]$ defined in Sect. 3.3.

First we observe that all the cyclic permutations of a product $\bar{P}$ have the same eigenvalues with the same multiplicities. Thus, if $\bar{P} = A_{i_{k^*}} \ldots A_{i_1}$ is an s.m.p. for a family $\mathscr{F}$, then each of its cyclic permutations

$$A_{i_s} \ldots A_{i_1} A_{i_{k^*}} \ldots A_{i_{s+1}}, \quad s = 1, \ldots, k^* - 1,$$

still is an s.m.p. for $\mathscr{F}$.

Definition 20 (Asymptotic simplicity). A nondefective bounded family $\mathscr{F}$ of complex $n \times n$-matrices is called asymptotically simple if it has a minimal s.m.p. $\bar{P}$ with only one leading eigenvector (modulo scalar nonzero factors) such that the set $\mathscr{E}$ of the leading eigenvectors of $\mathscr{F}$ is equal to the set of the leading eigenvectors of $\bar{P}$ and of its cyclic permutations.

We remark that the asymptotic simplicity of $\mathscr{F}$ means that the set $\mathscr{E}$ of the leading eigenvectors of $\mathscr{F}$ is $\mathscr{F}$-cyclic, i.e., for any pair $(x, y) \in \mathscr{E} \times \mathscr{E}$ there exist $\alpha, \beta \in \mathbf{C}$ with

$$|\alpha| \cdot |\beta| = 1$$

and two normalized products $\hat{P}, \hat{Q} \in \Sigma(\hat{\mathscr{F}})$ such that

$$y = \alpha \hat{P} x \quad \text{and} \quad x = \beta \hat{Q} y.$$

Theorem 10 (Small CPE Theorem). *Assume that a finite family $\mathscr{F}$ of complex $n \times n$-matrices is nondefective and asymptotically simple. Moreover, let $x \neq 0$ be a leading eigenvector of $\mathscr{F}$ and assume that (23) is satisfied. Then the set*

$$\partial \mathscr{S}[\hat{\mathscr{F}}, x] \bigcap \mathscr{T}[\hat{\mathscr{F}}, x] \tag{38}$$

is finite modulo scalar factors of unitary modulus. As a consequence, there exist a finite number of normalized products $\hat{P}^{(1)}, \ldots, \hat{P}^{(s)} \in \Sigma(\hat{\mathscr{F}})$ such that

$$\mathscr{S}[\hat{\mathscr{F}}, x] = \mathrm{absco}\Big(\{x, \hat{P}^{(1)} x, \ldots, \hat{P}^{(s)} x\}\Big), \tag{39}$$

so that $\mathscr{S}[\hat{\mathscr{F}}, x]$ is a b.c.p.

The next results give a better description of the structure of the b.c.p. $\mathscr{S}[\hat{\mathscr{F}}, x]$.

Theorem 11. *Let the hypotheses of Theorem 10 hold. Then each leading eigenvector ξ of $\mathscr{F}$ in the set $\varXi = \mathscr{E} \bigcap \partial \mathscr{S}[\hat{\mathscr{F}}, x]$ satisfies one of the following two properties:*

(a) *ξ is a vertex of the b.c.p. $\mathscr{S}[\hat{\mathscr{F}}, x]$;*
(b) *there exist $s \geq 2$ vertices $\xi_1, \ldots, \xi_s$ of the b.c.p. $\mathscr{S}[\hat{\mathscr{F}}, x]$ such that*

$$\xi_1, \ldots, \xi_s \in \varXi \quad \text{and} \quad \xi \in \text{absco}\Big(\{\xi_1, \ldots, \xi_s\}\Big). \tag{40}$$

The following particular subclass of asymptotically simple families allow us to refine Theorem 11.

Definition 21 (Absolute asymptotic simplicity). A nondefective bounded family $\mathscr{F}$ of complex $n \times n$-matrices is said to be absolutely asymptotically simple if it is asymptotically simple and has a unique minimal s.m.p. $\bar{P}$ (modulo cyclic permutations).

Proposition 7. *Let the hypotheses of Theorem 10 hold and, moreover, let the family $\mathscr{F}$ be absolutely asymptotically simple. Then all the leading eigenvectors of $\mathscr{F}$ in the set $\varXi = \mathscr{E} \bigcap \partial \mathscr{S}[\hat{\mathscr{F}}, x]$ are vertices of the b.c.p. $\mathscr{S}[\hat{\mathscr{F}}, x]$.*

Several examples showing the sharpness of the assumptions and counterexamples when one of the assumption is not fulfilled are given in [27].

In particular, whenever two or more spectrum maximizing products coexist, in general $\partial \mathscr{S}[\hat{\mathscr{F}}, x] \bigcap \mathscr{T}[\hat{\mathscr{F}}, x]$ is not finite, unless the eigenvectors are $\mathscr{F}$-cyclic. This situation, which appears to be non generic, occurs in several applications (see, e.g., the matrices associated to some Daubechies wavelets [13] or those appearing in the four-point subdivision scheme presented in [31], which we shall consider in Sect. 5.6), where all the matrices of the family have the same spectral radius which, in some cases, equals the joint spectral radius itself (see also the forthcoming Example 16).

Moreover, if the family $\mathscr{F}$ is *real*, it makes sense to consider the possibility that there exists a minimal s.m.p. with a pair of complex conjugate leading eigenvectors instead of a unique (necessarily real) leading eigenvector. Therefore, Guglielmi and Zennaro [25] modified the concept of asymptotic simplicity in the case of real families in order to include this possibility and also proved the small CPE theorem in this more general setting (under some suitable additional technical assumptions).

4.4 Polytope Extremality Results for Nonnegative Families

We conclude this section by considering again the special case of families $\mathscr{F}$ of nonnegative matrices (see Sects. 3.4 and 3.6).

It is clear that, if $x \in \mathbf{R}_+^n$, then

$$\mathscr{T}[\hat{\mathscr{F}}, x] \subset \mathbf{R}_+^n$$

as well. Therefore, we can define the set

$$\mathscr{Q}[\hat{\mathscr{F}}, x] = \text{clos}\left(\text{co}_-\left(\mathscr{T}[\hat{\mathscr{F}}, x]\right)\right).$$

Definition 22. A matrix $\bar{P} \in \Sigma_{k^*}(\mathscr{F})$ is called a dominant product for the family $\mathscr{F}$ if it is an s.m.p. (i.e., $\rho(P)^{1/k^*} = \rho(\mathscr{F})$) and there exists $\rho^* < 1$ such that $\rho(P) \leq \rho^*$ for all $P \in \Sigma(\hat{\mathscr{F}})$ which are neither a power of $\hat{P} = \bar{P}/\rho(\mathscr{F})$ nor a power of its cyclic permutations.

The following adaptation of the Small CPE Theorem has been proved by Guglielmi and Protasov [19].

Theorem 12 (Small positive polytope extremality (PPE) Theorem). *Assume that a finite family $\mathscr{F}$ of nonnegative $n \times n$-matrices is positively-irreducible and such that*

(i) there exists a dominant product $\bar{P}$;
(ii) $\bar{P}$ has a unique leading eigenvector $x \in \mathbf{R}_+^n$ (modulo positive scalar factors).

Then the set $\partial \mathscr{Q}[\hat{\mathscr{F}}, x] \cap \mathscr{T}[\hat{\mathscr{F}}, x]$ is finite. As a consequence, there exist a finite number of normalized products $\hat{P}^{(1)}, \ldots, \hat{P}^{(s)} \in \Sigma(\hat{\mathscr{F}})$ such that

$$\mathscr{Q}[\hat{\mathscr{F}}, x] = \text{co}_-\left(\{x, \hat{P}^{(1)}x, \ldots, \hat{P}^{(s)}x\}\right). \tag{41}$$

Moreover, all the leading eigenvectors of $\mathscr{F}$ in the set $\Xi = \mathscr{E} \cap \partial \mathscr{Q}[\hat{\mathscr{F}}, x]$ are vertices of the positive polytope $\mathscr{Q}[\hat{\mathscr{F}}, x]$.

5 Algorithms and Examples

The available methods for the approximation and computation of the joint spectral radius of a family $\mathscr{F}$ can be classified according to the following taxonomy.

(i) Branch and bound methods: based on the construction of a sequence of lower bounds (based on spectral radii) and upper bounds (based on norms) (see Daubechies and Lagarias [14] and Gripenberg [18]).
(ii) Optimal ellipsoid norm methods: in general, these methods only provide an approximation to $\rho(\mathscr{F})$ (see Ando and Shih [1] and Blondel et al. [8]).
(iii) Kronecker lifting methods: based on the transformation of the original problem into a higher dimensional one with special structure, that is, into an augmented

family with an invariant cone K (see e.g. Blondel and Nesterov [6], Parrillo and Jadbabaie [42] and Protasov [44]).

(iv) Extremal polytope norm methods: based on the theory presented in Sects. 3 and 4 (see Guglielmi et al. [27], Guglielmi and Zennaro [23–25], Guglielmi and Protasov [19], Kozyakin [36], Maesumi [38, 39] and Protasov [43, 44]).

For results about the complexity of the computation of the j.s.r. we refer the reader to the seminal paper by Blondel and Tsitsiklis [50].

In this section we discuss in some detail a class of algorithms derived by methods of type (iv) and we fix our attention on finite families $\mathscr{F} = \{A_i\}_{1 \leq i \leq d}$ which are assumed to have the finiteness property.

5.1 The General Procedure

The first step is to look for a candidate s.m.p. $\bar{P} \in \Sigma_{k^*}(\mathscr{F})$ for some $k^* \geq 1$ by means of some heuristic method (see, e.g., Sect. 6) and to define the scaled family $\mathscr{F}^* = \mathscr{F}/\rho(\bar{P})^{1/k^*}$, which satisfies $\rho(\mathscr{F}^*) \geq 1$.

The second step is to look for an invariant absolutely convex set for $\mathscr{F}^*$, i.e., the unit ball of an extremal norm.

If the procedure succeeds, then we can conclude that $\rho(\mathscr{F}^*) = 1$, that is $\rho(\mathscr{F}) = \rho(\bar{P})^{1/k^*}$ and $\bar{P}$ is actually an s.m.p. for $\mathscr{F}$.

The General Algorithm (Algorithm G)

---

1. Compute a candidate s.m.p. $\bar{P} \in \Sigma_{k^*}(\mathscr{F})$ for some $k^* \geq 1$ and a leading eigenvector $x \neq 0$ of $\bar{P}$.
2. Set $\vartheta = \rho(\bar{P})^{1/k^*}$ and define the scaled family

$$\mathscr{F}^* = \{\vartheta^{-1} A_i\}_{1 \leq i \leq d}.$$

3. Set $\mathscr{T}^{(0)} = \{x\}$.
4. For $m \geq 1$

   a. Set

   $$\mathscr{P}^{(m-1)} = \mathrm{absco}(\mathscr{T}^{(m-1)})$$

   and

   $$\mathscr{T}^{(m)} = \mathscr{F}^*(\mathscr{T}^{(m-1)}) \cup \mathscr{T}^{(m-1)}.$$

b. If $\text{span}(\mathscr{T}^{(m-1)}) = \mathbf{C}^n$ ($\mathbf{R}^n$ in the real case) and

$$\mathscr{T}^{(m)} \setminus \mathscr{T}^{(m-1)} \subset \mathscr{P}^{(m-1)}$$

   i. $\mathscr{P}^{(m-1)}$ is an invariant unit ball for $\mathscr{F}^*$ and the algorithm halts.

c. End If

5. End For

5.2 The Special Procedure for Families of Nonnegative Matrices

With respect to the general case, there are some peculiar differences, which are outlined hereafter.

First of all, remark that, by the Perron-Frobenius theorem, the candidate s.m.p. $\bar{P}$ has the positive real leading eigenvalue $\lambda_{\max} = \rho(\bar{P})$ and the corresponding leading eigenvector x may be chosen to belong to . Moreover, if the matrix $\bar{P}$ is primitive, than the leading eigenvector x is unique modulo scalar factors (see, e.g., Berman and Plemmons [4]).

A Special Algorithm for Nonnegative Matrices (Algorithm N)

1. Compute a candidate s.m.p. $\bar{P} \in \Sigma_{k^*}(\mathscr{F})$ for some $k^* \geq 1$ and a leading eigenvector $x \neq 0$ of $\bar{P}$, $x \in \mathbf{R}_+^n$.
2. Set $\vartheta = \rho(\bar{P})^{1/k^*}$ and define the scaled family

$$\mathscr{F}^* = \{\vartheta^{-1} A_i\}_{1 \leq i \leq d}.$$

3. Set $\mathscr{T}^{(0)} = \{x\}$.
4. For $m \geq 1$

   a. Set

   $$\mathscr{P}^{(m-1)} = \text{co}_-(\mathscr{T}^{(m-1)})$$

   and

   $$\mathscr{T}^{(m)} = \mathscr{F}^*(\mathscr{T}^{(m-1)}) \cup \mathscr{T}^{(m-1)}.$$

b. If

$$\mathcal{T}^{(m)} \setminus \mathcal{T}^{(m-1)} \subset \mathcal{P}^{(m-1)}$$

  i. $\mathcal{P}^{(m-1)}$ is an invariant positive polytope for $\mathcal{F}^*$ and the algorithm halts.

 c. End If

5. End For

Note that the condition $\mathrm{span}(\mathcal{T}^{(m-1)}) = \mathbf{C}^n$ is not required any more and that $\mathrm{absco}(\mathcal{T}^{(m-1)})$ is replaced by $\mathrm{co}_-(\mathcal{T}^{(m-1)})$.

The linear programming problem to be solved in order to check whether a vector $z \in \mathcal{P}^{(\ell)}$ for some ℓ is simpler with respect to the general real case.

We have also experimentally observed that, for a given family of nonnegative matrices, the polytope unit ball constructed by using the general Algorithm G has a number of vertices which is much higher than the positive polytope constructed by using the refined Algorithm N.

5.3 Termination Criteria

While implementing Algorithm G or N, we can use some criteria to detect whether the guess product $\bar{P}$ is not an s.m.p. They are based on the following results.

Theorem 13 (Cicone et al. [12]). *Let $\mathcal{F}$ be a finite irreducible family of matrices. Then $\bar{P}$ is not an s.m.p. if and only if, at some step m of the For cycle in Algorithm G (or N), the initial vector x lies strictly inside $\mathcal{P}^{(m-1)}$, that is*

$$x \in \overset{\circ}{\mathcal{P}}{}^{(m-1)}. \tag{42}$$

Theorem 14 (Guglielmi and Protasov [19]). *Let $\mathcal{F}$ be a finite irreducible family of matrices and assume that the leading eigenvalue of $\bar{P}$ is unique and simple and that $x \neq 0$ is the corresponding eigenvector. Moreover, let x' be the leading eigenvector of $\bar{P}^H$ normalized so that $(x', x) = 1$, where $(\cdot, \cdot)$ denotes the standard Euclidean scalar product. Then $\bar{P}$ is not an s.m.p. if and only if, at some step m of the For cycle in Algorithm G (or N), it holds that*

$$\left|(x', A_i^* v)\right| > 1 \quad \text{for some } v \in \mathcal{T}^{(m)} \text{ and } i \in \mathcal{I}. \tag{43}$$

Observe that we can replaced the vector x (and x') in (43) by any other leading eigenvector of $\mathcal{F}$, e.g., by leading eigenvectors of cyclic permutations of $\bar{P}$. Using more vectors makes the criterium more effective.

Fig. 2 Example 12. Step $m = 1$

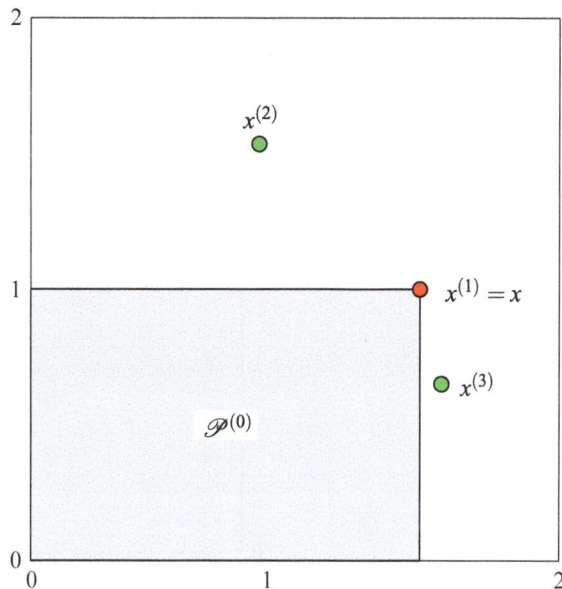

5.4 Some Illustrative Examples

Now we show some examples of application of Algorithms G and N.

Example 12. Consider the family of nonnegative matrices $\mathscr{F} = \{A_1, A_2\}$ defined by (36) for $\alpha = \frac{9}{10}$ and apply Algorithm N starting from the guess s.m.p. $\bar{P} = A_1 A_2$, whose unique leading eigenvector is

$$x = \left[(1+\sqrt{5})/2,\ 1\right]^{\mathrm{T}}.$$

Then we scale the family $\mathscr{F}$ to obtain

$$\mathscr{F}^* = \{A_1^*, A_2^*\} = \{A_1/\vartheta, A_2/\vartheta\},$$

where $\vartheta = \sqrt{\rho(P)} = \sqrt{\alpha}\,(1+\sqrt{5})/2$, and start the For cycle.

Step $m = 1$. We compute the vectors

$$x^{(2)} = A_2^* x^{(1)} \quad \text{and} \quad x^{(3)} = A_1^* x^{(1)}.$$

Both vectors lie externally to the initial polytope $\mathscr{P}^{(0)} = \mathrm{co}_-(\{x\})$ and, so, contribute to form the new positive polytope $\mathscr{P}^{(1)}$ (see Fig. 2, where they are indicated in green to represent alive leaves in the computational tree).

Fig. 3 Example 12. Step $m = 2$

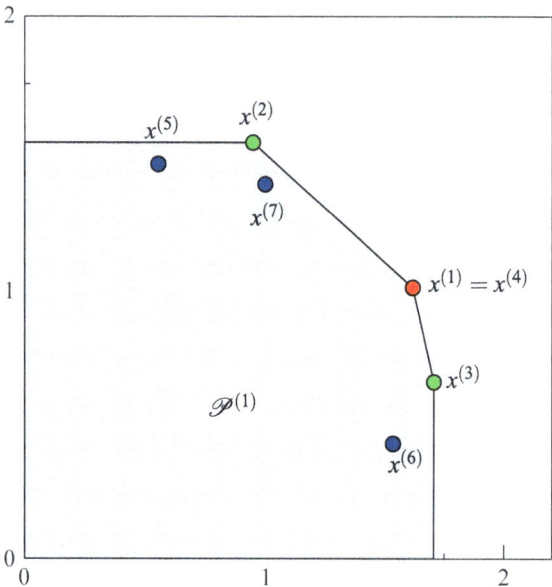

Step $m = 2$. We compute the vectors

$$x^{(4)} = A_1^* x^{(2)}, \quad x^{(5)} = A_2^* x^{(2)}, \quad x^{(6)} = A_1^* x^{(3)} \quad \text{and} \quad x^{(7)} = A_2^* x^{(3)}.$$

All vectors are included in the polytope $\mathscr{P}^{(1)} = \text{co}_-(\{x^{(1)}, x^{(2)}, x^{(3)}\})$ and, thus, the algorithm halts (see Fig. 3, where the new vertices are indicated in blue to represent dead leaves in the computational tree).

Note that, if we had applied the general Algorithm G, we would not have stopped at the second step because the vectors $x^{(5)}$ and $x^{(6)}$ would have been added to the previous vertices to form a new polytope. ◇

Example 13. Consider the family $\mathscr{F} = \{A_1, A_2\}$, where

$$A_1 = \begin{bmatrix} 1 & -1 \\ 0 & 1 \end{bmatrix}, \quad A_2 = \begin{bmatrix} 0 & -1 \\ 1 & 0 \end{bmatrix}.$$

We apply Algorithm G starting from the guess s.m.p. $\bar{P} = A_1 A_2 A_1^2 A_2$.

It turns out that the leading eigenvalue of $\bar{P}$ is real, so that we can choose a corresponding real eigenvector x. The algorithm halts after five steps of the For cycle determining the invariant unit ball $\mathscr{P}^{(4)} = \text{co}_s(\mathscr{T}^{(4)})$.

It is interesting to see what happens if a different initial vector x' is chosen in place of the leading eigenvector x. After 15 iterations the norm of the scaled family $\mathscr{F}^*$ associated to the computed polytope $\mathscr{P}^{(15)}$ is $\|\mathscr{F}^*\|_{\mathscr{P}^{(15)}} \approx 1.00455$. After 30 iterations it becomes $\|\mathscr{F}^*\|_{\mathscr{P}^{(30)}} \approx 1.0000022$ (the polytope $\mathscr{P}^{(30)}$ is shown in Fig. 5).

Fig. 4 Example 13. Invariant unit ball $\mathscr{P}^{(4)}$ and trajectory $\mathscr{T}^{(5)}$

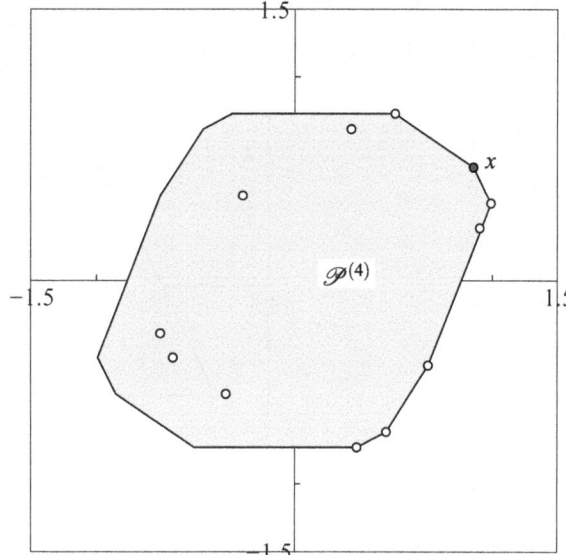

Fig. 5 Example 13. Polytope computed after 30 iterations

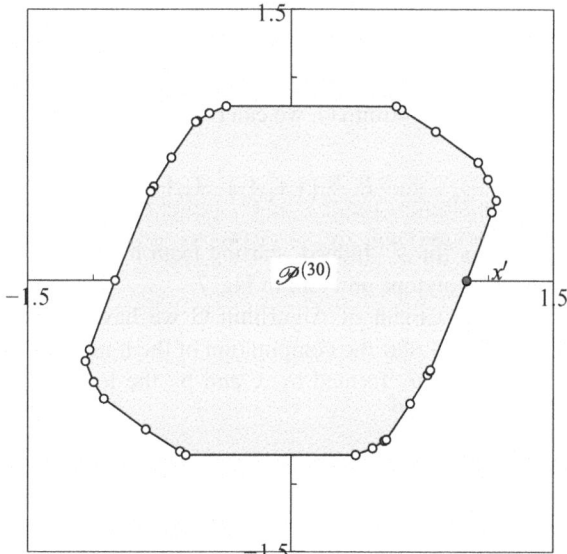

It is also interesting to see what happens if a wrong guess product $\bar{P}$ (which is not an s.m.p.) is used to implement Algorithm G. We choose $\bar{P} = A_2^2$ (which has real eigenvalues), whose leading eigenvector is $\tilde{x} = [1, \, 0]^T$. The behaviour of the algorithm is illustrated in Fig. 6, where we observe the polytope $\mathscr{P}^{(3)}$. The initial

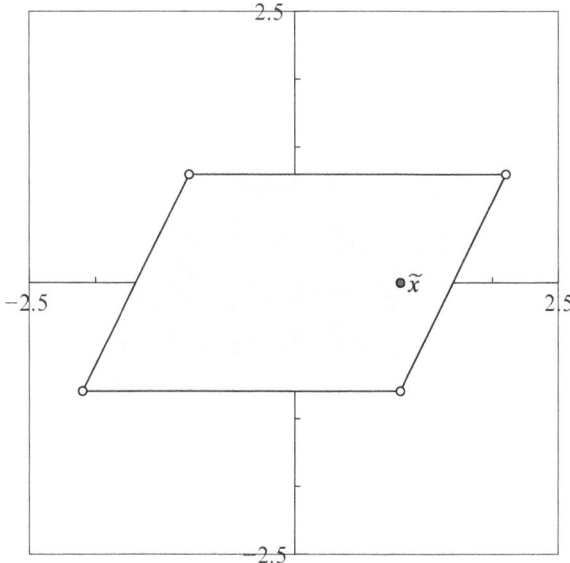

Fig. 6 Example 13. The effect of a wrong guess s.m.p. $\bar{P}$

vector $\tilde{x}$ appears to be internal to the polytope. This occurrence implies that $\bar{P}$ is not an s.m.p. (see Theorem 13). ◇

Example 14. Consider the family $\mathscr{F}$ of Examples 2 and 7.

By using Algorithm G, we can prove that the product

$$\bar{P} = \left(\left(A_1^2 A_2 \right)^2 A_1 A_2 \right)^3 A_2 A_1 A_2 \in \Sigma_{27}(\mathscr{F})$$

is an s.m.p. for $\mathscr{F}$. Indeed, starting from its unique real leading eigenvector x we obtain the polytope unit ball in Fig. 7.

A useful variant of Algorithm G we have adopted to reduce the number of iterations is to start the computation of the trajectory from $\mathscr{T}^{(0)}$ equal to the whole set of 27 vectors formed by x and by the leading eigenvectors of all the cyclic permutations of $\bar{P}$. ◇

Example 15. Consider the family $\mathscr{F} = \{A_1, A_2\}$, where

$$A_1 = \begin{bmatrix} \cos(1) & -\sin(1) \\ \sin(1) & \cos(1) \end{bmatrix}, \quad A_2 = \begin{bmatrix} \cos(1) & \sin(1) \\ -\sin(1) & \cos(1) \end{bmatrix}.$$

Since $\|A_1\|_2 = \|A_2\|_2 = 1$, it is immediate to observe that $\rho(\mathscr{F}) = 1$. Moreover, it is also easy to check that the b.c.p.

$$\mathscr{P} = \text{absco}\left(\{[1, i]^T, [1, -i]^T\} \right)$$

is invariant for $\mathscr{F}$ and, thus, is the unit ball of an extremal norm.

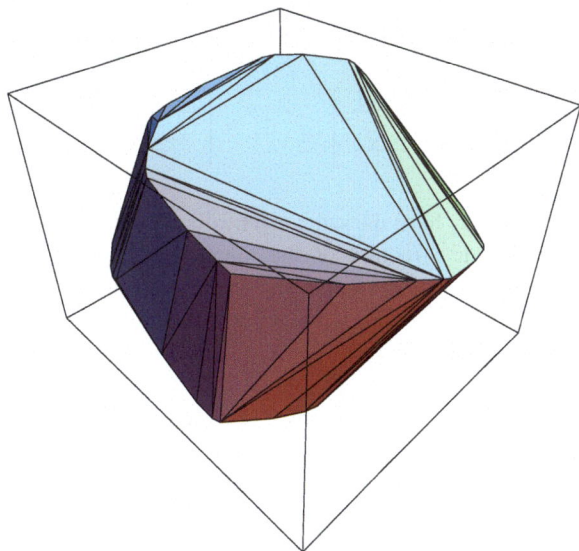

Fig. 7 Example 14. Extremal unit ball for the family $\mathscr{F}$

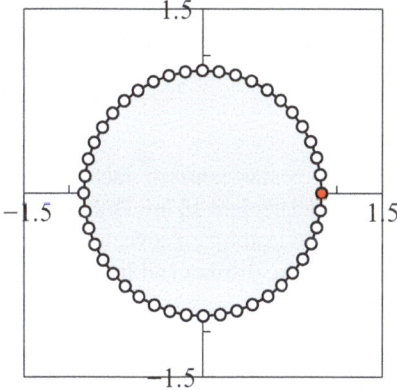

Fig. 8 Example 15. Polytope obtained after $m = 22$ steps

However, implementing Algorithm G starting from the initial vector $x = [1, 0]^T$, in the limit we get a trajectory whose closure is the unit circle in the real plane, which is the boundary of an invariant unit ball for $\mathscr{F}$, but is not a polytope. Indeed, after any finite number m of steps in the For cycle, we obtain a polytope $\mathscr{P}^{(m-1)}$ as in Fig. 8. ◇

Example 16. Here we address the 2 × 2-matrix family $\mathscr{F} = \{A_1, A_2\}$ from [27], where

$$A_1 = \begin{bmatrix} 3-\sqrt{5} & \frac{3-\sqrt{5}}{2} \\ \frac{3-\sqrt{5}}{2} & \frac{3-\sqrt{5}}{2} \end{bmatrix}, \quad A_2 = \begin{bmatrix} \frac{3-\sqrt{5}}{2} & \frac{3-\sqrt{5}}{2} \\ \frac{3-\sqrt{5}}{2} & 3-\sqrt{5} \end{bmatrix}.$$

The matrices A_1 and A_2 have the same eigenvalues 1 and $\frac{7-3\sqrt{5}}{2}$.

Given the palindromic structure of the matrices, by showing that they are simultaneously similar to symmetric matrices (see Guglielmi et al. [28]), it can be proved that $\rho(\mathscr{F}) = 1$, that A_1 and A_2 are the only minimal s.m.p.'s of $\mathscr{F}$, and that the only l.s.m.p.'s of $\mathscr{F}$ are just

$$A_1^\infty = \lim_{k\to\infty} A_1^k \quad \text{and} \quad A_2^\infty = \lim_{k\to\infty} A_2^k.$$

Therefore, the set of the leading eigenvectors of $\mathscr{F}$ is given by the leading eigenvectors of A_1 and A_2, namely,

$$\alpha_1 x_1, \quad x_1 = [\frac{1+\sqrt{5}}{2}, 1]^T, \quad \alpha_1 \neq 0,$$

and

$$\alpha_2 x_2, \quad x_2 = [1, \frac{1+\sqrt{5}}{2}]^T, \quad \alpha_2 \neq 0,$$

respectively.

This set is finite (modulo scalar nonzero factors) but not $\mathscr{F}$-cyclic. As a consequence, the hypotheses of Theorem 10 are violated.

Now set $x = x_1$, which is the leading eigenvector of the s.m.p. A_1. It turns out that all the vectors $A_1^k x$, $k \geq 1$, are distinct and lie on the segment of $\mathbf{R}^2$ that joins x and

$$A_2^\infty x = \frac{14+6\sqrt{5}}{15+7\sqrt{5}} \left[1, \frac{1+\sqrt{5}}{2}\right]^T,$$

which is the leading eigenvector of the s.m.p. A_2, and that all vectors $A_1 A_2^k x$, $k \geq 1$, lie inside the b.c.p. $\mathrm{absco}\left(\{x, A_2^\infty x\}\right)$ (see Fig. 9). Therefore we can conclude that $\mathscr{S}[\mathscr{F}, x] = \mathrm{absco}\left(\{x, A_2^\infty x\}\right)$.

Although $\mathscr{S}[\mathscr{F}, x]$ is a b.c.p., the auspicable finiteness property of the set $\partial\mathscr{S}[\hat{\mathscr{F}}, x] \cap \mathscr{T}[\hat{\mathscr{F}}, x]$ stated by Theorem 10 does not hold and Algorithm G would not halt in a finite time.

Fig. 9 Real section of $\mathscr{S}[\mathscr{F}, x]$ in Example 16

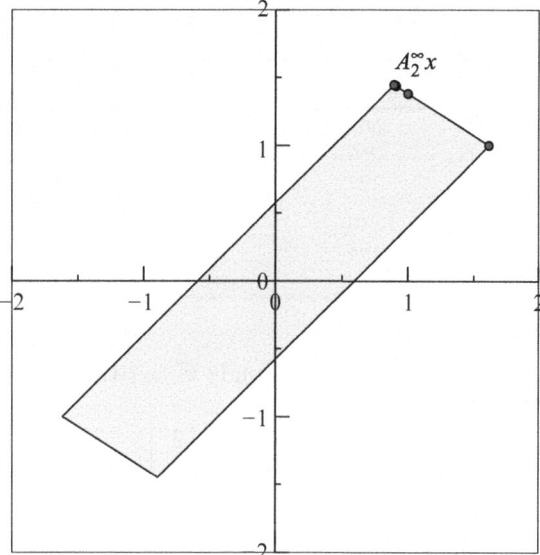

Fig. 10 Real section of the polytope absco($\{x_1, x_2\}$) in Example 16. The *red points* indicate x_1 and x_2, the *blue* ones indicate $A_1 x_2$ and $A_2 x_1$. This shows that the drawn polytope defines an extremal real norm for $\mathscr{F}$

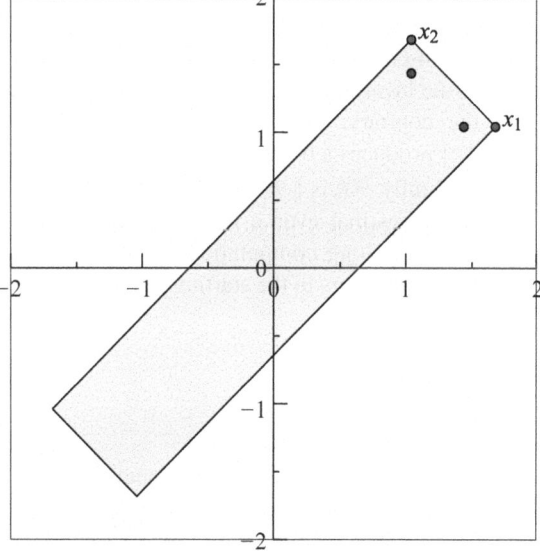

Nevertheless, it is interesting to observe that choosing $\alpha_1 = \alpha_2 = 1$ provides the extremal b.c.p. absco($\{x_1, x_2\}$) (see Fig. 10), whereas other choices of α_1 and α_2 do not necessarily provide an extremal norm.

This gives an indication that when the set of leading eigenvectors is not $\mathscr{F}$-cyclic, the balancing of them may play a fundamental role. ◇

Table 1 Statistics for Algorithms G and N for families of two real random matrices of dimension n

	General real case				Nonnegative case		
n	m	# vertices	s.m.p.	n	m	# vertices	s.m.p.
10	14	565	$A_1^4 A_2$	10	4	11	A_1
10	11	266	$A_1 A_2$	10	3	5	$A_1 A_2$
10	8	122	$A_1 A_2^3$	10	5	16	$A_1^2 A_2$
25	21	1,240	$A_1^2 A_2 A_1 A_2$	25	6	18	$A_1 A_2$
25	24	3,194	$A_1^2 A_2^3$	25	5	11	$A_1^2 A_2$
25	15	1,610	$A_1 A_2$	25	6	18	$A_1^2 A_2 A_1 A_2$

Example 17. Consider the family $\mathscr{F} = \{A_1, A_2\}$, where

$$A_1 = \begin{bmatrix} -3 & -2 & 1 & 2 \\ -2 & 0 & -2 & 1 \\ 1 & 3 & -1 & -5 \\ -3 & -3 & -2 & -1 \end{bmatrix}, \quad A_2 = \begin{bmatrix} 1 & 0 & -3 & -1 \\ -4 & -2 & -1 & -4 \\ -1 & 0 & -1 & 2 \\ -1 & -2 & -1 & 2 \end{bmatrix}$$

(see Guglielmi and Zennaro [25].

The s.m.p. is $\bar{P} = A_1 A_2$, which has two complex conjugate leading eigenvectors x and $\bar{x}$.

Algorithm G with initial set $\mathscr{T}^{(0)} = \{x, \bar{x}\}$ halts after $m = 7$ steps of the For cycle and the invariant b.c.p. $\mathscr{P}^{(6)}$ has 38 essential vertices, pairwise conjugate.

On the contrary, after $m = 20$ steps of the For cycle, Algorithm G with $\mathscr{T}^{(0)} = \{x\}$ produces a b.c.p. $\mathscr{P}^{(20)}$ with 98 essential vertices such that the norm of the scaled family $\mathscr{F}^*$ is $\|\mathscr{F}^*\|_{\mathscr{P}}^{(20)} \approx 1.037$.

This indicates that even a not so accurate approximation of the joint spectral radius requires a long computation if one does not include both leading complex conjugate eigenvectors in the starting set $\mathscr{T}^{(0)}$. ◇

5.5 Statistics

In order to get an idea of the computational performances of the presented algorithm when applied to an arbitrary family (in its general form G) and to a nonnegative family (in its specialized form N), in Table 1 we consider families $\mathscr{F} = \{A_1, A_2\}$ of two real random matrices of different dimensions n. For each considered family, we report the number m of steps of the For cycle needed to find the invariant unit ball $\mathscr{P}^{(m-1)}$, the number of essential vertices of $\mathscr{P}^{(m-1)}$ and the s.m.p. $\bar{P}$.

It clearly seems that the nonnegative case has a much smaller complexity than the general case. Intuitively, the difference might be due to the fact that in the general case the vertices are distributed in all orthants, whereas in the nonnegative case they are confined to the sole orthant $\mathbf{R}_+^n$.

5.6 An Application in Interpolatory Subdivision

We consider now the interpolatory four-point subdivision scheme, introduced by Dubuc [16] and recently studied by Hechler et al. [31]. In order to analyze the convergence of the scheme, we have to consider the family of matrices $\mathscr{F}_\omega = \{A_1, A_2\}$, where

$$A_{1,\omega} = \begin{pmatrix} 4\omega & 4\omega & 0 & 0 \\ -2\omega & 1-4\omega & -2\omega & 0 \\ 0 & 4\omega & 4\omega & 0 \\ 0 & -2\omega & 1-4\omega & -2\omega \end{pmatrix}, \quad A_{2,\omega} = \begin{pmatrix} -2\omega & 1-4\omega & -2\omega & 0 \\ 0 & 4\omega & 4\omega & 0 \\ 0 & -2\omega & 1-4\omega & -2\omega \\ 0 & 0 & 4\omega & 4\omega \end{pmatrix},$$

where $\omega \in \mathbf{R}$ is a nonnegative parameter of the scheme.

As is shown in [31], $\rho(\mathscr{F}) \le 1$ if and only if $\omega \le \omega^*$, ω^* being the only real solution of the cubic equation $32\omega^3 + 4\omega - 1 = 0$. It turns out that

$$\omega^* = \frac{1}{12}\left(27 + 3\sqrt{105}\right)^{1/3} - \frac{1}{2}\left(27 + 3\sqrt{105}\right)^{-1/3} \approx 0.19273.$$

By applying Algorithm G to the scaled family $\mathscr{F}_{\omega^*}^*$, we find that the product

$$\bar{P} = A_{1,\omega^*} A_{2,\omega^*}$$

is spectrum maximizing. This provides part of the proof of the results given in [31].

Let x be the *real* leading eigenvector of $\bar{P}$. Then, with $\mathscr{T}^{(0)} = \{x\}$, the real extremal polytope generated by Algorithm G turns out to be

$$\mathscr{P}^{(4)} = \text{absco}\left(\{x^{(1)}, x^{(2)}, \ldots, x^{(16)}\}\right),$$

where

$x^{(1)} = x,$ $\quad x^{(2)} = A_{2,\omega^*} x^{(1)},$ $\quad x^{(3)} = A_{1,\omega^*} x^{(1)},$ $\quad x^{(4)} = A_{2,\omega^*} x^{(2)},$
$x^{(5)} = A_{1,\omega^*} x^{(3)},$ $\quad x^{(6)} = A_{1,\omega^*} x^{(4)},$ $\quad x^{(7)} = A_{2,\omega^*} x^{(3)},$ $\quad x^{(8)} = A_{2,\omega^*} x^{(4)},$
$x^{(9)} = A_{1,\omega^*} x^{(6)},$ $\quad x^{(10)} = A_{1,\omega^*} x^{(7)},$ $\quad x^{(11)} = A_{2,\omega^*} x^{(6)},$ $\quad x^{(12)} = A_{2,\omega^*} x^{(7)},$
$x^{(13)} = A_{1,\omega^*} x^{(8)},$ $\quad x^{(14)} = A_{1,\omega^*} x^{(12)},$ $\quad x^{(15)} = A_{2,\omega^*} x^{(9)},$ $\quad x^{(16)} = A_{2,\omega^*} x^{(12)}.$

Numerical experiments suggest that formally analogous extremal polytopes, with the same number of vertices obtained in the same sequence from the leading eigenvector $x(\omega)$ of the spectrum maximizing product $\bar{P} = A_{1,\omega} A_{2,\omega}$, are obtained for $0.14 \le \omega \le \omega^*$.

An interesting value of the parameter is $\tilde{\omega} = \frac{1}{16}$. In fact, for $\omega \le \tilde{\omega}$ the family $\mathscr{F}_\omega$ has both $A_{1,\omega}$ and $A_{2,\omega}$ as spectrum maximizing products, each of which has a unique real leading eigenvector.

Starting from the initial set $\mathscr{T}^{(0)} = \{x(\omega), \hat{x}(\omega)\}$ of both such leading eigenvectors, scaled to be unitary with respect to the same norm, Algorithm G converges in only two steps. Although the assumptions of Theorem 10 are violated, we believe that the finite convergence occurs because of the symmetry in the problem, which is consistent with the same scaling of the two leading eigenvectors.

In all experimented cases Algorithm G provides the extremal real polytope

$$\mathscr{P}^{(1)}(\omega) = \text{absco}\left(\{x^{(1)}(\omega), x^{(2)}(\omega), x^{(3)}(\omega), x^{(4)}(\omega)\}\right),$$

where

$$x^{(1)}(\omega) = x(\omega), \quad x^{(2)}(\omega) = \hat{x}(\omega)$$

and

$$x^{(3)}(\omega) = A_{1,\omega} x^{(2)}(\omega), \quad x^{(4)}(\omega) = A_{2,\omega} x^{(1)}(\omega).$$

6 Fast Heuristic Methods for s.m.p. Searching

In this last section we present some techniques, based on heuristic considerations, which can be used to detect an s.m.p. $\bar{P}$ of a finite family $\mathscr{F} = \{A_i\}_{1 \leq i \leq d}$ that has the finiteness property.

Such techniques consist in *smart* recursive selections of products in the sets $\Sigma_k(\mathscr{F})$, $1 \leq k \leq k^*$, for given $k^* \geq 1$, of which to evaluate the spectral radius, that include the s.m.p. $\bar{P}$. The term smart has to be intended in the sense that the cardinality of the set of the selected products is a function of k^* of order much lower than the exponential order d^{k^*} (corresponding to a brute force exhaustive investigation).

6.1 General Structure of the Methods

In all cases, the simple inspiring fact is that every product $P \in \Sigma_k(\mathscr{F})$, $k \geq 2$, has one of the forms

- $P = RS^2$
- $P = S^2R$
- $P = SRS$

where $R \in \Sigma_r(\mathscr{F})$ and $S \in \Sigma_s(\mathscr{F})$ for $r \geq 0$, $s \geq 1$ with $k = r + 2s$, or, if none the above holds, one of the forms

- $P = AR$
- $P = RA$

where $R \in \Sigma_{k-1}(\mathscr{F})$ and $A \in \mathscr{F}$. Here we understand that $\Sigma_0(\mathscr{F}) = \{I\}$, where I is the identity matrix.

On the basis of such observation, we define recursively product sequences $\{P^{(i)}\}_{i \geq 1}$ with $P^{(i)} \in \Sigma_{k_i}(\mathscr{F})$, $k_i < k_{i+1}$, such that

- $P^{(1)} = A$, where $A \in \mathscr{F}$
- $P^{(i+1)}$ has one of the forms

  o $P^{(i+1)} = P^{(i)} S^{(i)}$
  o $P^{(i+1)} = S^{(i)} P^{(i)}$
  o $P^{(i+1)} = P^{(i)} A$
  o $P^{(i+1)} = A P^{(i)}$

where $A \in \mathscr{F}$ and $S^{(i)} \in \Sigma_{s_i}(\mathscr{F})$, $1 \leq s_i \leq k_i$, is such that $P^{(i)} = R^{(i)} S^{(i)}$ or $P^{(i)} = S^{(i)} R^{(i)}$ with $R^{(i)} \in \Sigma_{r_i}(\mathscr{F})$, $r_i \geq 0$, $k_i = r_i + s_i$.

The most critical part of the method is the choice of the criterion for selecting the new elements $P^{(i+1)}$ of the sequences at each step within the *preliminary* sets $\tilde{\Sigma}'_{k_{i+1}}(\mathscr{F}) \subseteq \Sigma_{k_{i+1}}(\mathscr{F})$ of all the products which are generated by the rule above.

It is just the case to observe that, without any selection mechanism, the method would generate all the possible products of $\Sigma(\mathscr{F})$.

There are three *selection criteria* we consider:

- (*increasing path* criterion) We only select products such that:

$$\rho(P^{(i+1)})^{\frac{1}{k_{i+1}}} \geq \rho(P^{(i)})^{\frac{1}{k_i}}.$$

- (*maximal path* criterion) We only select products such that $\rho(P^{(i+1)})$ is not less than the maximum $\rho(P)$ of those products $P \in \tilde{\Sigma}'_{k_{i+1}}(\mathscr{F})$ which contain all the matrices of the family $\mathscr{F}$ as factors. It is understood that this maximum is zero if there are no such products P, as it clearly happens as long as $k_{i+1} \leq d - 1$, where d is the cardinality of the family.
- (*maximal increasing path* criterion) We apply both the previous selection criteria, the latter in sequence after the former.

Remark that, in the last two criteria, selecting $P^{(i+1)}$ after comparison with the sole products P which contain all the matrices of the family $\mathscr{F}$ as factors prevents us from cutting off in the very first steps some *good* product sequences including an s.m.p. $\bar{P}$. In general, the bigger is d and the larger is the difference among the spectral radii of the various matrices of the family, the higher is the probability that this could occur.

This more cautious selection strategy will be referred to as the *safety device* and will be illustrated in Sect. 6.4 (see Examples 19 and 20).

6.2 Outline of the Algorithm

We define the various sequences of products advancing on the degree k (one degree per step).

We start from $\tilde{\Sigma}_1(\mathscr{F}) = \mathscr{F}$ and, for each $k \geq 2$, we define recursively the sets of products $\tilde{\Sigma}_k(\mathscr{F}) \subseteq \Sigma_k(\mathscr{F})$ until a given degree k_{max} is riched.

First we define the preliminary (in general, larger) set $\tilde{\Sigma}'_k(\mathscr{F})$ of products $\tilde{P}$ as follows:

⋄ For $h = [\frac{k+1}{2}], \ldots, k-2$

- $\forall P \in \tilde{\Sigma}_h(\mathscr{F})$ compute factorizations

$$P = R_{left} S_{right} \quad \text{and} \quad P = S_{left} R_{right}$$

with $S_{left}, S_{right} \in \Sigma_{k-h}(\mathscr{F})$;
- define the products of degree k

$$\tilde{P}^{left}_{left} = S_{left} P, \quad \tilde{P}^{right}_{left} = S_{right} P, \quad \tilde{P}^{left}_{right} = P S_{left}, \quad \tilde{P}^{right}_{right} = P S_{right};$$

⋄ $\forall P \in \tilde{\Sigma}_{k-1}(\mathscr{F})$ and $\forall A \in \mathscr{F}$ define

$$\tilde{P}_{left} = AP, \quad \tilde{P}_{right} = PA.$$

Then we compute the spectral radius $\rho(\tilde{P})$ for all $\tilde{P} \in \tilde{\Sigma}'_k(\mathscr{F})$ and select those products $\tilde{P}$ which satisfy the chosen *selection criterion* in order to form the set $\tilde{\Sigma}_k(\mathscr{F})$.

As for the implementation of the algorithm, we obviously exploit the fact that the spectral radius of a product of matrices is invariant by cyclic permutations.

Moreover, in order to avoid useless repeated computations of products and spectral radii, we use a one-to-one representation of products by integer numbers in base $d+1$, where d is the cardinality of the family $\mathscr{F}$. More precisely, to each product $P = A_{i_1} \cdots A_{i_k} \in \Sigma_k(\mathscr{F})$ we associate the *identifier*

$$\mathrm{id}(P) = \sum_{j=1}^{k} i_j (d+1)^{k-j}$$

and suitably operate on these identifiers to handle the factorizations of each product P and to search for the factors R and S and for the new products $\tilde{P}$ in the data base of all the products that have already been computed during the implementation of the algorithm (even if they have not been included in the sets $\tilde{\Sigma}_h(\mathscr{F})$).

6.3 Computational Complexity

In order to make a rough estimation of the computational complexity of the algorithm with respect to the degree k, we define the following quantities:

- $\nu(k)$ is the cardinality of $\tilde{\Sigma}_k(\mathscr{F})$;
- $C(k)$ is the computational complexity of the kth step to compute $\tilde{\Sigma}_k(\mathscr{F})$;
- $TC(k) = \sum_{h=2}^{k} C(h)$ is the total computational complexity up to the kth step to compute $\tilde{\Sigma}_h(\mathscr{F})$ for $2 \le h \le k$.

For each $k \ge 2$, in order to determine the preliminary set $\tilde{\Sigma}'_k(\mathscr{F})$ and to use any of the three selection criteria, the previous algorithm is such that:

◇ For $h = [\frac{k+1}{2}], \ldots, k-2$ at most $4\nu(h)$ new products $\tilde{P}$ of the type PS and SP and at most $2\nu(h)$ spectral radii are computed;
◇ At most $2d\nu(k-1)$ new products $\tilde{P}$ of the type PA and AP and at most $d\nu(k-1)$ spectral radii are computed.

Therefore, if γ is the cost for computing two matrix products and a spectral radius, apart from minor overhead we have

$$C(k) \le \left(d\nu(k-1) + 2 \sum_{h=[\frac{k+1}{2}]}^{k-2} \nu(h) \right) \gamma$$

and, consequently,

$$TC(k) \le \left(\sum_{h=2}^{k-1} (d + 2\mu_{k,h})\nu(h) \right) \gamma, \qquad (44)$$

where $\mu_{k,h} = \min\{h-1, k-h-1\}$.

It is evident that the computational complexity depends directly on the numbers $\nu(h)$ of products selected at each previous step. Thus, the selection criterion must reduce $\nu(h)$ as much as possible without cutting off all *good* sequences $\{P^{(i)}\}_{i \ge 1}$ that contain an s.m.p. $\bar{P}$.

With the *increasing path* selection criterion, experiments show that, in average, $\nu(h)$ is not significantly smaller than d^h, which corresponds to a full acceptance of products $\tilde{P}$ (i.e., no selection mechanism). Therefore, this strategy is too slow and must be discarded.

With the *maximal path* selection criterion, in average, $\nu(h)$ is drastically reduced. In fact, apart from the first few steps where the safety device often works, in most cases the maximum spectral radius in the preliminary set $\tilde{\Sigma}'_h(\mathscr{F})$ is attained by one product only, modulo cyclic permutations. This suggests that, in most cases, $\nu(h) \le h$. Furthermore, in practice, not all the cyclic permutations of the maximal product are selected by the algorithm and, indeed, there exists a constant $\bar{\nu}$ such that

$$\nu(h) \leq \bar{\nu} \quad \text{for all } h.$$

In conclusion, with this selection criterion we often have

$$C(k) = \mathcal{O}(k)$$

and, consequently,

$$TC(k) = \mathcal{O}(k^2),$$

that is a quite satisfactory computational complexity.

The *maximal increasing path* selection criterion may be seen as a more stringent variant of the maximal path selection criterion in that it also requires that the selected sequences of products be monotonic in the spectral radius. This fact slightly increases the probability that all the good sequences containing an s.m.p. are discarded and does not reduce significantly the computational complexity, which remains of the same order.

6.4 Some Illustrative Examples

We analyzed the algorithms based on the maximal path and on the maximal increasing path selection criteria on a certain number of test families with various characteristics, mainly with very small cardinality, say $d = 2$ and $d = 3$. In any case, they never failed to find out an s.m.p. $\bar{P}$.

Here we give three particular examples, which resume some significant situations we may face using the proposed algorithms.

The computations have been carried out by using MATLAB. However, since we did not optimize the programs under the point of view of the computational efficiency, we do not give the experimental results regarding the computational complexity and the CPU time. However, in this respect, we confine ourselves to give, as rough indicators, the cardinalities $\nu(k)$ of the sets $\tilde{\Sigma}_k(\mathscr{F})$, $2 \leq k \leq k_{max}$, and the upper bound to $TC(k_{max})$ suggested by (44). Observe that all these quantities are independent of the particular computational tricks that one could use to increase the efficiency of the algorithm.

We always choose the degree k^* of the s.m.p. $\bar{P}$ as the maximum number k_{max} of steps of the algorithm.

Example 18. We consider the family $\mathscr{F} = \{A_1, A_2\}$, where

$$A_1 = \begin{bmatrix} -1 & 1 & -1 \\ -1 & -1 & 1 \\ 0 & 1 & 1 \end{bmatrix}, \quad A_2 = \begin{bmatrix} -1 & 1 & -1 \\ -1 & -1 & 0 \\ 1 & 1 & 1 \end{bmatrix}.$$

which was already introduced in Examples 2 and 7.

The dimension of the matrices is $n = 3$, the cardinality of the family is $d = 2$, $\rho(A_1) = 1.64089\ldots$ and $\rho(A_2) = 1.57082\ldots$. Moreover, as we already saw in Sect. 5.4, Example 14, a minimal s.m.p. is

$$\bar{P} = \left(\left(A_1^2 A_2 \right)^2 A_1 A_2 \right)^3 A_2 A_1 A_2 \in \Sigma_{27}(\mathscr{F})$$

and $\rho(\mathscr{F}) = \rho(\bar{P})^{\frac{1}{27}} = 1.78893\ldots$.

Both algorithms detect $\bar{P}$ and the same set of other six cyclic permutations of it. Remark that, as we already anticipated in Sect. 6.3, not all the cyclic permutations of $\bar{P}$ are detected.

There are two good sequences (obtained by using both algorithms) that contain $\bar{P}$. The first six elements of both sequences are:

$P^{(1)} = A_1$, $\quad\rho(P^{(1)}) = 1.64089\ldots$
$P^{(2)} = A_1 A_2$, $\quad\rho(P^{(2)})^{\frac{1}{2}} = 1.68179\ldots$
$P^{(3)} = (A_1 A_2)^2$, $\quad\rho(P^{(3)})^{\frac{1}{4}} = 1.68179\ldots$
$P^{(4)} = A_1^2 A_2 A_1 A_2$, $\quad\rho(P^{(4)})^{\frac{1}{5}} = 1.70501\ldots$
$P^{(5)} = \left(A_1^2 A_2 \right)^2 A_1 A_2$, $\quad\rho(P^{(5)})^{\frac{1}{8}} = 1.78308\ldots$
$P^{(6)} = \left(\left(A_1^2 A_2 \right)^2 A_1 A_2 \right)^2$, $\quad\rho(P^{(6)})^{\frac{1}{16}} = 1.78308\ldots$

Then the last but first element of the first good sequence is

$$P^{(7)} = \left(\left(A_1^2 A_2 \right)^2 A_1 A_2 \right)^2 A_2 A_1 A_2, \qquad \rho(P^{(7)})^{\frac{1}{19}} = 1.78843\ldots$$

and the last but first element of the second good sequence is

$$P^{(7)} = \left(\left(A_1^2 A_2 \right)^2 A_1 A_2 \right)^3, \qquad \rho(P^{(7)})^{\frac{1}{24}} = 1.78308\ldots$$

Eventually, for both good sequences we have $P^{(8)} = \bar{P}$.

As for the cardinalities $\nu(k)$ of the sets $\tilde{\Sigma}_k(\mathscr{F})$, for the maximal path criterion method we have: $\nu(1) = 2$, $\nu(2) = 2$, $\nu(3) = 3$, $\nu(4) = 2$, $\nu(5) = 3$, $\nu(6) = 2$, $\nu(7) = 7$, $\nu(8) = 6$, $\nu(9) = 9$, $\nu(10) = 5$, $\nu(11) = 3$, $\nu(12) = 12$, $\nu(13) = 5$, $\nu(14) = 14$, $\nu(15) = 4$, $\nu(16) = 6$, $\nu(17) = 17$, $\nu(18) = 11$, $\nu(19) = 5$, $\nu(20) = 18$, $\nu(21) = 3$, $\nu(22) = 3$, $\nu(23) = 4$, $\nu(24) = 6$, $\nu(25) = 4$, $\nu(26) = 12$, $\nu(27) = 7$.

Therefore, by (44), we obtain $TC(27) \leq 2672\gamma$.

For the maximal increasing path criterion method, instead we have: $\nu(1) = 2$, $\nu(2) = 2$, $\nu(3) = 0$ (i.e., the set $\tilde{\Sigma}_3(\mathscr{F})$ is empty), $\nu(4) = 2$, $\nu(5) = 3$, $\nu(6) = 2$, $\nu(7) = 5$, $\nu(8) = 6$, $\nu(9) = 3$, $\nu(10) = 3$, $\nu(11) = 2$, $\nu(12) = 12$, $\nu(13) = 0$ (i.e., the set $\tilde{\Sigma}_{13}(\mathscr{F})$ is empty), $\nu(14) = 6$, $\nu(15) = 5$, $\nu(16) = 6$, $\nu(17) = 3$,

$\nu(18) = 3$, $\nu(19) = 3$, $\nu(20) = 12$, $\nu(21) = 3$, $\nu(22) = 2$, $\nu(23) = 6$, $\nu(24) = 6$, $\nu(25) = 10$, $\nu(26) = 4$, $\nu(27) = 7$.

This time, by (44), we obtain $TC(27) \leq 1616\gamma$. ◇

Example 19. We consider the family $\mathscr{F} = \{A_1, A_2\}$, where

$$A_1 = \begin{bmatrix} -1 & 1 & -1 \\ -1 & -1 & 1 \\ 0 & 1 & 1 \end{bmatrix}, \qquad A_2 = \begin{bmatrix} -1 & 1 & -1 \\ -1 & -1 & 0 \\ 0 & 1 & 1 \end{bmatrix}.$$

The dimension of the matrices is $n = 3$, the cardinality of the family is $d = 2$, $\rho(A_1) = 1.64089\ldots$ and $\rho(A_2) = 1.59817\ldots$. A minimal s.m.p. is

$$\bar{P} = \left(A_1^2 A_2^2\right)^9 A_2 A_1 A_2^3 \in \Sigma_{41}(\mathscr{F})$$

and $\rho(\mathscr{F}) = \rho(\bar{P})^{\frac{1}{41}} = 1.68418\ldots$.

Both algorithms detect $\bar{P}$ along with a set of other 11 cyclic permutations of it in the case of the maximal path criterion and a set of other 5 cyclic permutations of it in the case of the maximal increasing path criterion.

Again, not all the cyclic permutations of $\bar{P}$ are detected in either case.

A good sequence that contains $\bar{P}$, obtained by using the maximal path criterion, is:

$$\begin{aligned}
P^{(1)} &= A_1, & \rho(P^{(1)}) &= 1.64089\ldots \\
P^{(2)} &= A_1 A_2, & \rho(P^{(2)})^{\frac{1}{2}} &= 1.60573\ldots \\
P^{(3)} &= A_1^2 A_2, & \rho(P^{(3)})^{\frac{1}{3}} &= 1.63754\ldots \\
P^{(4)} &= A_1^2 A_2^2, & \rho(P^{(4)})^{\frac{1}{4}} &= 1.68179\ldots \\
P^{(5)} &= \left(A_1^2 A_2^2\right)^2, & \rho(P^{(5)})^{\frac{1}{8}} &= 1.68179\ldots \\
P^{(6)} &= \left(A_1^2 A_2^2\right)^3, & \rho(P^{(6)})^{\frac{1}{12}} &= 1.68179\ldots \\
P^{(7)} &= \left(A_1^2 A_2^2\right)^4, & \rho(P^{(7)})^{\frac{1}{16}} &= 1.68179\ldots \\
P^{(8)} &= A_1 \left(A_1^2 A_2^2\right)^4, & \rho(P^{(8)})^{\frac{1}{17}} &= 1.66681\ldots \\
P^{(9)} &= A_1^3 A_2 A_1 \left(A_1^2 A_2^2\right)^4, & \rho(P^{(9)})^{\frac{1}{21}} &= 1.67961\ldots \\
P^{(10)} &= A_1^3 A_2 A_1 \left(A_1^2 A_2^2\right)^6, & \rho(P^{(10)})^{\frac{1}{29}} &= 1.68335\ldots \\
P^{(11)} &= A_1^3 A_2 A_1 \left(A_1^2 A_2^2\right)^7, & \rho(P^{(11)})^{\frac{1}{33}} &= 1.68389\ldots \\
P^{(12)} &= A_1^3 A_2 A_1 \left(A_1^2 A_2^2\right)^8, & \rho(P^{(12)})^{\frac{1}{37}} &= 1.68412\ldots
\end{aligned}$$

and, eventually, $P^{(13)} = \bar{P}$.

Observe that $\rho(P^{(i)})^{\frac{1}{k_i}} < \rho(A_1)$ for $i = 2, 3$. This means that the *safety device* worked. Indeed, without using it, this good sequence would not have been obtained.

Moreover, as an example of non-monotonicity, observe that $\rho(P^{(8)})^{\frac{1}{k_8}} < \rho(P^{(7)})^{\frac{1}{k_7}}$.

A good sequence that contains $\bar{P}$, obtained by using the maximal increasing path criterion, is:

$$P^{(1)} = A_2, \qquad \rho(P^{(1)}) = 1.59817\ldots$$
$$P^{(2)} = A_1 A_2, \qquad \rho(P^{(2)})^{\frac{1}{2}} = 1.60573\ldots$$
$$P^{(3)} = A_1^2 A_2, \qquad \rho(P^{(3)})^{\frac{1}{3}} = 1.63754\ldots$$
$$P^{(4)} = A_1^2 A_2^2, \qquad \rho(P^{(4)})^{\frac{1}{4}} = 1.68179\ldots$$
$$P^{(5)} = \left(A_1^2 A_2^2\right)^2, \qquad \rho(P^{(5)})^{\frac{1}{8}} = 1.68179\ldots$$
$$P^{(6)} = \left(A_1^2 A_2^2\right)^3, \qquad \rho(P^{(6)})^{\frac{1}{12}} = 1.68179\ldots$$
$$P^{(7)} = \left(A_1^2 A_2^2\right)^4, \qquad \rho(P^{(7)})^{\frac{1}{16}} = 1.68179\ldots$$
$$P^{(8)} = \left(A_1^2 A_2^2\right)^5, \qquad \rho(P^{(8)})^{\frac{1}{20}} = 1.68179\ldots$$
$$P^{(9)} = \left(A_1^2 A_2^2\right)^6, \qquad \rho(P^{(9)})^{\frac{1}{24}} = 1.68179\ldots$$
$$P^{(10)} = \left(A_1^2 A_2^2\right)^7, \qquad \rho(P^{(10)})^{\frac{1}{28}} = 1.68179\ldots$$
$$P^{(11)} = A_1 \left(A_1^2 A_2^2\right)^7, \qquad \rho(P^{(11)})^{\frac{1}{29}} = 1.68233\ldots$$
$$P^{(12)} = A_1^3 A_2 A_1 \left(A_1^2 A_2^2\right)^7, \qquad \rho(P^{(12)})^{\frac{1}{33}} = 1.68389\ldots$$
$$P^{(13)} = A_1^3 A_2 A_1 \left(A_1^2 A_2^2\right)^8, \qquad \rho(P^{(13)})^{\frac{1}{37}} = 1.68412\ldots$$

and, eventually, $P^{(14)} = \bar{P}$.

Observe that, also in this case, the *safety device* worked, since $\rho(P^{(i)})^{\frac{1}{k_i}} < \rho(A_1)$ for $i = 1, 2, 3$.

As for the cardinalities $\nu(k)$ of the sets $\tilde{\Sigma}_k(\mathscr{F})$, for the maximal path criterion method we have: $\nu(1) = 2$, $\nu(2) = 3$, $\nu(3) = 4$, $\nu(4) = 3$, $\nu(5) = 4$, $\nu(6) = 4$, $\nu(7) = 5$, $\nu(8) = 3$, $\nu(9) = 4$, $\nu(10) = 9$, $\nu(11) = 6$, $\nu(12) = 3$, $\nu(13) = 4$, $\nu(14) = 10$, $\nu(15) = 4$, $\nu(16) = 3$, $\nu(17) = 6$, $\nu(18) = 5$, $\nu(19) = 11$, $\nu(20) = 3$, $\nu(21) = 4$, $\nu(22) = 4$, $\nu(23) = 9$, $\nu(24) = 3$, $\nu(25) = 6$, $\nu(26) = 6$, $\nu(27) = 14$, $\nu(28) - 3$, $\nu(29) = 4$, $\nu(30) = 8$, $\nu(31) - 17$, $\nu(32) = 3$, $\nu(33) = 8$, $\nu(34) = 10$, $\nu(35) = 8$, $\nu(36) = 3$, $\nu(37) = 10$, $\nu(38) = 4$, $\nu(39) = 10$, $\nu(40) = 3$, $\nu(41) = 12$.

Therefore, by (44), we obtain $TC(41) \leq 5034\gamma$.

For the maximal increasing path criterion method, instead we have: $\nu(1) = 2$, $\nu(2) = 3$, $\nu(3) = 4$, $\nu(4) = 3$, $\nu(5) = 1$, $\nu(6) = 4$, $\nu(7) = 1$, $\nu(8) = 3$, $\nu(9) = 4$, $\nu(10) = 1$, $\nu(11) = 6$, $\nu(12) = 3$, $\nu(13) = 4$, $\nu(14) = 4$, $\nu(15) = 4$, $\nu(16) = 3$, $\nu(17) = 3$, $\nu(18) = 4$, $\nu(19) = 1$, $\nu(20) = 3$, $\nu(21) = 4$, $\nu(22) = 6$, $\nu(23) = 6$, $\nu(24) = 3$, $\nu(25) = 5$, $\nu(26) = 4$, $\nu(27) = 4$, $\nu(28) = 3$, $\nu(29) = 10$, $\nu(30) = 4$, $\nu(31) = 1$, $\nu(32) = 3$, $\nu(33) = 4$, $\nu(34) = 3$, $\nu(35) = 6$, $\nu(36) = 3$, $\nu(37) = 4$, $\nu(38) = 4$, $\nu(39) = 4$, $\nu(40) = 3$, $\nu(41) = 6$.

This time, by (44), we obtain $TC(41) \leq 3176\gamma$. ◇

Example 20. We consider the family $\mathscr{F} = \{A_1, A_2, A_3\}$, where

$$A_1 = \begin{bmatrix} -1 & 2 & -1 \\ -1 & -1 & 1 \\ 0 & 1 & 1 \end{bmatrix}, \quad A_2 = \begin{bmatrix} -1 & 1 & -1 \\ -1 & -1 & 0 \\ 1 & 1 & 2 \end{bmatrix}, \quad A_3 = \begin{bmatrix} -1 & 1 & 0 \\ -1 & -1 & 2 \\ -1 & 0 & 1 \end{bmatrix}.$$

The dimension of the matrices is $n = 3$, the cardinality of the family is $d = 3$, $\rho(A_1) = 1.86765\ldots$, $\rho(A_2) = 1.79632\ldots$ and $\rho(A_3) = 1$. A minimal s.m.p. is

$$\bar{P} = A_1 A_3 (A_2 A_3 A_2)^2 \in \Sigma_8(\mathscr{F})$$

and $\rho(\mathscr{F}) = \rho(\bar{P})^{\frac{1}{8}} = 2.37745\ldots$.

Both algorithms detect $\bar{P}$ along with a set of other two cyclic permutations of it, which are the same ones.

Again, not all the cyclic permutations of $\bar{P}$ are detected.

The following good sequence that contains $\bar{P}$ is obtained by both algorithms:

$$P^{(1)} = A_2, \qquad \rho(P^{(1)}) = 1.79632\ldots$$
$$P^{(2)} = A_3 A_2, \qquad \rho(P^{(2)})^{\frac{1}{2}} = 1.86120\ldots$$
$$P^{(3)} = A_1 A_3 A_2, \qquad \rho(P^{(3)})^{\frac{1}{3}} = 2.24790\ldots$$
$$P^{(4)} = A_1 (A_3 A_2)^2, \qquad \rho(P^{(4)})^{\frac{1}{5}} = 2.37279\ldots$$

and, eventually, $P^{(5)} = \bar{P}$.

Observe that, since $\rho(P^{(i)})^{\frac{1}{k_i}} < \rho(A_1)$ for $i = 1, 2$, the *safety device* worked. More precisely, it is worth remarking that, since A_1^2 is the only product that maximizes the spectral radius in $\Sigma_2(\mathscr{F})$ and since it is not a factor of $\bar{P}$, without using the safety device it is not possible to generate $\bar{P}$. In fact, all the products produced by both algorithms should include the factor A_1^2.

As for the cardinalities $\nu(k)$ of the sets $\tilde{\Sigma}_k(\mathscr{F})$, for the maximal path criterion method we have: $\nu(1) = 3$, $\nu(2) = 9$, $\nu(3) = 3$, $\nu(4) = 9$, $\nu(5) = 3$, $\nu(6) = 3$, $\nu(7) = 4$, $\nu(8) = 3$.

Therefore, by (44), we obtain $TC(8) \leq 195\gamma$.

For the maximal increasing path criterion method, instead we have: $\nu(1) = 3$, $\nu(2) = 7$, $\nu(3) = 3$, $\nu(4) = 7$, $\nu(5) = 3$, $\nu(6) = 3$, $\nu(7) = 12$, $\nu(8) = 3$.

This time, by (44), we obtain $TC(8) \leq 191\gamma$. ◇

As one could reasonably guess a priori, in all examples we considered the computational complexity of the maximal increasing path criterion method is lower than that of the maximal path criterion method, even if not dramatically.

By carefully analyzing some further outputs of our algorithms (that we did not report here), another remarkable fact, not so surprising, is that, for some values of k, the quantity

$$\tilde{\rho}_k(\mathscr{F}) = \max_{P \in \tilde{\Sigma}_k(\mathscr{F})} \rho(P)^{1/k} \qquad (45)$$

is smaller for the maximal increasing path criterion method than for the maximal path criterion method. On the other hand, in Example 18 we see that, for $k = 3$ and $k = 13$, the maximal increasing path criterion method is such that $\nu(k) = 0$. This is due to the fact that it produces an empty preliminary set $\tilde{\Sigma}'_k(\mathscr{F})$. Therefore, $\tilde{\rho}_k(\mathscr{F}) = 0$.

Moreover, for some values of k, it happens that $\tilde{\rho}_k(\mathscr{F})$ is strictly less than $\bar{\rho}_k(\mathscr{F})$ (see (9)) for both methods. However, this was never observed for $k = k^*$, the degree of the minimal s.m.p. $\bar{P}$.

6.5 Conclusive Remarks on the Methods

Although our experiments with the heuristic methods proposed in this section were always successful, there is no proof yet that they always work and, indeed, this is not even our guess.

Another problem which is not satisfactorily solved yet is that of finding a reliable stopping criterion to the iterative generation of the sets $\tilde{\Sigma}'_k(\mathscr{F})$, other than observing a stagnation of the values $\tilde{\rho}_k(\mathscr{F})$ defined by (45). However, apart from being computationally expensive, even this strategy might reveal to be misleading since it is not clear a priori for how long the stagnation should last to make us sure that the maximum has been reached.

The conclusion is that, so far, we need to use also another method which is able to check whether a given product $\bar{P}$ is an s.m.p. or not. Therefore, we currently propose to use the heuristic methods introduced in this section to provide reasonable candidate s.m.p.'s for the initialization of Algorithms G and N presented in Sect. 5.

Acknowledgements We thank C.I.M.E. for the excellent support in the organization of the Summer School and INdAM-GNCS for partial funding.

References

1. T. Ando, M.-H. Shih, Simultaneous contractibility. SIAM J. Matrix Anal. Appl. **19**, 487–498 (1998)
2. N.E. Barabanov, Lyapunov indicator for discrete inclusions, I–III. Autom. Rem. Contr. **49**, 152–157 (1988)
3. M.A. Berger, Y. Wang, Bounded semigroups of matrices. Lin. Algebra Appl. **166**, 21–27 (1992)
4. A. Berman, R.J. Plemmons, *Nonnegative Matrices in the Mathematical Sciences*. Classics in Applied Mathematics, vol. 9 (SIAM, Philadelphia, 1994), xx+340 pp.
5. J. Berstel, Growth of repetition-free words – A review. Theor. Comput. Sci. **340**, 280–290 (2005)
6. V.D. Blondel, Y. Nesterov, Computationally efficient approximations of the joint spectral radius. SIAM J. Matrix Anal. Appl. **27**, 256–272 (2005)
7. V.D. Blondel, J. Theys, A.A. Vladimirov, An elementary counterexample to the finiteness conjecture. SIAM J. Matrix Anal. Appl. **24**, 963–970 (2003)
8. V.D. Blondel, Y. Nesterov, J. Theys, On the accuracy of the ellipsoid norm approximation of the joint spectral radius. Lin. Algebra Appl. **394**, 91–107 (2005)
9. V.D. Blondel, R. Jungers, V.Y. Protasov, On the complexity of computing the capacity of codes that avoid forbidden difference patterns. IEEE Trans. Inf. Theor. **52**, 5122–5127 (2006)
10. T. Bousch, J. Mairesse, Asymptotic height optimization for topical IFS, Tetris heaps and the finiteness conjecture. J. Am. Math. Soc. **15**, 77–111 (2002)

11. S. Boyd, L. Vandenberghe, *Convex Optimization* (Cambridge University Press, Cambridge, 2004), xiv+716 pp.
12. A. Cicone, N. Guglielmi, S. Serra-Capizzano, M. Zennaro, Finiteness property of pairs of 2×2 sign-matrices via real extremal polytope norms. Lin. Algebra Appl. **432**, 796–816 (2010)
13. I. Daubechies, *Ten Lectures on Wavelets*. CBMS-NSF Regional Conference Series in Applied Mathematics, vol. 61 (SIAM, Philadelphia, 1992), xx+357 pp.
14. I. Daubechies, J. Lagarias, Two-scale difference equations. II. Local regularity, infinite products of matrices and fractals. SIAM. J. Math. Anal. **23**, 1031–1079 (1992)
15. I. Daubechies, J.C. Lagarias, Sets of matrices all infinite products of which converge. Lin. Algebra Appl. **161**, 227–263 (1992)
16. S. Dubuc, Interpolation through an iterative scheme. J. Math. Anal. Appl. **114**, 185–204 (1986)
17. L. Elsner, The generalized spectral-radius theorem: An analytic-geometric proof. Lin. Algebra Appl. **220**, 151–159 (1995)
18. G. Gripenberg, Computing the joint spectral radius. Lin. Algebra Appl. **234**, 43–60 (1996)
19. N. Guglielmi, V.Y. Protasov, Exact computation of joint spectral characteristics of linear operators. Found. Comput. Math. **13**, 37–97 (2013)
20. N. Guglielmi, M. Zennaro, On the asymptotic properties of a family of matrices. Lin. Algebra Appl. **322**, 169–192 (2001)
21. N. Guglielmi, M. Zennaro, On the zero-stability of variable stepsize multistep methods: The spectral radius approach. Numer. Math. **88**, 445–458 (2001)
22. N. Guglielmi, M. Zennaro, On the limit products of a family of matrices. Lin. Algebra Appl. **362**, 11–27 (2003)
23. N. Guglielmi, M. Zennaro, Balanced complex polytopes and related vector and matrix norms. J. Convex Anal. **14**, 729–766 (2007)
24. N. Guglielmi, M. Zennaro, An algorithm for finding extremal polytope norms of matrix families. Lin. Algebra Appl. **428**, 2265–2282 (2008)
25. N. Guglielmi, M. Zennaro, Finding extremal complex polytope norms for families of real matrices. SIAM J. Matrix Anal. Appl. **31**, 602–620 (2009)
26. N. Guglielmi, M. Zennaro, On the asymptotic regularity of a family of matrices. Lin. Algebra Appl. **436**, 2093–2104 (2012)
27. N. Guglielmi, F. Wirth, M. Zennaro, Complex polytope extremality results for families of matrices. SIAM J. Matrix Anal. Appl. **27**, 721–743 (2005)
28. N. Guglielmi, C. Manni, D. Vitale, Convergence analysis of C^2 Hermite interpolatory subdivision schemes by explicit joint spectral radius formulas. Lin. Algebra Appl. **434**, 784–902 (2011)
29. L. Gurvits, Stability of discrete linear inclusions. Lin. Algebra Appl. **231**, 47–85 (1995)
30. K.G. Hare, N. Sidorov, I. Morris, J. Theys, An explicit counterexample to the Lagarias-Wang finiteness conjecture. Adv. Math. **226**, 4667–4701 (2011)
31. J. Hechler, B. Mößner, U. Reif, C^1-continuity of the generalized four-point scheme. Lin. Algebra Appl. **430**, 3019–3029 (2009)
32. R. Jungers, V. Blondel, On the finiteness properties for rational matrices. Lin. Algebra Appl. **428**, 2283–2295 (2008)
33. R.M. Jungers, *The Joint Spectral Radius: Theory and Applications*. Lecture Notes in Control and Information Sciences, vol. 385 (Springer, Berlin, 2009), xiv+144 pp.
34. R.M. Jungers, V.Y. Protasov, Counterexamples to the complex polytope extremality conjecture. SIAM J. Matrix Anal. Appl. **31**, 404–409 (2009)
35. R.M. Jungers, V.Y. Protasov, V.D. Blondel, Overlap-free words and spectra of matrices. Theor. Comput. Sci. **410**, 3670–3684 (2009)
36. V.S. Kozyakin, On the computational aspects of the theory of joint spectral radius. Dokl. Math. **80**, 487–491 (2009)
37. J.C. Lagarias, Y. Wang, The finiteness conjecture for the generalized spectral radius of a set of matrices. Lin. Algebra Appl. **214**, 17–42 (1995)

38. M. Maesumi, Optimum unit ball for joint spectral radius: An example from four-coefficient MRA, in *Approximation Theory VIII: Wavelets and Multilevel Approximation*, ed. by C.K. Chui, L.L. Schumaker, vol. 2 (World Scientific, Singapore, 1995), pp. 267–274
39. M. Maesumi, Calculating joint spectral radius of matrices and Hölder exponent of wavelets, in *Approximation Theory IX*, ed. by C.K. Chui, L.L. Schumaker (World Scientific, Singapore, 1998), pp. 1–8
40. O. Mason, R.N. Shorten, Quadratic and copositive Lyapunov functions and the stability of positive switched linear systems, in *Proceedings of the American Control Conference (ACC 2007)* (2007), pp. 657–662
41. B.E. Moision, A. Orlitsky, P.H. Siegel, On codes that avoid specified differences. IEEE Trans. Inf. Theor. **47**, 433-422, (2001)
42. P.A. Parrilo, A. Jadbabaie, Approximation of the joint spectral radius using sum of squares. Lin. Algebra Appl. **428**, 2385–2402 (2008)
43. V.Y. Protasov, The joint spectral radius and invariant sets of linear operators. Fundam. Prikl. Mat. **2**(1), 205–231 (1996)
44. V.Y. Protasov, The generalized spectral radius. A geometric approach. Izv. Math. **61**, 995–1030 (1997)
45. M.H. Shih, Simultaneous Schur stability. Lin. Algebra Appl. **287**, 323–336 (1999)
46. M.H. Shih, J.W. Wu, C.T. Pang, Asymptotic stability and generalized Gelfand spectral radius formula. Lin. Algebra Appl. **252**, 61–70 (1997)
47. R. Shorten, F. Wirth, O. Mason, K. Wulff, C. King, Stability criteria for switched and hybrid systems. SIAM Rev. **49**, 545–592 (2007)
48. G.C. Rota, G. Strang, A note on the joint spectral radius. Kon. Nederl. Acad. Wet. Proc. **63**, 379–381 (1960)
49. G. Strang, The joint spectral radius, Commentary by Gilbert Strang. Collected works of Gian-Carlo Rota (2000)
50. J.N. Tsitsiklis, V.D. Blondel, The Lyapunov exponent and joint spectral radius of pairs of matrices are hard—when not impossible—to compute and to approximate. Math. Contr. Signals Syst. **10**, 31–40 (1997)
51. C. Vagnoni, M. Zennaro, The analysis and the representation of balanced complex polytopes in 2D. Found. Comput. Math. **9**, 259–294 (2009)
52. J.S. Vandergraft, Spectral properties of matrices which have invariant cones. SIAM J. Appl. Math. **16**, 1208–1222 (1968)
53. L. Villemoes, Wavelet analysis of refinement equations. SIAM J. Math. Anal. **25**, 1433–1460 (1994)
54. A.N. Willson, A stability criterion for nonautonomous difference equations with application to the design of a digital FSK oscillator. IEEE Trans. Circ. Syst. **21**, 124–130 (1974)

LECTURE NOTES IN MATHEMATICS

Edited by J.-M. Morel, B. Teissier; P.K. Maini

Editorial Policy (for Multi-Author Publications: Summer Schools / Intensive Courses)

1. Lecture Notes aim to report new developments in all areas of mathematics and their applications - quickly, informally and at a high level. Mathematical texts analysing new developments in modelling and numerical simulation are welcome. Manuscripts should be reasonably selfcontained and rounded off. Thus they may, and often will, present not only results of the author but also related work by other people. They should provide sufficient motivation, examples and applications. There should also be an introduction making the text comprehensible to a wider audience. This clearly distinguishes Lecture Notes from journal articles or technical reports which normally are very concise. Articles intended for a journal but too long to be accepted by most journals, usually do not have this "lecture notes" character.

2. In general SUMMER SCHOOLS and other similar INTENSIVE COURSES are held to present mathematical topics that are close to the frontiers of recent research to an audience at the beginning or intermediate graduate level, who may want to continue with this area of work, for a thesis or later. This makes demands on the didactic aspects of the presentation. Because the subjects of such schools are advanced, there often exists no textbook, and so ideally, the publication resulting from such a school could be a first approximation to such a textbook. Usually several authors are involved in the writing, so it is not always simple to obtain a unified approach to the presentation.

   For prospective publication in LNM, the resulting manuscript should not be just a collection of course notes, each of which has been developed by an individual author with little or no coordination with the others, and with little or no common concept. The subject matter should dictate the structure of the book, and the authorship of each part or chapter should take secondary importance. Of course the choice of authors is crucial to the quality of the material at the school and in the book, and the intention here is not to belittle their impact, but simply to say that the book should be planned to be written by these authors jointly, and not just assembled as a result of what these authors happen to submit.

   This represents considerable preparatory work (as it is imperative to ensure that the authors know these criteria before they invest work on a manuscript), and also considerable editing work afterwards, to get the book into final shape. Still it is the form that holds the most promise of a successful book that will be used by its intended audience, rather than yet another volume of proceedings for the library shelf.

3. Manuscripts should be submitted either online at www.editorialmanager.com/lnm/ to Springer's mathematics editorial, or to one of the series editors. Volume editors are expected to arrange for the refereeing, to the usual scientific standards, of the individual contributions. If the resulting reports can be forwarded to us (series editors or Springer) this is very helpful. If no reports are forwarded or if other questions remain unclear in respect of homogeneity etc, the series editors may wish to consult external referees for an overall evaluation of the volume. A final decision to publish can be made only on the basis of the complete manuscript; however a preliminary decision can be based on a pre-final or incomplete manuscript. The strict minimum amount of material that will be considered should include a detailed outline describing the planned contents of each chapter.

   Volume editors and authors should be aware that incomplete or insufficiently close to final manuscripts almost always result in longer evaluation times. They should also be aware that parallel submission of their manuscript to another publisher while under consideration for LNM will in general lead to immediate rejection.

4. Manuscripts should in general be submitted in English. Final manuscripts should contain at least 100 pages of mathematical text and should always include

    – a general table of contents;
    – an informative introduction, with adequate motivation and perhaps some historical remarks: it should be accessible to a reader not intimately familiar with the topic treated;
    – a global subject index: as a rule this is genuinely helpful for the reader.

    Lecture Notes volumes are, as a rule, printed digitally from the authors' files. We strongly recommend that all contributions in a volume be written in the same LaTeX version, preferably LaTeX2e. To ensure best results, authors are asked to use the LaTeX2e style files available from Springer's webserver at
    ftp://ftp.springer.de/pub/tex/latex/svmonot1/ (for monographs) and
    ftp://ftp.springer.de/pub/tex/latex/svmultt1/ (for summer schools/tutorials).
    Additional technical instructions, if necessary, are available on request from:
    lnm@springer.com.

5. Careful preparation of the manuscripts will help keep production time short besides ensuring satisfactory appearance of the finished book in print and online. After acceptance of the manuscript authors will be asked to prepare the final LaTeX source files and also the corresponding dvi-, pdf- or zipped ps-file. The LaTeX source files are essential for producing the full-text online version of the book. For the existing online volumes of LNM see:
    http://www.springerlink.com/openurl.asp?genre=journal&issn=0075-8434.
    The actual production of a Lecture Notes volume takes approximately 12 weeks.

6. Volume editors receive a total of 50 free copies of their volume to be shared with the authors, but no royalties. They and the authors are entitled to a discount of 33.3 % on the price of Springer books purchased for their personal use, if ordering directly from Springer.

7. Commitment to publish is made by letter of intent rather than by signing a formal contract. Springer-Verlag secures the copyright for each volume. Authors are free to reuse material contained in their LNM volumes in later publications: a brief written (or e-mail) request for formal permission is sufficient.

Addresses:
Professor J.-M. Morel, CMLA,
École Normale Supérieure de Cachan,
61 Avenue du Président Wilson, 94235 Cachan Cedex, France
E-mail: morel@cmla.ens-cachan.fr

Professor B. Teissier, Institut Mathématique de Jussieu,
UMR 7586 du CNRS, Équipe "Géométrie et Dynamique",
175 rue du Chevaleret,
75013 Paris, France
E-mail: teissier@math.jussieu.fr

For the "Mathematical Biosciences Subseries" of LNM:

Professor P. K. Maini, Center for Mathematical Biology,
Mathematical Institute, 24-29 St Giles,
Oxford OX1 3LP, UK
E-mail : maini@maths.ox.ac.uk

Springer, Mathematics Editorial I,
Tiergartenstr. 17,
69121 Heidelberg, Germany,
Tel.: +49 (6221) 4876-8259
Fax: +49 (6221) 4876-8259
E-mail: lnm@springer.com

The manufacturer's authorised representative in the EU is Springer Nature Customer Service Centre GmbH, Europaplatz 3, 69115 Heidelberg, Germany. If you have any concerns regarding our products, please contact ProductSafety@springernature.com

Printed and bound by CPI Group (UK) Ltd, Croydon, CR0 4YY

23/03/2026

02076665-0001